12th International Conference on Excitonic and Photonic Processes in Condensed Matter and Nano Materials (EXCON 2018)

Journal of Physics: Conference Series Volume 1220

Nara City, Japan
8 – 13 July 2018

ISBN: 978-1-5108-8820-3
ISSN: 1742-6588

Printed from e-media with permission by:

Curran Associates, Inc.
57 Morehouse Lane
Red Hook, NY 12571

Some format issues inherent in the e-media version may also appear in this print version.

Institute of Physics
Dirac House, Temple Back
Bristol BS1 6BE UK

Phone: 44 1 17 929 7481
Fax: 44 1 17 920 0979

techtracking@iop.org

Additional copies of this publication are available from:

Curran Associates, Inc.
57 Morehouse Lane
Red Hook, NY 12571 USA
Phone: 845-758-0400
Fax: 845-758-2633
Email: curran@proceedings.com
Web: www.proceedings.com

TABLE OF CONTENTS

IOP Conf. Series: Journal of Physics: Conf. Series **1220** (2019) 011001 doi:10.1088/1742-6596/1220/1/011001

Preface

The 12th International Conference on Excitonic and Photonic Processes in Condensed Matter and Nano Materials

EXCON2018

July 8-13, 2018, NARA, JAPAN

This volume contains a collection of papers of the 12th International Conference on Excitonic and Photonic Processes in Condensed Matter and Nano Materials (EXCON 2018).

EXCON 2018 was held in the historic city of Nara, Japan, on July 8-13, 2018. The previous eleven EXCON conferences were held in Darwin, Australia (1994), Kurort Gohrisch, Germany (1996), Boston, USA (1998), Osaka, Japan (2000), Darwin, Australia (2002), Kraków, Poland (2004), Winston-Salem, USA (2006), Kyoto, Japan (2008), Brisbane, Australia (2010), Groningen, The Netherlands (2012) and Montréal, Canada (2015). The EXCON series of conferences has provided an interdisciplinary forum for mutual research communications among solid-state physicists, photo-physicists, photo-chemists, photo-biologists, material scientists, and device-oriented researchers from academia and industry. A variety of excitonic and photonic phenomena in diverse materials from bulk materials to micro- and nanostructures have been covered in the conferences, such as quantum dots, metallic antennas, atomic-layer materials, bio-materials and molecular materials. The complete list of topics covered by EXCON 2018 is as follows:

- Coherent excitonic processes, including multi-excitonic processes;
- Excitonic and photonic phenomena in nanostructures;
- Excitonic and photonic phenomena in atomic-layer materials;
- Strong light-matter coupling through microcavities, photonic crystals, and plasmonic structures;

IOP Conf. Series: Journal of Physics: Conf. Series **1220** (2019) 011001 doi:10.1088/1742-6596/1220/1/011001

- Exciton-lattice interaction and photo-induced phase transition;
- Excited states in disordered and amorphous systems;
- Excitonic processes in organic materials and polymers;
- High-density excitonic systems and condensation;
- Light-matter interaction in metamaterials;
- Excitonic processes in photovoltaic materials and devices;
- Ultrafast dynamics of excitonic relaxation, energy transport, and charge transfer;
- Applications of excitonic and photonic processes in quantum information processing;
- Optical devices: LEDs and solar cells in inorganic and organic systems;
- Optical manipulation of nanostructures;
- New experimental techniques applied to excitonic and photonic processes;
- New theoretical and computational techniques applied to excitonic and photonic processes.

At EXCON 2018, two outstanding researchers were invited for the plenary lectures to view the front line of the research field covered by this conference. Prof. Chihaya Adachi from Kyushu University presented a talk entitled “Control of excitonic processes in donor-acceptor molecules aiming for high-performance light-emitting devices”, and Prof. Alfred Leitenstorfer from the University of Konstanz gave a talk on “Spatio-temporal quantum physics of light and matter.” There were more than 250 attendees at EXCON 2018, including 16 invited speakers from 13 countries. All the presentations and interdisciplinary discussions greatly contributed to the exchange of new ideas and the definite success of the conference. One notable aspect was the participation of a great number of young researchers. In particular, we had many interesting presentations by students. This conference honored impressive presentations of students with the Best Student Award. This award was won by Ms. Tsugumi Umanodan (Tokyo Institute of Technology, Japan) and Mr. Kenta Ushiro (Osaka City University, Japan).

EXCON 2018 was sponsored by several parties. One co-sponsor was the project of the Grant in Aid for Scientific Research on Innovative Areas, MEXT, Japan: "Nano-Material Manipulation and Structural Order Control with Optical Forces". This project aims to realize optical force technologies for the mechanical manipulation of individual nano-objects, leading to scientific principles of a novel scheme for creating structural order through the optical manipulation of microscopic substances. This conference had a special session of optical force science. Exciting presentations, including five invited talks, presented cutting-edge research in this area, whose origin was the invention of optical tweezers by Dr. Ashkin in 1986. We are greatly pleased that Dr. Ashkin won the Nobel prize in the same year as EXCON 2018. During the conference, a prize ceremony was held by the International Consultative Committee for Condensed Matter Photo-Science, which is officially supported by Spectra-Physics Corporation. The Spectra-Physics Prize has been established to honor an outstanding exciton researcher who achieved original and novel experimental research results with pioneering features, and it was awarded to Prof. Kazunari Matsuda, Kyoto University. This prize greatly encourages young scientists in the fields covered by this conference.

EXCON 2018 received kind support from several sources. We would like to thank the International Advisory Committee and Program Advisory Board for their precious advice throughout the conference's preparation. Furthermore, we should mention that this conference could not have achieved its success without the dedicated work of the members of the local organizing committee, staff members of Nara Visitors Bureau and Polaris secretary office, and students and secretaries in the committee member groups. We would also like to thank the companies that presented exhibition booths, which were indispensable for this conference.

Finally, it is a pleasure to announce that the next conference of the EXCON series will be held in 2021 in Lund, Sweden. We believe in the success of the next EXCON and are very much looking forward to EXCON 2021.

IOP Conf. Series: Journal of Physics: Conf. Series **1220** (2019) 011001 doi:10.1088/1742-6596/1220/1/011001

Hajime Ishihara: Chair of EXCON 2018

Takuya Iida: Guest Editor

Osaka, March 2019

Committees

Conference Chair

Hajime Ishihara: Osaka University/Osaka Prefecture University, Japan

International Advisory Committee

Hajime Ishihara: EXCON2018 Chair; Osaka, Japan

Tadashi Itoh: Osaka, Japan

Safa Kasap: Saskatoon, Canada

Jasper Knoester: Groningen, The Netherlands

Paul van Loosdrecht: Cologne, Germany

Atsuo Matsui: Kobe, Japan

Piotr Petelenz: Kraków, Poland

Gregory Scholes: Princeton, USA

IOP Conf. Series: Journal of Physics: Conf. Series **1220** (2019) 011001 doi:10.1088/1742-6596/1220/1/011001

Yong-Gu Shim: Osaka Prefecture University
Nobuhiko Yokoshi: Osaka Prefecture University

IOP Conf. Series: Journal of Physics: Conf. Series **1220** (2019) 011002 doi:10.1088/1742-6596/1220/1/011002

Peer review statement

All papers published in this volume of *Journal of Physics: Conference Series* have been peer reviewed through processes administered by the proceedings Editors. Reviews were conducted by expert referees to the professional and scientific standards expected of a proceedings journal published by IOP Publishing.

Electron–hole collisions in an atomically thin semiconductor

F Langer[1], C P Schmid[1], S Schlauderer[1], P Nagler[1], C Schüller[1], T Korn[1], M Gmitra[1], J Fabian[1], P G Hawkins[2], U Huttner[2], J T Steiner[2], S W Koch[2], M Kira[3] and R Huber[1]

[1] *University of Regensburg, 93040 Regensburg, Germany*
[2] *University of Marburg, 35032 Marburg, Germany*
[3] *University of Michigan, Ann Arbor, Michigan 48109, USA*

Corresponding author email address: fabian1.langer@ur.de

Abstract. Strong-field biasing of a solid with intense lightwaves leads to simultaneous interband excitation and intraband acceleration of electron–hole pairs. These coupled dynamics result in high-harmonic emission from the bulk solid. For a controlled acceleration of quasiparticles with well-defined initial conditions, we prepare coherent electron–hole pairs by a resonant near-infrared pulse before a strong multi-terahertz field accelerates these entities. The ballistic dynamics manifests itself as high-order sidebands to the near-infrared excitation spectrum. This mechanism allows for the implementation of a quasiparticle collider in order to study those entities in close analogy to conventional collision experiments. Accelerating electrons and holes in a monolayer of a transition metal dichalcogenide extends this scheme to internal quantum degrees of freedom. We show how a strong lightwave can transport electron–hole pairs from one valley to the other faster than one oscillation of the carrier wave, effectively switching the valley pseudospin on a sub-cycle scale. This scheme paves the way to ultimately fast valleytronics.

1. Introduction

Light is an electromagnetic wave and consists of an oscillating electric and magnetic field. Owing to the transient nature of the fields, they may be harnessed as an ultrafast bias for solid state materials. While the photon energy may be well below the fundamental band gap of the semiconductor, the electric field can still promote carriers via Landau-Zener tunnelling [1]. Since the field is only applied for an optical half cycle, large peak amplitudes may be used without damaging the crystal by excessive carrier excitation [2]. These large peak fields can accelerate the carriers throughout a significant portion of the Brillouin zone once they reach a mobile state. If the corresponding Bloch wave even hits the edge of the Brillouin zone, it is Bragg reflected by the periodic lattice and undergoes a Bloch oscillation cycle [3]. Such dynamics can only be observed, however, as long as the coherence of the Bloch state is preserved. Ultrafast lightwave biasing accelerates charges faster than any scattering can occur and, hence, opens the door to studying and exploiting coherent strong-field processes in condensed matter.

Starting in the early 2000s, extremely nonlinear optical phenomena have been demonstrated in the terahertz and mid-infrared domain [4]. The first observation of non-perturbative high-harmonic generation in a solid was reported in 2011 [5]. These experiments have been modelled by pure intraband dynamics, which could explain salient features of the observed characteristics. By excluding the process of interband excitation, however, a major part of the microscopic light-matter matter interaction is left out. Not only the initial carrier distribution is set by the interband process, but it will also contribute to the macroscopic emission [6]. Subsequently, more refined models have been developed [7]-[10].

We have studied high-harmonic emission from bulk gallium selenide resulting from irradiation with intense multi-terahertz (THz) waveforms. The emitted harmonics cover the entire THz to ultraviolet spectral domain of almost 13 optical octaves in a single, phase-stable spectrum [7]. Gaining insight into the microscopic mechanism necessitates resolving the sub-cycle process on its natural timescale. To this end, we developed a cross-correlation technique capable of clocking the high-harmonic emission to the fundamental driving field [9]. These experiments reveal a distinctly different dynamics as compared to gaseous high-harmonic sources. Here, high-harmonic bursts are emitted synchronized to the field crests of one polarity only. The observed dynamics originate from a non-perturbative quantum interference of different excitation pathways, underpinning the influence of interband processes on high-harmonic emission [9]. The intricate mechanism can even be exploited to set the polarization state, the sub-cycle repetition rate, and to shape the high-harmonic carrier field by making use of crystal symmetry [10].

Although the relative contributions of interband excitation and intraband acceleration to high-harmonic emission can be controlled, they should be separated for a controlled acceleration of electrons and holes within a solid. In the following, we will show, how this can be achieved to realize coherent quasiparticle collisions on a sub-cycle timescale and how the scope of lightwave acceleration can be extended to quantum mechanical attributes such as the valley pseudospin in monolayer transition metal dichalcogenides (TMDCs).

2. Coherent electron–hole collisons

The processes of interband excitation and intraband acceleration are most elegantly separated by utilizing two different electromagnetic pulses. While a visible or near-infrared pulse creates coherent electron–hole pairs by resonant interband transitions, a strong lightwave can be used to accelerate these carriers within their bands. This scheme is illustrated in figure 1. When the lightwave drives only intraband currents, it will ballistically accelerate the wave packets during a half cycle (figure 1b). When the polarity of the field changes, however, electrons and holes are decelerated and set on a recollision path. If they reach spatial overlap again, they may annihilate and recombine (figure 1c), emitting their excess energy in the form of electromagnetic radiation. Since this process can occur periodically throughout the duration of the accelerator field, the light emission will be modulated and additional peaks to the resonance spectrum appear. This process is called high-order sideband generation and has been first observed for excitons in quantum wells driven by low-frequency terahertz fields [11].

Here, we demonstrate coherent electron–hole collisions on a sub-cycle timescale in the bulk semiconductor tungsten diselenide (WSe_2) [12]. We inject coherent carriers into this material by addressing the excitonic resonance of bulk WSe_2 with a 100-fs near-infrared pulse. The exciton binding energy amounts to 60 meV [13], which provides stable excitons even at room temperature. Intense THz pulses are collinearly focused on the crystal and are used to ionize, accelerate and recollide the coherent electron–hole pairs. The combined interaction of the near-infrared and the THz pulses results in the emission of a broadband high-order sideband spectrum (figure 2a). The sideband peaks exhibit a spacing of twice the driving THz frequency. By Fourier transformation, this translates into a temporal periodicity

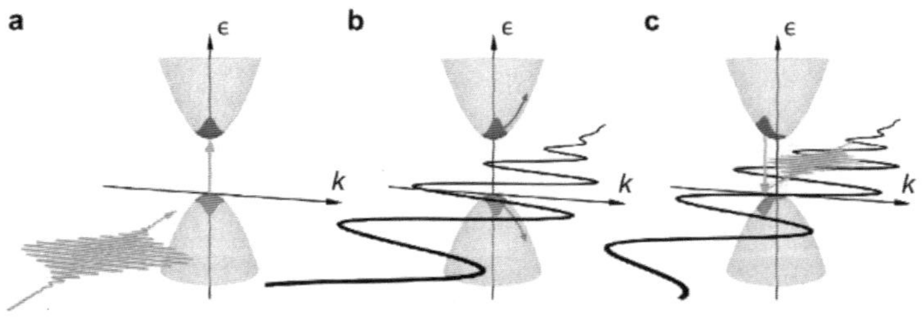

Figure 1. (a) A resonant optical pulse (orange) creates coherent electron–hole pairs (red and blue wave packets) in the conduction and valence band (parabolas), respectively. (b) A strong lightwave (black) accelerates the wave packets within the bands. (c) In the following half cycle, electrons and holes are recollided and emit high-order sideband radiation.

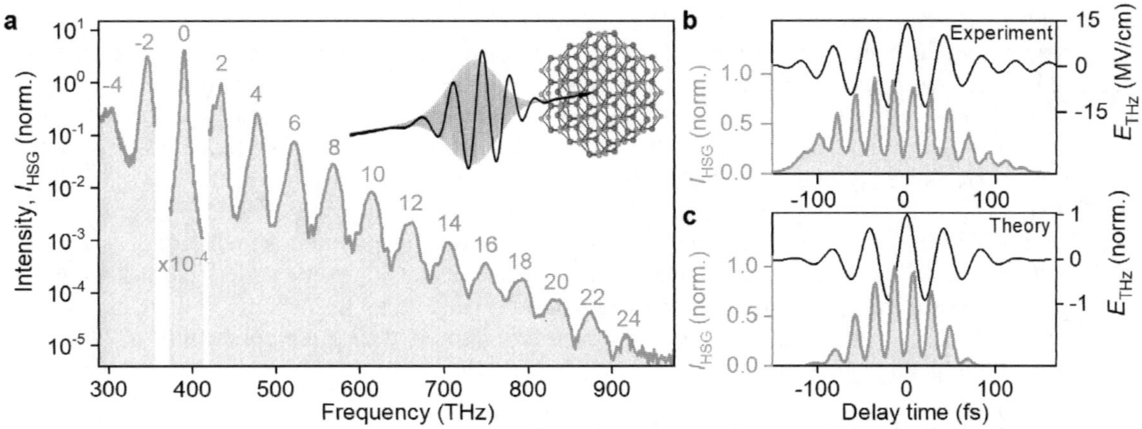

Figure 2. (a) High-order sideband spectrum obtained from bulk tungsten diselenide. The interband resonance is addressed with a 100-fs near-infrared pulse in resonance with the excitonic absorption peak (order 0). When the THz bias is applied even-order sidebands appear, spanning the entire visible-to-ultraviolet spectrum. Inset: Pictogram of the experimental scheme. (b) A sub-cycle injection of excitons with an 8-fs optical pulse allows for tracking the ballistic dynamics. We observe good and bad excitation times (high and low sideband intensity). (c) The experimental results are well reproduced by a quantum mechanical calculation.

of half the driving period. Hence, coherent electron–hole collisions occur for both field polarities equally, in accordance with the inversion symmetry of bulk WSe_2.

For a better understanding of the microscopic mechanism, we prepare coherent electron-hole pairs with an 8-fs near-infrared pulse and track the sub-cycle dynamics of this process. When we delay the ultrashort injection pulse with respect to the THz field, we observe certain delay times t_{ex} of high and low sideband intensity and, hence, good and bad injection times, respectively (figure 2b). Interestingly, the sideband intensity only peaks after the crest of a THz half cycle, evidencing a ballistic origin of the sidebands: electrons and holes have to be injected after the field maximum, so that they are accelerated at first, but the following half cycle is strong enough to invert their motion and to recollide them at high velocity. With a full quantum-mechanical calculation of the light-matter interaction based on the

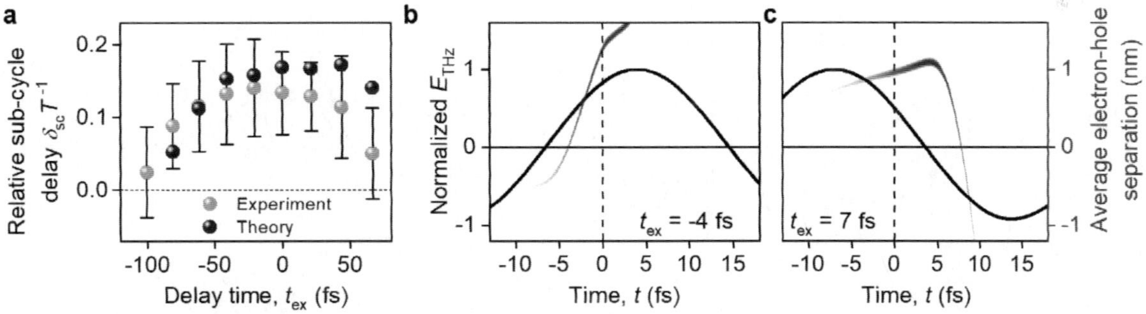

Figure 3. (a) Relative sub-cycle delay of individual sideband peaks with respect to the nearest THz field crest, shown for experiment (orange) and calculation (blue). The error bars depict the standard deviation obtained from 25 consecutive measurements. (b) Calculated trajectory (blue) of an electron with respect to the hole residing at position 0 for a bad excitation time of t_{ex} = -4 fs. The black line shows the THz field and time t = 0 is set to mark the peak intensity of the injection pulse (dashed line). The intensity of the blue curve encodes the density of coherent electron–hole pairs, which are driven apart by the THz field. (c) Same as (b) for a good excitation time of t_{ex} = 7 fs. The trajectory crosses position 0 again, indicating a collision between electrons and holes.

semiconductor Bloch equations, we are able to reproduce the dynamics of the experiment (figure 2c) [12]. Intricate details such as the relative sub-cycle delay of the sideband peaks with respect to the nearest THz field crest, are reproduced in great detail (figure 3a). A cluster expansion approach allows us to include many-body correlations and to track the average electron–hole separation during lightwave acceleration by their mutual correlation [12]. This allows for an extraction of relative electron–hole trajectories, which directly underpin the ballistic acceleration for bad excitation times (figure 3b, no recollision) and good excitation times (figure 3c, recollision).

This approach offers a completely novel way to study many-body excitations in the condensed phase. It brings the powerful tools developed for high-energy particle physics to the solid state and allows for an implementation of the same principles in studying quasiparticles. This scheme may be extended to other correlated phenomena such as Cooper pairs in high-temperature superconductors or Dirac-like fermions in topological insulators.

3. Excitons in monolayer transition metal dichalcogenides

Particularly interesting quasiparticles for future nano-electronic applications are excitons in monolayer transition metal dichalcogenides. This bound state is defined by its centre of mass momentum K and discrete quantum numbers, describing the relative electron–hole motion. Due to the reduced dimensionality and the non-uniform, weak screening in these materials, the excitons exhibit a large binding energy in excess of 0.2 eV and a non-hydrogenic Rydberg series in their excitation spectrum [14]. However, only the optically bright excitons with vanishing centre of mass momentum can be accessed by optical spectroscopy. Here, we study the intra-excitonic transitions, irrespective of the electron–hole pair's centre of mass momentum, by employing phase-locked THz probe pulses, tracking the 1*s*-2*p* resonance. This technique reveals unique and direct insight into the entire exciton ensemble, its many-body interactions, and ultrafast dynamics [15].

In a pump-probe geometry, we resonantly create excitons within the light cone by a 90-fs near-infrared pulse (figure 4a). A THz probe pulse then addresses intra-excitonic transitions, which are tracked by its absorption. With electro-optical detection, we resolve pump-induced changes in the electric field of the THz waveform, which allow us to extract the real part of the optical conductivity $\Delta\sigma_1$ and the dielectric function ε_1. We observe a clear absorption peak in $\Delta\sigma_1$ and a zero-crossing in ε_1, which are caused by the 1*s*-2*p* transition (figure 4b-c). A two-dimensional hydrogen model of the Wannier excitons including nonlocal dielectric screening confirms this assignment and allows for an extraction of the

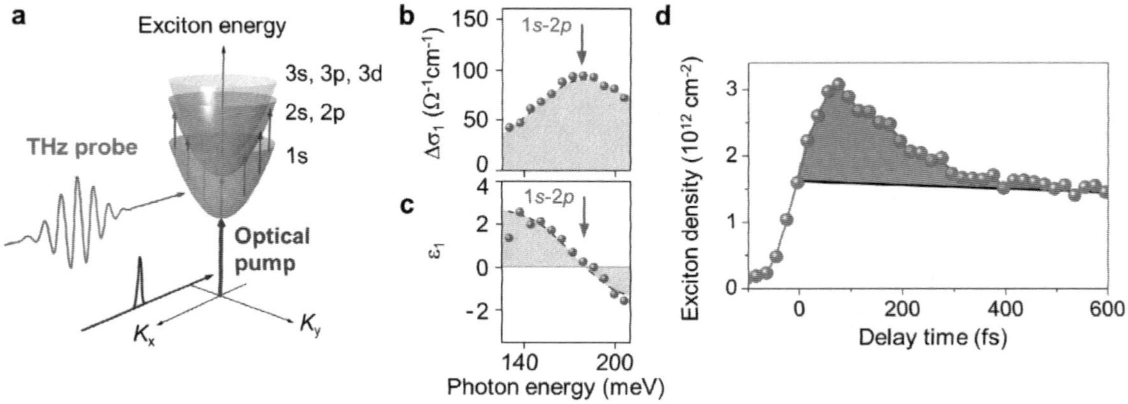

Figure 4. (a) Schematic of the optical-pump-THz-probe experiment exploring internal exciton quantum transitions. (b) Real part of the optical conductivity $\Delta\sigma_1$ measured in the THz region, 275 fs after the optical pump pulse. The excitonic 1*s*-2*p* resonance is clearly observed. (c) Real part of the dielectric function ε_1, evidencing the intra-excitonic resonance. (d) Temporal evolution of the exciton density after optical excitation. The initial ultrafast decay (grey area) stems from radiative recombination, while the slower decay is attributed to non-radiative Auger recombination.

oscillator strength [15]. Furthermore, we directly observe the temporal evolution of the exciton density upon photo-excitation (figure 4d): the exciton density rises quickly, but features an ultrafast decay afterwards, followed by a more moderate decay dynamics. This behaviour is caused by the large dipole moment of excitons in monolayer TMDCs, resulting in an ultrafast radiative decay. On longer timescales, excitons may scatter to optically dark states, where the main mechanism is Auger recombination [15].

These results represent the first-ever direct observation of the ultrafast radiative decay of excitons in monolayer TMDCs, highlighting their strong coupling to light and new possibilities to utilize internal orbital resonances for future quantum control of these materials.

4. Lightwave control of the valley pseudospin

Excitons in monolayer TMDCs also feature an additional degree of freedom, hitherto untapped by lightwave electronics: the valley pseudospin. This quantum mechanical attribute is a spin-like quantity describing whether an electron resides in one of two distinguishable energy minima of the band structure. It can be used as a binary quantum information carrier, opening the field of valleytronics [16]. Here, we demonstrate a sub-cycle control scheme for the valley pseudospin. To this end, we utilize atomically strong THz fields to accelerate and collide electrons and holes in monolayer WSe_2. We read out changes in the pseudospin by the polarization of the emitted high-order sidebands [17].

As a first step, we compare high-order sideband generation in bulk and monolayer WSe_2. Figure 5a illustrates typical spectra. While the bulk crystal (orange curve) exclusively shows even orders, also odd orders appear in the monolayer spectrum, when the semiconductor is biased along its zigzag direction (red curve). In the armchair case (blue curve), virtually no odd orders are observed. The electronic structure in reciprocal space reveals that the zigzag direction connects K and K' valleys (inset to figure 5a), while the armchair direction only connects valleys of the same type. The odd orders, therefore, directly reflect the influence of the valley pseudospin. Additionally, the optical selection rules connected with the valley degree of freedom selectively rotate the polarisation state of the odd orders perpendicular to the driving field (figure 5a, arrows) [17].

To determine, if we can switch the valley pseudospin on a sub-cycle timescale, we prepare coherent excitons selectively in the K valley by a right-handed circularly polarized light pulse (figure 5b, orange

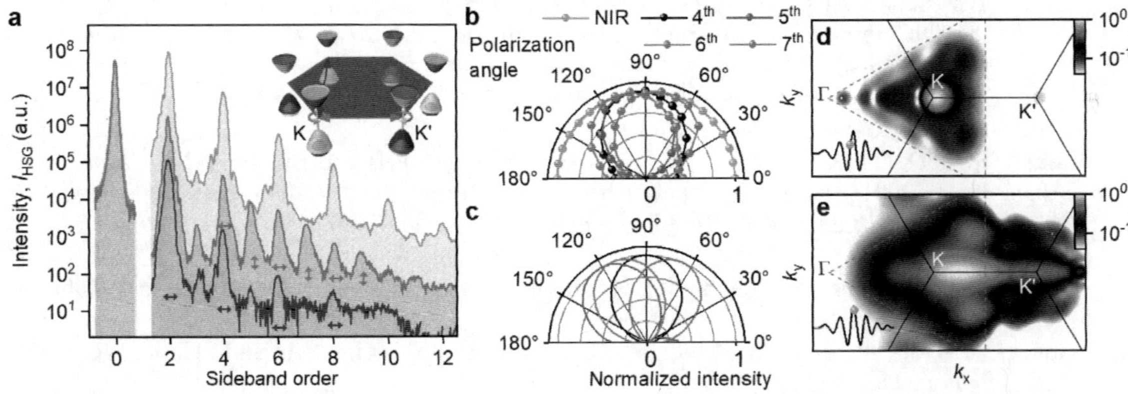

Figure 5. (a) High-order sideband spectra generated from bulk (orange) and monolayer WSe_2 (red and blue, polarization indicated by arrows). Inset: Hexagonal Brillouin zone of WSe_2 with K and K' valleys, as well as the zigzag (red) and the armchair direction (blue). (b) Sideband polarization after valley-selective excitation using a circularly polarized injection pulse (orange). All sidebands are elliptically polarized (black and red spheres). (c) A quantum mechanical calculation models and reproduces the experiment. (d) Snapshot of the coherent electron–hole polarization p_k near a zero crossing of the field (indicated in the inset). The electrons and holes are created in the K valley. (e) After a few femtoseconds, the electric field has accelerated a pronounced ledge to the K' valley, effectively switching the valley pseudospin of those electron–hole pairs.

spheres). The sidebands, however, exhibit a strongly elliptical polarization, resulting from admixtures of the non-excited valley with opposite helicity. We quantify the intervalley transport by modelling the microscopic dynamics quantum mechanically. The calculations reproduce the polarization state of the sidebands (figure 5c). The coherent electron–hole polarization p_k is proportional to the density of electrons and holes and allows us to track their excursion in the band structure. Figure 5d illustrates a snapshot near a zero-crossing of the THz field. Here, all excitons remain in the K valley. After a few femtoseconds, the THz field has accelerated a significant portion of electrons and holes towards the K' valley (figure 5e). The peak prevailing in the K valley originates from continued excitation by the 100-fs injection pulse. We predict, however, a transfer fidelity as high as 96% for a 5-fs excitation pulse [17].

5. Summary

In conclusion, we demonstrate that THz pulses can not only probe many-body excitations in solids extremely sensitively, but they can also control their dynamics and quantum properties. We accelerate and collide coherent excitons in bulk WSe_2, demonstrating quasiparticle collisions on a sub-cycle scale. We expect this to be a useful tool in the investigation of emergent and correlated effects in condensed matter. Going to the monolayer limit of WSe_2, we study the internal structure and ultrafast radiative decay dynamics of optically excited excitons using ultrasensitive THz probing. In a second step, we extend the scope of lightwave acceleration in solids to quantum degrees of freedom and introduce a sub-cycle control scheme of the valley pseudospin in monolayer TMDCs. This paves the way to ultimately fast quantum information processing in nano-materials.

References

[1] Landau L 1932 *Phys. Z. Sowjetunion* **2** 46-51, Zener C 1932 *Proc. R. Soc.* A **137** 696-702
[2] Zener C 1934 *Proc. R. Soc.* A **145** 523-9
[3] Bloch F 1929 *Z. Phys.* **52** 555-600
[4] Chin A H, Calderón A G and Kono J 2001 *Phys Rev. Lett.* **86** 3292-5, Leinß S Kampfrath T, v Volkmann K, Wolf M, Steiner J T, Kira M, Koch S W Leitenstorfer A and Huber R *et al.* 2008 *Phys. Rev. Lett.* **101** 246401, Kampfrath T *et al.* 2011 *Nat. Photon.* **5** 31-4, Liu M *et al.* 2012 Nature **487** 345-8, Junginger F, Mayer B, Schmidt C, Schubert O, Mährlein S, Leitenstorfer A, Huber R and Pashkin A 2012 *Phys. Rev. Lett.* **109** 147403
[5] Ghimire S, DiChiara A D, Sistrunk E, Agostini P, DiMauro L F and Reis D A 2011 *Nat. Phys.* **7** 138-41
[6] Golde D, Meier T and Koch S W 2008 *Phys. Rev.* B **77** 075330
[7] Schubert O *et al.* 2014 *Nat. Photon.* **8** 119-23
[8] Vampa G, McDonald C R, Orlando G, Klug D D, Corkum P B and Brabec T 2014 *Phys. Rev. Lett.* **113** 073091, Higuchi T, Stockman M I and Hommelhoff P 2014 *Phys. Rev. Lett.* **113** 213901, Wu M, Ghimire S, Reis D A, Schafer K J and Gaarde M B 2015 *Phys. Rev.* A **91** 043839
[9] Hohenleutner M, Langer F, Schubert O, Knorr M, Huttner U, Koch S W, Kira M and Huber R 2015 *Nature* **523** 572-5
[10] Langer F, Hohenleutner M, Huttner U, Koch S W, Kira M and Huber R 2017 *Nat. Photon.* **11** 227-31
[11] Zaks B, Liu R B and Sherwin M S 2012 *Nature* **483** 580-3
[12] Langer F *et al.* 2016 *Nature* **533** 225-9
[13] Beal A R, Knights J C and Liang W Y 1972 *J Phys* C **5** 3540-51
[14] Chernikov A, Berkelbach T C, Hill H M, Rigosi A, Li Y, Aslan O B, Reichmann D R, Hybertsen M S and Heinz T F 2014 *Phys. Rev. Lett.* **113** 076802
[15] Poellmann C *et al.* 2015 *Nat. Mater.* **14** 889-93
[16] Xu X, Yao W, Xiao D and Heinz T F 2014 *Nat. Phys.* **10** 343-50, Schaibley J R, Yu H, Clark G, Rivera P, Ross J S, Seyler K L, Yao W and Xu X 2016 *Nat. Rev. Mat.* **1** 16055
[17] Langer F *et al.* 2018 *Nature* **557** 76-80

IOP Conf. Series: Journal of Physics: Conf. Series **1220** (2019) 012002 doi:10.1088/1742-6596/1220/1/012002

Giant-dipole excitons in cuprous oxide

M Kurz and S Scheel

Institut für Physik, Universität Rostock, Albert-Einstein-Straße 23, 18059 Rostock, Germany

E-mail: markus.kurz@uni-rostock.de

Abstract. In this work we present the interaction potentials and eigenenergies of a novel species of Wannier excitons when exposed to crossed electric and magnetic fields. More precisely, we present the theory of giant-dipole excitons in Cu_2O. We derive an exact formulation of the field-dressed excitonic Hamiltonian and exemplarily calculate the excitonic eigenenergies in an exact diagonalization approach for external field strengths of $B = 4\,\mathrm{T}$ and $E = 1\,\mathrm{kV/cm}$. For this particular field configuration, we obtain level spacings between $1.14\,\mu\mathrm{eV}$ and $77.6\,\mu\mathrm{eV}$.

1. Introduction

In semiconductor physics, a bound state between an excited electron in the conduction band and a positively charged hole in the valence band is known as exciton. As the interaction between both constituents can be modeled by a screened Coulomb interaction, an exciton is often considered to be a solid state analogue to the hydrogen atom [1, 2]. For instance, recent measurements of highly excited Rydberg excitons up to principal quantum numbers $n = 25$ have attracted quite some attentions [3]. However, the simple hydrogen model is incapable in reproducing the experimentally observed level-splitting due to fine- and hyperfine splitting, thus it has been expanded by considering both the complex level structure and the cubic symmetry O_h of Cu_2O [4]. In the presence of external electric and magnetic fields the symmetry of the excitonic states is further reduced, leading to level structure providing numerous complex splittings of excitonic absorption lines [5, 6].

In atomic physics, an exotic state of highly excited Rydberg states in crossed external fields are so-called giant-dipole states [7]. When the separation of relative and center-of-mass degrees of freedom are treated correctly, the effect of the center-of-mass motion on the internal degrees of freedom is an effective potential that gives rise to an outer potential well for certain applied field strengths. Bound states in these outer potential wells are characterized be large spatial electron-hole separations up to several micrometers, leading to huge permanent electric dipole moments. However, the concept of giant-dipole states is not only restricted to atomic systems but, in principle, to any neutral bound quasi-particle species such as excitons as well. In a recently published work [8], a sufficient theoretical description of excitons exposed to external fields in Cu_2O has been derived. In this work the exact separation between gauge-dependent and gauge-independent terms was performed, giving the possibility of identifying both the kinetic as well as the exact single-particle potential for internal degrees of freedom.

In this paper, we present a brief derivation of the irreducible tensor representation of the excitonic Hamiltonian in external electric and magnetic fields. Following this, we derive giant-dipole potential surfaces for various external field strengths. Furthermore, within an exact diagonalization approach we calculate excitonic eigenspectra [9].

IOP Conf. Series: Journal of Physics: Conf. Series **1220** (2019) 012002 doi:10.1088/1742-6596/1220/1/012002

2. The excitonic Hamiltonian in external fields

We consider excitons in Cu_2O which are formed by an electron in the energetically lowest Γ_6^+-conduction band and a positively charged hole in the uppermost Γ_5^+-valence band, where the latter is triply degenerate. These three bands are deformed due to interband interactions and non-spherical symmetry properties of the solid. However, an effective description can be achieved by an $I = 1$ quasi-spin representation in the hole degrees of freedom [10]. For the electron one finds a nearly parabolic dispersion relation, while the kinetic energy Hamiltonian for the hole is a more complex expression due to the three coupled valence bands

$$H_e(\boldsymbol{p_e}) = \frac{\boldsymbol{p}_e^2}{2m_e}, \quad H_h(\boldsymbol{p_h}) = \frac{\boldsymbol{p}_h^2}{2m_0}(\gamma_1 + 4\gamma_2) - \frac{3\gamma_2}{m_0}(p_{h,x}^2 I_x^2 + \text{c.p.}) - \frac{6\gamma_3}{m_0}[p_{h,x}p_{h,y}\{I_x I_y\} + \text{c.p.}]. \quad (1)$$

The hole Hamiltonian is characterized by three Luttinger parameters $\gamma_i,\ i = 1, 2, 3$ [11]. The mapping $\{ab\} \equiv (ab + ba)/2$ is called the symmetric product and c.p. denotes the cyclic permutations of terms. Furthermore, the quasi-spin $\boldsymbol{I}$ couples to the hole-spin $\boldsymbol{S}_h$ via

$$H_{\text{so}} = \frac{2}{3}\Delta\left(1 + \boldsymbol{S}_h \cdot \boldsymbol{I}\right), \quad (2)$$

while the electron spin $\boldsymbol{S}_e$ is not considered in this work.

In the case that an external magnetic field is applied, the canonical momenta $\boldsymbol{p}_{e,h}$ are replaced by $\boldsymbol{p}_{e,h} \rightarrow \boldsymbol{p}_{e,h} \pm \boldsymbol{A}(\boldsymbol{r}_{e,h})$, where $\boldsymbol{A}(\boldsymbol{r})$ is the magnetic vector potential. The coupling of $\boldsymbol{S}_h$ and $\boldsymbol{I}$ to the external magnetic field is given by

$$H_{\text{B}} = \mu_B[(3\kappa + \frac{g_s}{2})\boldsymbol{I} \cdot \boldsymbol{B} - g_s \boldsymbol{S}_h \cdot \boldsymbol{B}]. \quad (3)$$

If an external homogeneous electric field is applied, the stark terms $\mp\boldsymbol{E}\cdot\boldsymbol{r}_{e,h}$ have to be added to the Hamiltonian. If one introduces the relative coordinate $\boldsymbol{r} = \boldsymbol{r}_e - \boldsymbol{r}_h$ and center-of-mass vector $\boldsymbol{R}$, one can perform a pseudoseparation of both degrees of freedom using the eigenstates and eigenvalues $\boldsymbol{K}$ of the so-called pseudomomentum $\hat{\boldsymbol{K}} = \boldsymbol{P} - (\boldsymbol{B} \times \boldsymbol{r})/2$ [7]. As it has been shown in previous studies, the excitonic Hamiltonian can then be transformed into a single-particle Hamiltonian $H_{\text{ex}} = T + V_{\text{gd}}$ with

$$\begin{aligned} T &= H_e(\boldsymbol{\pi}) + H_h(\boldsymbol{\pi}), \quad \pi_i = p_i - qA_i(\boldsymbol{r}) - \sum_k \left(\frac{m_h}{M}\delta_{ki} - \Omega_{ik}\right)\tilde{K}_k, \quad I_{ij} = 3\{I_i I_j\} - 2\delta_{ij}, \\ V_{\text{gd}}(\boldsymbol{r}) &= \left(\Omega_1 \tilde{K}^2 + \boldsymbol{E}\cdot\boldsymbol{r} - \frac{1}{r}\right)1_I - \Omega_2 \sum_i \tilde{K}_i^2 I_{ii} - \frac{2}{3}\Omega_3 \sum_{ij,j<i} \tilde{K}_i \tilde{K}_j I_{ij} + H_{\text{so}} + H_{\text{B}}, \end{aligned} \quad (4)$$

$q = (m_e - m_h)/M$, $M = m_e + m_h$ and $\tilde{\boldsymbol{K}} = \boldsymbol{K} + \boldsymbol{B} \times \boldsymbol{r}$ [8]. Here, the term V_{gd} represents a single-particle potential for the internal motion of the field-dressed excitons.

3. Excitonic giant-dipole potential surfaces and eigenenergies

Due to the spin-1 and spin-1/2 degrees of freedom, the potential V_{gd} can be expressed as a 6×6 matrix, where the matrix elements are functions of the external field parameters $\boldsymbol{B}||[100]$, $\boldsymbol{E}||[001]$ and the spatial coordinate $\boldsymbol{r}$, respectively. The eigenvalues $V_i(\boldsymbol{r}),\ i = 1, ..., 6$ are three-dimensional potential surfaces which parametrically depend on the external field strengths.

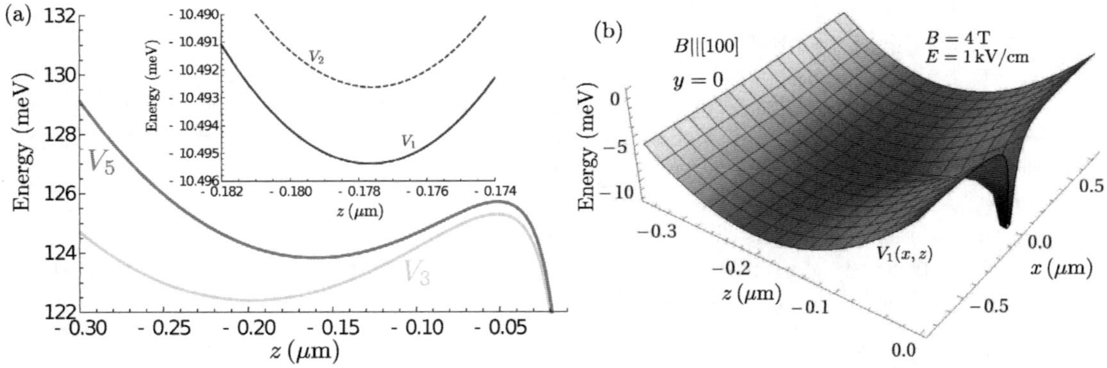

Figure 1. (a) One-dimensional potential curves for $x = y = 0$. The potentials V_3 and V_5 clearly possess local minima at $z \approx -0.2$ μm and $z \approx -0.16\,\mu$m. The inset shows the curves V_1 and V_2, respectively. For both curves we find minima near $z \approx -0.1775\,\mu$m. In (b) the two-dimensional surface $V_1(x, z)$ is shown for $y = 0$.

In Fig. 1(a) we show a one dimensional cut through the six potential curves along the negative z-direction for applied field strengths of $B = 4$ T and $E = 1$ kV/cm. The main figure shows the potential curves $V_3(z)$ (green solid curve) and $V_5(z)$ (red solid curve), respectively. Obviously, both potential curves possess local minima at $z \approx -0.2\,\mu$m and $z \approx -0.16\,\mu$m. This large electron-hole separation results in a huge electric dipole moment, justifying the expression excitonic giant-dipole states. The potential curves $V_4(z)$ and $V_6(z)$ are not shown in the main figure as they are not resolved on this energy scale. However, in the inset we present both $V_1(z)$ (solid blue line) and the adjacent curve $V_2(z)$ (dashed blue line). As one can see, these two curves possess local minima at $z \approx -0.1775\,\mu$m with an energetic separation of around $3\,\mu$eV. In Fig. 1(b) the two-dimensional potential surface $V_1(x, z)$ is shown for fixed $y = 0$. In this plot one clearly sees the local potential minimum at $x = 0, z = -0.1775\,\mu$m.

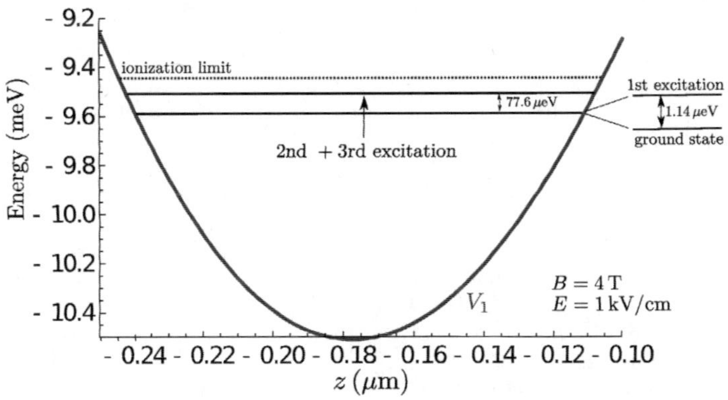

Figure 2. Excitonic eigenenergies for $B = 4$ T and $E = 1$ kV/cm. We find four bound states with an energetic separation between $1.14\,\mu$eV and $77.6\,\mu$eV.

For the calculation of the excitonic eigenenergies, we apply an exact diagonalization approach. In particular, in case the couplings among the single potential surfaces are neglected, one can derive analytic solutions for the bound states in the outer potential wells. Together with the basis function $|1, m\rangle$, $m = -1, 0, 1$ and $|1/2, \pm 1/2\rangle$ for the spin-1 and spin-1/2 Hilbert spaces, these functions can be used to define a basis set for the exact diagonalization of the excitonic Hamiltonian given by Eq. (4). As we are mostly interested in the energetically low lying states, we have used the giant-dipole functions of the potential surface V_1 for the exact diagonalization.

In Fig. 2 we present the excitonic eigenenergies for applied field strengths of $B = 4\,\mathrm{T}$ and $E = 1\,\mathrm{kV/cm}$. For comparison, we show a one-dimensional cut through the energetically lowest potential surface V_1 as well. For this surface we also indicate the ionization limit, i.e. the energy above which all excitonic states lie in the continuum. For the applied field configuration this energy is given by $-9.43\,\mathrm{meV}$, whereby the ground state energy is given by $-9.58\,\mathrm{meV}$. In total, we find four bound excitonic giant-dipole states. Their energetic spacing is determined by two energy scales. The first one is given by $1.14\,\mu\mathrm{eV}$ which is the spacing between the ground state and the first excited state as well as between the second and third excitation. This spacing is not resolved in the main figure but presented in the inset where we indicate the energetic separation between the ground state and the first excited state. The second energy scale is given by $77.6\,\mu\mathrm{eV}$ and is much larger than the first one. This is the spacing between the first and second excited state and is clearly visible in Fig. 2. As all bound states are quite close to the ionization limit, no more bound states are found for this particular field configuration.

In general, we find that bound excitonic states are only found for sufficiently strong magnetic fields as in this case the outer potential minima are deep enough to provide bound states. For instance, for $E = 1\,\mathrm{kV/cm}$ giant-dipole excitons are only found for $3.4\,\mathrm{T} \leq B \leq 4.4\,\mathrm{T}$ where we find between one and six bound states with energetic level spacings in the range of 0.6 ... 100 μeV.

4. Summary and conclusions

In the present article, we have presented a theoretical description of field-dressed excitons in Cu_2O. In particular, we have shown the exact kinetic and potential energy terms which can be derived from a pseudoseparation of the relative and center-of-mass degrees of freedom. We obtain potential energy surfaces for the internal motion, possessing local minima providing bound excitonic giant-dipole states. In future studies one might address the experimental preparation and verification of excitonic giant-dipole states. One possible approach may be to use time-dependent external fields to apply an adiabatic state transfer of highly excited Rydberg excitons into giant-dipole states. The experimental verification could be achieved via microwave spectroscopy and via the direct measurement of the large electric dipole moment.

Acknowledgments

We acknowledge support by the Focus Programme SPP 1929 GiRyd by the Deutsche Forschungsgemeinschaft (DFG).

References

[1] Frenkel J 1931 *Phys. Rev.* **37** 1276
[2] Mott N F 1938 *Trans. Faraday Soc.* **34** 500
[3] Kazimierczuk T *et al.* 2014 *Nature (London)* **514** 343
[4] Schöne F *et al.* 2016 *Phys. Rev. B* **93** 075203
[5] Heckötter J *et al.* 2017 *Phys. Rev. B* **95** 035210
[6] Schweiner F *et al.* 2016 *Phys. Rev. B* **95** 035202
[7] Dippel O, Schmelcher P, and Cederbaum L S 1994 *Phys. Rev. A* **49** 4415
[8] Kurz M Grünwald P, and Scheel S 2017 *Phys. Rev. B* **95** 245205
[9] Kurz M and Scheel S *(to be published)*
[10] Suzuki K, and Hensel J C 1974, *Phys. Rev. B* **9** 2731
[11] Luttinger J M 1956 *Phys. Rev.* **102** 1030

IOP Conf. Series: Journal of Physics: Conf. Series **1220** (2019) 012003 doi:10.1088/1742-6596/1220/1/012003

Optical observation of fermionic partons in Kitaev spin balls

Shoji Yamamoto and Taku Kimura

Department of Physics, Hokkaido University, Sapporo 060-0810, Japan

E-mail: yamamoto@phys.sci.hokudai.ac.jp

Abstract. We demonstrate a Kitaev spin liquid in a polyhedral cluster and propose detecting its fractional excitations—Majorana spinons—by magnetic Raman scattering. While little polarization dependence of the Raman spectra at sufficiently low temperatures is usual with quantum spin liquids, the present observations are primarily of geometric origin. The Raman scattering intensity peaks melt with increasing temperature, to be more precise, with increasing number of background gauge-flux excitations.

1. Introduction

Quantum spin liquids (QSLs) [1] have no conventional magnetic order but feature fractional excitations—partons—and emergent gauge degrees of freedom [2]. The Kitaev model [3] is exactly solvable to have a QSL ground state and its elementary excitations consist of itinerant Majorana fermions and local gauge vortices, called spinons and visons, respectively. While Kitaev-type interactions are found in some honeycomb iridates [4], they can be defined on any lattice with coordination number three. We consider a Kitaev spin liquid in "zero dimension" in an attempt to identify and characterize Majorana spinons on a rigorous basis. Magnetic Raman scattering is an excellent probe to detect Majorana spinons separately from visons.

2. Kitaev Hamiltonian

We consider the Kitaev Hamiltonian on a dodecahedral lattice of $L \equiv 20$ sites

$$\mathcal{H} = -\sum_{\lambda=x,y,z}\sum_{<i,j>_\lambda} J_\lambda \sigma_i^\lambda \sigma_j^\lambda;\ [\sigma_i^\lambda, \sigma_j^\mu] = 2\mathrm{i}\delta_{ij}\varepsilon_{\lambda\mu\nu}\sigma_i^\nu, \tag{1}$$

where $<i,j>_\lambda$ runs over nearest neighbors with λ depending on the (i,j) pair [Fig. 1(a)]. We set J_λ's all equal to J in what follows. Let

$$\sigma_i^\lambda = \mathrm{i}\eta_i^\lambda c_i;\ c_i^\dagger = c_i,\ (\eta_i^\lambda)^\dagger = \eta_i^\lambda;\ \{\eta_i^\lambda, \eta_j^\mu\} = 2\delta_{ij}\delta_{\lambda\mu},\ \{c_i, c_j\} = 2\delta_{ij},\ \{\eta_i^\lambda, c_j\} = 0. \tag{2}$$

Then the Hamiltonian (1) reads

$$\mathcal{H} = \mathrm{i}\sum_{\lambda=x,y,z}\sum_{<i,j>_\lambda} J_\lambda \hat{u}_{<i,j>_\lambda} c_i c_j;\ \hat{u}_{<i,j>_\lambda} \equiv \mathrm{i}\eta_i^\lambda \eta_j^\lambda. \tag{3}$$

Since the bond operators $\hat{u}_{<i,j>_\lambda}$ commute with the Hamiltonian as well as each other, they are treatable as classical $\mathbb{Z}_2$ variables, $u_{<i,j>_\lambda} = \pm 1$. Every anticlockwise multiplication—viewed from the outside of the dodecahedron—of the spin operators around pentagonal plaquette p,

$$\hat{W}_p \equiv \prod_{<i,j>_\lambda \in \partial p} \sigma_i^\lambda \sigma_j^\lambda = (-\mathrm{i})^5 \prod_{<i,j>_\lambda \in \partial p} \hat{u}_{<i,j>_\lambda}\ (p = 1, 2, \cdots, 3L/5), \tag{4}$$

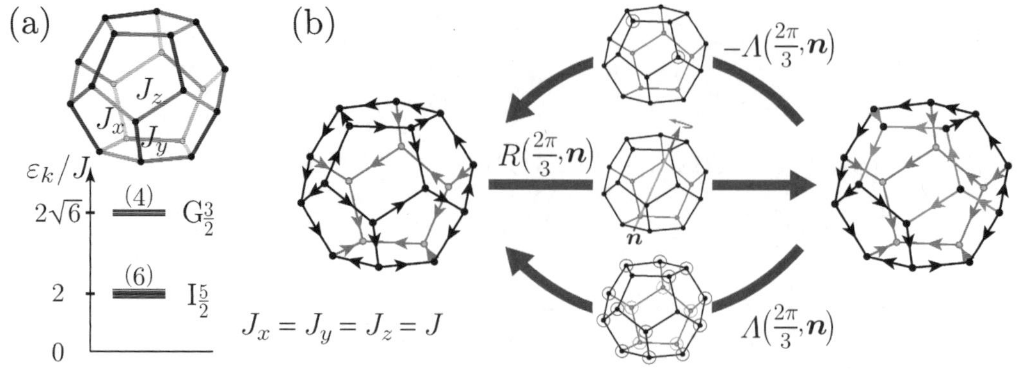

Figure 1. (a) Kitaev model on a dodecahedron and its quasiparticle eigenvalues ε_k, each labeled its degeneracy and belonging representation. (b) A projective symmetry group operation on the gauge-ground Kitaev dodecahedron, where $u_{<i,j>_\lambda}$ reads ± 1 for the $i \rightleftharpoons j$ arrows, respectively.

is also commutable with the Hamiltonian and thus treatable as a $\mathbb{Z}_2$ classical variable with eigenvalues $W_p = \pm\mathrm{i}$. A matter fermion c_i "feels" a gauge field $\{W_p\}$. A local flux excitation against the ground-state W_p configuration is called a vison.

While the original Hilbert space is block-diagonalized with every W_p configuration, there remains a variety of $u_{<i,j>_\lambda}$ configurations within each block. This is because we introduce four Majorana fermions at each site at the cost of redundant degrees of freedom. The ground state of the Kitaev dodecahedron is given by a uniform set of W_p's. Given a W_p configuration, all the possible sets of $\mathbb{Z}_2$ bond variables are degenerate in energy and therefore we may choose any of them. Given a $u_{<i,j>_\lambda}$ configuration, we can diagonalize the Majorana Hamiltonian into

$$\mathcal{H} = \sum_{k=1}^{L/2} \frac{\varepsilon_k}{2}\left(\alpha_k^\dagger \alpha_k - \alpha_k \alpha_k^\dagger\right) = \sum_{k=1}^{L/2} \varepsilon_k \left(n_k - \frac{1}{2}\right);\ n_k = \alpha_k^\dagger \alpha_k,\ c_i = \sum_{k=1}^{L/2} (\psi_{ik}\alpha_k + \psi_{ik}^* \alpha_k^\dagger), \quad (5)$$

where c_i's are recomplexified into α_k's to obtain the eigenvalues ε_k which are positive definite.

3. Projective symmetry group

Before proceeding to optical observation, we make a symmetry argument of this Kitaev spin ball [Fig. 1(b)]. While a dodecahedron belongs to the point group $\mathbf{I}_\mathrm{h}$, the dodecahedral Kitaev spin model (3) is not invariant under $\mathbf{I}_\mathrm{h}$ group operations. Even if we choose bond variables so as to give the uniform W_p configuration, neither simple rotation nor inversion keeps these local gauge fields invariant. Then we extend a simple point group by gauge transformations [5].

Suppose we rotate the gauge-ground Kitaev Hamiltonian anticlockwise about a three-fold axis, say $\boldsymbol{n}$, of the dodecahedron by an angle of $\frac{2\pi}{3}$, which is denoted by $R(\frac{2\pi}{3},\boldsymbol{n})$. Then, bond variables may be changed but the ground W_p configuration remains unchanged. Any two sets of bond variables giving the same W_p configuration can be converted to each other by local gauge transformations. The gauge transformation $c_i \to -c_i$ reads $u_{<i,j>_\lambda} \to -u_{<i,j>_\lambda}$ with j running over the nearest neighbor sites of site i. $\Lambda(\frac{2\pi}{3},\boldsymbol{n})$ operates on 18 sites, while $-\Lambda(\frac{2\pi}{3},\boldsymbol{n})$ reads operating on all the other sites. Both indeed recover the initial bond configuration. Every rotation $R \subset \mathbf{I}$ can be followed by two sets of local gauge transformations to keep any ground-state bond-variable configuration invariant. Since the inversion reverses all the pentagonal plaquette fluxes W_p, any gauged inversion cannot restore the initial bond-variable configuration. The symmetry group of the Kitaev dodecahedron is thus given by a gauge extension of the pure rotation subgroup of the icosahedral group $\mathbf{I}_\mathrm{h}$, which is denoted by $\widetilde{\mathbf{I}}$. This extended group is double valued, hence called a projective symmetry group (PSG) [6]. If we apply the gauged $\frac{2\pi}{3}$ rotation three times to any bond-variable configuration, it is multiplied by -1, which is reminiscent of the case with a half-odd integer spin. Note that $\widetilde{\mathbf{I}} \subset \mathrm{SU}(2)$ and $\mathbf{I} \subset \mathrm{SO}(3)$.

We can characterize every eigenstate of the Hamiltonian (5) in terms of this symmetry group unless the uniform W_p configuration is broken. Every eigenstate of the gauge-ground Kitaev dodecahedron belongs to any of the irreducible representations of $\widetilde{\mathbf{I}}$ and the degeneracy of each eigenvalue is given by the dimension of its belonging representation [Fig. 1(a)].

4. Magnetic Raman scattering

Within the Loudon-Fleury perturbation theory [7,8], magnetic Raman intensity reads

$$I(\omega)=\frac{1}{2\pi\hbar L}\int_{-\infty}^{\infty}\mathrm{d}t\,\mathrm{e}^{\mathrm{i}\omega t}\langle \mathrm{e}^{\frac{\mathrm{i}\mathcal{H}t}{\hbar}}\mathcal{R}\mathrm{e}^{-\frac{\mathrm{i}\mathcal{H}t}{\hbar}}\mathcal{R}\rangle_T;\ \mathcal{R}=-\sum_{\lambda=x,y,z}\sum_{<i,j>_\lambda}J_\lambda(\boldsymbol{e}_{\mathrm{in}}\cdot\boldsymbol{d}_{ij})(\boldsymbol{e}_{\mathrm{sc}}\cdot\boldsymbol{d}_{ij})\sigma_i^\lambda\sigma_j^\lambda, \quad (6)$$

where $\boldsymbol{e}_{\mathrm{in}}$ ($\boldsymbol{e}_{\mathrm{sc}}$) is the polarization vector of incident (scattered) light, while $\boldsymbol{d}_{ij}$ is the vector connecting the ith and jth sites. Having in mind that $[\mathcal{R},\hat{u}_{<i,j>_\lambda}]=0$, the magnetic Raman operator is expressed in terms of the fermionic quasiparticles as

$$\mathcal{R}=\frac{1}{2}\sum_{k,l=1}^{L/2}\left[X_{kl}(2\alpha_k^\dagger\alpha_l-\delta_{kl})+Y_{kl}\alpha_k^\dagger\alpha_l^\dagger+Y_{lk}^*\alpha_k\alpha_l\right];$$
$$X_{kl}=\sum_{\lambda=x,y,z}\sum_{<i,j>_\lambda}\mathrm{i}J_\lambda(\boldsymbol{e}_{\mathrm{in}}\cdot\boldsymbol{d}_{ij})(\boldsymbol{e}_{\mathrm{sc}}\cdot\boldsymbol{d}_{ij})u_{<i,j>_\lambda}[\psi_{ik}^*\psi_{jl}-\psi_{jk}^*\psi_{il}],$$
$$Y_{kl}=\sum_{\lambda=x,y,z}\sum_{<i,j>_\lambda}\mathrm{i}J_\lambda(\boldsymbol{e}_{\mathrm{in}}\cdot\boldsymbol{d}_{ij})(\boldsymbol{e}_{\mathrm{sc}}\cdot\boldsymbol{d}_{ij})u_{<i,j>_\lambda}[\psi_{ik}^*\psi_{jl}^*-\psi_{jk}^*\psi_{il}^*]. \quad (7)$$

Note that $[\mathcal{R},\mathcal{H}]\neq 0$ in general. Neglecting the Rayleigh and anti-Stokes terms, the scattering intensity mediated by matter fermions with a particular W_p configuration is calculated as

$$I_{\{W_p\}}(\omega)=\frac{1}{2L}\sum_{k,l=1}^{L/2}\Big[2|X_{kl}|^2n_k(1-n_l)\delta(\hbar\omega+\varepsilon_k-\varepsilon_l)+|Y_{kl}|^2(1-n_k)(1-n_l)\delta(\hbar\omega-\varepsilon_k-\varepsilon_l)\Big]. \quad (8)$$

Then we take a thermal average to obtain the spectrum,

$$I(\omega)=\frac{1}{Z}\sum_{\{W_p\}}\sum_{\{n_k\}}{}_{\{W_p\}}\langle\{n_k\}|\mathrm{e}^{-\beta\mathcal{H}}I_{\{W_p\}}(\omega)\mathcal{P}|\{n_k\}\rangle_{\{W_p\}}\equiv\langle I_{\{W_p\}}(\omega)\rangle_T, \quad (9)$$

where $Z=\sum_{\{W_p\}}\sum_{\{n_k\}}{}_{\{W_p\}}\langle\{n_k\}|\mathrm{e}^{-\beta\mathcal{H}}\mathcal{P}|\{n_k\}\rangle_{\{W_p\}}$, W_p's each run over $\pm\mathrm{i}$, n_k's each run over 0 and 1, and $\mathcal{P}$ projects out all the unphysical combinations of $\{W_p\}$ and $\{n_k\}$ [9].

Figure 2(a) shows $I(\omega)$ with various polarization vectors in the xy-plane at absolute zero, where Eq. (8) consists only of the second term and therefore every Raman-active mode is characterized by a product representation of two quasiparticle eigenstates belonging to either $\mathrm{I}_{\frac{5}{2}}$ or $\mathrm{G}_{\frac{3}{2}}$ representation. On the other hand, the Raman spectra can be characterized by irreducible representations of the lattice point group $\mathbf{I}_{\mathrm{h}}$. The Kitaev dodecahedron has only one Raman-active mode belonging to the five-dimensional $\mathrm{H_g}$ representation of $\mathbf{I}_{\mathrm{h}}$. When we decompose the Raman operator as $\mathcal{R}=\sum_{i=1}^5 E_{\mathrm{H_g}(i)}\mathcal{R}^{\mathrm{H_g}(i)}$ neglecting the irrelevant Rayleigh component of $\mathrm{A_g}$ symmetry, the contributions $\langle\mathrm{e}^{\frac{\mathrm{i}\mathcal{H}t}{\hbar}}\mathcal{R}^{\mathrm{H_g}(i)}\mathrm{e}^{-\frac{\mathrm{i}\mathcal{H}t}{\hbar}}\mathcal{R}^{\mathrm{H_g}(i)}\rangle_{T=0}$ are all degenerate with each other as long as the ground state is a QSL [10]. Then $I(\omega)$ reads

$$I(\omega)=\frac{1}{2\pi\hbar L}\int_{-\infty}^{\infty}\mathrm{d}t\,\mathrm{e}^{\mathrm{i}\omega t}\langle\mathrm{e}^{\frac{\mathrm{i}\mathcal{H}t}{\hbar}}\mathcal{R}^{\mathrm{H_g}(1)}\mathrm{e}^{-\frac{\mathrm{i}\mathcal{H}t}{\hbar}}\mathcal{R}^{\mathrm{H_g}(1)}\rangle_{T=0}\sum_{i=1}^{5}E_{\mathrm{H_g}(i)}^2, \quad (10)$$

hence comes very little polarization dependence originating merely in the coefficients $E_{\mathrm{H_g}(i)}$. There is no polarization dependence at all in a planar honeycomb Kitaev spin liquid [8], where the Raman operator consists only of the two-dimensional $\mathrm{E_2}$ representation of the point group $\mathbf{C}_{6\mathrm{v}}$ as $\mathcal{R}=\sum_{i=1}^2 E_{\mathrm{E_2}(i)}\mathcal{R}^{\mathrm{E_2}(i)}$ with $E_{\mathrm{E_2}(1)}=(e_{\mathrm{in}}^x e_{\mathrm{sc}}^x-e_{\mathrm{in}}^y e_{\mathrm{sc}}^y)/\sqrt{2}$ and $E_{\mathrm{E_2}(2)}=(e_{\mathrm{in}}^x e_{\mathrm{sc}}^y+e_{\mathrm{in}}^y e_{\mathrm{sc}}^x)/\sqrt{2}$, and satisfy $\sum_{i=1}^2 E_{\mathrm{E_2}(i)}^2=1/2$ as well as $\langle\mathrm{e}^{\frac{\mathrm{i}\mathcal{H}t}{\hbar}}\mathcal{R}^{\mathrm{E_2}(1)}\mathrm{e}^{-\frac{\mathrm{i}\mathcal{H}t}{\hbar}}\mathcal{R}^{\mathrm{E_2}(1)}\rangle_{T=0}=\langle\mathrm{e}^{\frac{\mathrm{i}\mathcal{H}t}{\hbar}}\mathcal{R}^{\mathrm{E_2}(2)}\mathrm{e}^{-\frac{\mathrm{i}\mathcal{H}t}{\hbar}}\mathcal{R}^{\mathrm{E_2}(2)}\rangle_{T=0}$. However, no or little polarization dependence of the magnetic Raman spectra is not necessarily

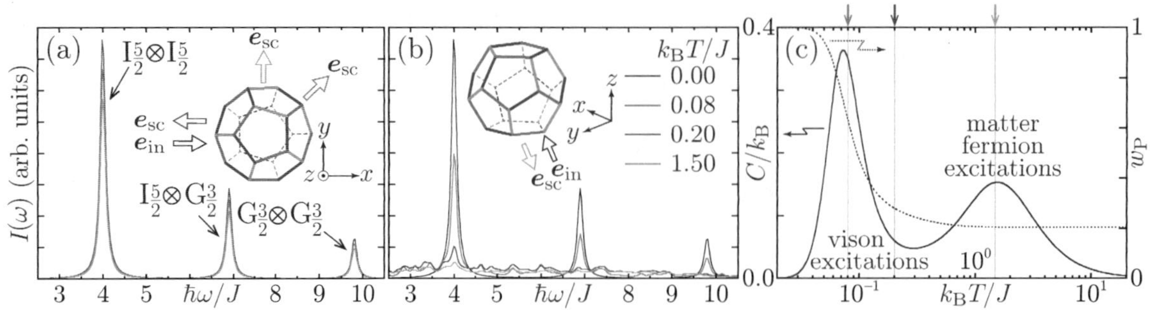

Figure 2. (a) $I(\omega)$ at $T=0$ with $(\vartheta_{\rm in},\varphi_{\rm in})=(\frac{\pi}{2},0)$ and $\vartheta_{\rm sc}=\frac{\pi}{2}$, $\varphi_{\rm sc}=\frac{\pi}{4},\frac{\pi}{2},\pi$; (b) $I(\omega)$ at various T with $(\vartheta_{\rm in},\varphi_{\rm in})=(\frac{\pi}{4},0)$ and $(\vartheta_{\rm sc},\varphi_{\rm sc})=(\frac{3\pi}{4},\pi)$; where $\mathbf{e}_{\rm in}\equiv(1,\vartheta_{\rm in},\varphi_{\rm in})$ and $\mathbf{e}_{\rm sc}\equiv(1,\vartheta_{\rm sc},\varphi_{\rm sc})$. (c) Specific heat C and the flux-sum magnitude $w_{\rm P}$ as functions of T.

the case with QSLs. It is primarily of geometric origin. We will demonstrate elsewhere strongly polarization-dependent magnetic Raman spectra in Kitaev spin clusters of octahedral symmetry. A strong polarization dependence is observed in three-dimensional Kitaev QSLs [10] as well.

Figure 2(b) shows $I(\omega)$ as a function of temperature. As soon as visons are thermally excited, the Hamiltonian (3) no longer belongs to $\widetilde{\mathbf{I}}$. Vison excitations can be measured by the parameter

$$w_{\rm P}=\frac{1}{Z}\sum_{\{W_p\}}\sum_{\{n_k\}}{}_{\{W_p\}}\langle\{n_k\}|{\rm e}^{-\beta\mathcal{H}}\hat{w}_{\rm P}\mathcal{P}|\{n_k\}\rangle_{\{W_p\}};\ \hat{w}_{\rm P}\equiv\frac{5}{3L}\left|\sum_{p=1}^{3L/5}\hat{W}_p\right| \tag{11}$$

and observed through specific heat. Figure 2(c) shows their temperature dependences. $w_{\rm P}=1$ without any vison at $T=0$, while $w_{\rm P}=5040/2048/12=0.205078125$ with visons emerging at random in the $T\to\infty$ limit. At $k_{\rm B}T/J=0.2$, visons are almost fully excited and therefore the symmetry-definite Raman scattering intensity peaks melt away. Matter fermions are excited at much higher temperatures. Specific heat is thus doubly peaked with increasing temperature.

5. Concluding remarks

Gapless algebraic quantum spin liquids [11] such as critical Heisenberg antiferromagnets in one dimension are also attracting much interest in the context of fractional spinon excitations, where Jordan-Wigner fermions behave as elementary spin-$\frac{1}{2}$ entities and they are indeed detected in pairs by inelastic neutron scattering [12]. Our proposal consists of observing fractional magnetic excitations—Majorana spinons—by inelastic light scattering rather than usual magnetic probes. PSG characterization of magnetic Raman spectra mediated by fractional quasiparticles may be available to other QSL nanostructures such as hexagonal fragments. We hope this study will stimulate further chemical explorations of Kitaev spin systems, say, nanoribbons and nanotubes.

Refereneces

[1] Savary L and Balents L 2017 *Rep. Prog. Phys.* **80** 016502
[2] Misguich G 2011 in *Introduction to Frustrated Magnetism* ed C Lacroix *et al.* (Heidelberg: Springer) chapter 16
[3] Kitaev A 2006 *Ann. Phys.* **321** 2
[4] Jackeli G and Khaliullin G 2009 *Phys. Rev. Lett.* **102** 017205
[5] Mellado P, Petrova O and Tchernyshyov O 2015 *Phys. Rev.* B **91** 041103(R)
[6] Wen X G 2002 *Phys. Rev.* B **65** 165113
[7] Fleury P A and Loudon R 1968 *Phys. Rev.* **166** 514
[8] Knolle J, Chern G W, Kovrizhin D L, Moessner R and Perkins N B 2014 *Phys. Rev. Lett.* **113** 187201
[9] Pedrocchi F L, Chesi S and Loss D 2011 *Phys. Rev.* B **84** 165414
[10] Perreault B, Knolle J, Perkins N B and Burnell F J 2015 *Phys. Rev.* B **92** 094439
[11] Giamarchi T 2004 *Quantum Physics in One Dimension* (Oxford: Oxford University Press) chapter 3
[12] Mourigal M, Enderle M, Klöpperpieper A, Caux J S, Stunault A and Rønnow H M 2013 *Nature Phys.* **9** 435

IOP Conf. Series: Journal of Physics: Conf. Series **1220** (2019) 012004 doi:10.1088/1742-6596/1220/1/012004

Circularly polarized two-photon-induced luminescence from plasmonic two-dimensional chiral Au nanostructures

Khai Q. Le[1] and H. Okamoto[1,2]
[1] Institute for Molecular Science, Okazaki, Aichi 4444-8585, Japan
[2] The Graduate School for Advanced Studies (Sokendai), Okazaki, Aichi 4444-8585, Japan
E-mail: aho@ims.ac.jp

Abstract. A chiral nanostructure, which exhibits optical activity, absorbs different amounts of left-handed circularly polarized (LCP) and right-handed circularly polarized (RCP) light. In this work, we report the observation of dissymmetry between two-photon-induced LCP and RCP photoluminescence from plasmonic two-dimensional (2D) chiral Au nanostructures. Under excitation by linearly polarized femtosecond pulses from a mode-locked Ti:sapphire laser with a low incident power of 3 mW, the 2D chiral plasmonic nanostructure yields circularly polarized two-photon-induced photoluminescence (TPIPL) due to resonance with a chiral multipolar plasmon mode of the nanostructure. The handedness of the circularly polarized TPIPL was dependent on the handedness of the chiral plasmonic nanostructure. The chiral nature of TPIPL may find potential applications in optical devices, sensing of chiral molecular environments in biological systems, and so forth.

1. Introduction

Excitation of surface plasmon resonance is associated with strong local electric fields near the surface of the metal nanostructure. The strongly confined fields enhance optical emission phenomena such as Raman scattering, photoluminescence (PL), second-harmonic generation, two-photon induced photoluminescence (TPIPL), and so forth. In a typical TPIPL process, two photons are absorbed by the material at the same time to generate electrons in high energy state, followed by emission of a photon with higher energy than the incident ones.

Control of polarization states of PL from molecules and nanomaterials are occasionally desired. In particular, direct light sources of circularly polarized luminescence (CPL) are essential in application of on-chip integrated devices [1]. Some organic dyes with chiral structures achieve direct CPL, although the syntheses are in general complex. We recently analyzed CPL from a hybrid nanomaterial consists of a two-dimensional chiral plasmonic nanostructure and achiral dye molecules as an alternative. The handedness of the PL from the material was found to depend on the handedness of the nanostructure, and high ellipticity was observed (dissymmetry factor $g>0.1$; $g = 2(I_L - I_R)/(I_L + I_R)$ is the difference between left- and right-circularly polarized PL intensities, I_L and I_R, respectively) at wavelengths resonant with some specific plasmon mode [2].

Despite the low quantum efficiency, PL from metal nanostructures are advantageous because of lower photobleaching under some circumstances in bioimaging and other applications [3,4]. Plasmonic nanoparticles arranged along helical templates were reported to yield circularly polarized TPIPL [5]. In the present study, we observed two-photon induced CPL from gold in the visible wavelength region by pulsed excitation of chiral gold nanostructures. In this case, dye molecules are not hybridized. We

IOP Conf. Series: Journal of Physics: Conf. Series **1220** (2019) 012004 doi:10.1088/1742-6596/1220/1/012004

investigated correlation between the dissymmetry factors of the two-photon induced CPL and the circular dichroism (CD) spectrum of the chiral nanostructure. Through the electromagnetic field analysis of the plasmon mode, we elucidate the physics behind the generation of highly circularly polarized TPIPL. We report that the dissymmetry factor was enhanced in resonance with a multipolar plasmon mode having a chiral geometrical arrangement of induced oscillating dipoles.

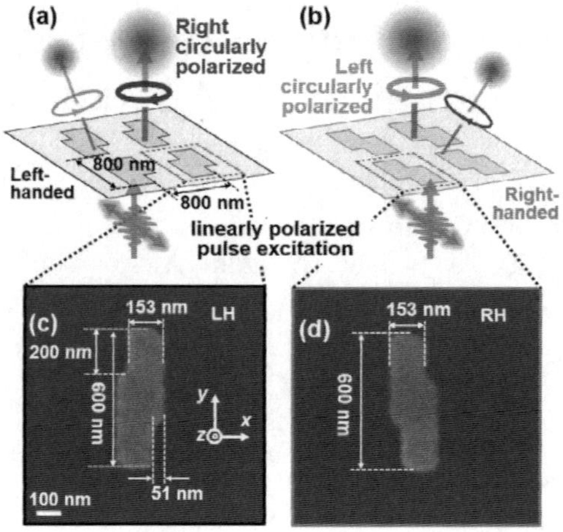

Figure 1. (a,b) Sketch of the chiral nanostructures and circularly polarized TPIPL upon linearly polarized femtosecond pulse irradiation, (c,d) SEM images of the fabricated Au nanostructures.

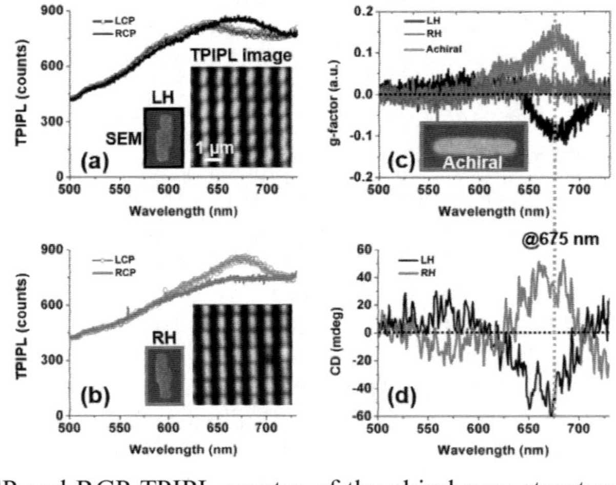

Figure 2. (a,b) LCP and RCP TPIPL spectra of the chiral nanostructures. Insets: SEM and TPIPL microscopic images of the chiral nanostructures. (c) Dissymmetry factor (g) spectra of TPIPL for the chiral and achiral nanostructures. (d) CD spectra of the chiral nanostructures. The dotted line indicates the wavelength giving dissymmetry maximum (675 nm).

2. Results and discussion

The plasmonic nanostructure arrays were designed as inducing CD over the wavelength range of interest. The constituents of the arrays were enantiomeric Au nanostructures consisting of two overlapped rectangles forming lightning-bolt-like shaped left-handed (LH) and right-handed (RH) structures as seen in Fig. 1(a,b). The scanning electron micrographs (SEM) of the fabricated individual constituents are shown in Fig. 1(c,d) with the measured dimensions. These enantiomeric nanostructures were fabricated on a glass substrate forming two arrayed patterns with a total area of 200 × 200 μm^2 and a lattice constant of 800 nm, using a typical nanofabrication procedure based on electron beam lithography and lift-off techniques. The two nanostructured arrays were separated by a 2 mm distance to avoid their mutual interaction which might cause unexpected diffraction and scattering by these neighboring nanostructures. The thickness of the fabricated Au nanostructures was 50 nm, which was produced over a 2-nm chromium adhesion layer. The insets of Figs. 2(a,b) show far-field TPIPL microscopic images of the chiral nanostructures. The TPIPL images look consistent with the SEM images, although the spatial resolution is limited. Under linearly polarized femtosecond pulsed laser irradiation (average power of 3 mW) onto the LH nanostructure at approximately 800 nm, we observed a RCP TPIPL at ~675 nm with stronger intensity than that of left-handed circularly polarized (LCP) TPIPL, as seen in Fig. 2(a). In contrast, at ~570 nm, the right-handed circularly polarized (RCP) TPIPL was slightly weaker than the LCP TPIPL. Figure 2(b) shows the TPIPL spectra for LCP and RCP luminescence from the RH nanostructure. The handedness of the TPIPL for the RH nanostructure was the same as that for the LH one but in the reversed manner. The dissymmetry factor $g = 2(TPIPL_{LCP} - TPIPL_{RCP})/(TPIPL_{LCP} + TPIPL_{RCP})$ for the LH nanostructure was negative at ~675 nm and positive at ~570 nm whereas that for the RH nanostructure was positive at ~675 nm and negative at ~570 nm. They were approximately equal to each other in amplitude and of opposite signs as seen in Fig. 2 (c). The schematic illustration of the handedness dependent TPIPL from the chiral nanostructures is given in Fig. 1(a,b). The precision of the optical measurement setup is justified by measuring the dissymmetry factor for TPIPL from an achiral nanostructured surface consisting of periodic Au nanorectangles, which is expected to be null due to its symmetric geometry. The width, length and thickness of the Au nanorectangle were 170, 510, and 50 nm, respectively. The lattice constant of the achiral nanostructured array was 700 nm. As expected, the dissymmetry factor showed nearly null values (Fig. 2(c)) over the wavelength range of interest for the achiral nanostructure, which justifies our measurement.

Interestingly, the TPIPL dissymmetry spectra, which were defined as the difference between the LCP and RCP TPIPL for the chiral Au nanostructures shown in Fig. 2 (c), are correlated to the CD spectra of the respective enantiomers as seen in Fig. 2 (d). The mirrored images of the CD spectra correspond to those of the TPIPL dissymmetry spectra of the chiral nanostructures. At ~675 nm, the LH nanostructure absorbs more RCP light than LCP one and emits stronger RCP TPIPL than LCP TPIPL. The same scenarios happen for the RH nanostructure but in a reversed manner.

In order to elucidate the physics behind the two-photon induced CPL, we conducted an electromagnetic model analysis of the plasmon mode at 675 nm, where the strongest magnitude of the dissymmetry factor g was observed. We numerically solved the light-matter interaction problem by the full-wave Maxwell's equations solver based on the MATLAB-based toolbox MNPBEM [6]. The Maxwell solver was developed based on the three-dimensional boundary element method. We simulated a unit cell consisting of the chiral nanostructure on a glass substrate (the dimensions were taken from the measured ones in Fig. 1(c)). Periodic boundary conditions were applied to mimic the entire periodic array of the nanostructure. The unit cell of the nanostructure used for simulation is illustrated in Fig. 3 (a). The light incident on the nanostructure was assumed as a plane wave polarized in the direction along the x- or y-axis, as defined in the inset of Fig. 3 (a). The complex refractive index of Au with realistic material losses needed for the numerical simulation was given in the refractive index library of MNPBEM.

We found that the TPIPL dissymmetry of the chiral nanostructures showed a resonance with a two-dimensional chiral multipolar plasmon mode. Figure 3 (b) shows the local electric field snapshot at 20 nm above the nanostructure at 675 nm. Figure 3 (c) shows the electric field intensity at the same wavelength. The arrows indicate the electric polarization directions of the corresponding multipolar plasmon mode, which shows a chiral spatial arrangement of the induced local polarizations. This mode may strongly interact with chiral electromagnetic fields (i.e., CPL), and this is supposed to be the reason of the strong TPIPL dissymmetry.

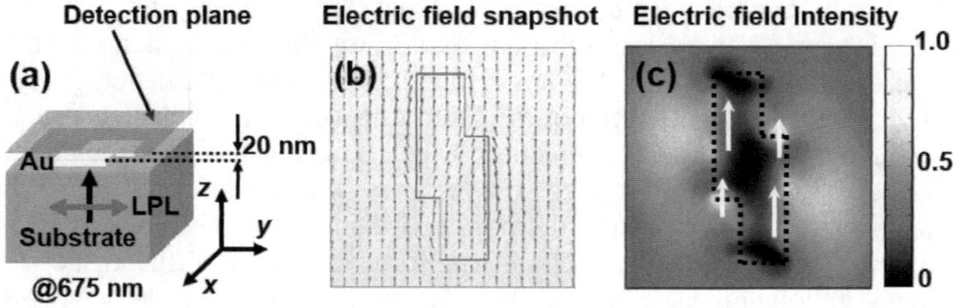

Figure 3. (a) Unit cell geometry of the chiral nanostructure used for electromagnetic modelling. (b) A simulated snapshot of electric field induced by the incident light at 675 nm. (b) Electric field intensity and schematic electric fields (white arrows) associated with the plasmon modes resonant with 675-nm light.

3. Conclusion

In this work, we observed circularly polarized TPIPL directly from chiral plasmonic nanostructures upon linearly polarized ultrashort pulsed laser irradiation. The dissymmetry factor spectra of circularly polarized TPIPL showed resonance with a multipolar chiral plasmon mode of the nanostructure and were correlated to the CD spectra of the chiral nanostructures. Our findings show that a proper design of chiral plasmonic nanostructure provides a chiral light source that directly emits CPL. The chiral nature of TPIPL may find potential applications in optical devices, sensing of chiral molecular environments in biological systems, and so forth.

Acknowledgments

The authors thank Ms. A. Ishikawa (IMS) for the nanostructured sample fabrication, and Drs. S. Hashiyada and T. Narushima (IMS) for their technical assistance. This work was supported by Grants-in-Aid for Scientific Research (KAKENHI) (Nos. JP22225002, JP15H02161, JP15K13683, JP16H06505 to H.O.) and by the Core-to-Core Program (A. Advanced Research Networks) from the Japan Society for the Promotion of Science (JSPS), and by the Photon Frontier Network Program of the Ministry of Education, Culture, Sports, Science, and Technology (MEXT), Japan.

References

[1] He L, Li M 2014 *Opt. Lett.* **39** 2553
[2] Le K Q, Hashiyada S, Kondo M and Okamoto H 2018 *J. Phys. Chem. C* **122** 24924
[3] Wang H, Huff T B, Zweifel D A, He W, Low P S, Wei A and Cheng J-X 2005 *Proc. Natl. Acad. Sci. USA* **102** 15752
[4] Jiang Y, Horimoto N N, Imura K, Okamoto H, Matsui K and Shigemoto R 2009 *Adv. Mater.* **21** 2309
[5] Singh H J, Saumitra, Singh V R, Sikdar S K and Jayaprakash B 2016 *ACS Photon.* **3** 863
[6] Hohenester U and Trügler A 2012 *Comput. Phys. Commun.* **183** 370

IOP Conf. Series: Journal of Physics: Conf. Series **1220** (2019) 012005 doi:10.1088/1742-6596/1220/1/012005

Ultrafast electronic relaxation dynamics in a valence fluctuation material $Sm_{0.83}Y_{0.17}S$

Ryohei Ikeda[1], Hiroshi Watanabe[2,1], Yusuke Negoro[1], Yusuke Takeno[1], Keiichiro Imura[3], Hiroyuki S. Suzuki[4,5], Noriaki K. Sato[3], Shin-ichi Kimura[2,1]

[1]Department of Physics, Osaka University, Toyonaka 560-0043, Japan
[2]Graduate School of Frontier Biosciences, Osaka University, Suita 567-0871, Japan
[3]Department of Physics, Nagoya University, Nagoya 464-8602, Japan
[4]Institute for Materials Science, Tsukuba 305-0047, Japan
[5]Institute for Solid State Physics, The University of Tokyo, Kashiwa 277-8581, Japan

E-mail: hwata@fbs.osaka-u.ac.jp

Abstract. We report ultrafast electronic relaxation dynamics of yttrium-doped samarium mono-sulfide, $Sm_{0.83}Y_{0.17}S$, which is one of valence fluctuating compounds, by pump-probe measurements. We observed a large increase of the Drude weight in the reflectivity spectrum by the photo-excitation and a double exponential decay of the relaxation time to a metastable state. This suggests that the photo-induced effect can be explained as the change of carrier density. The metastable state has a long lifetime (>1 ns) and the carrier density is slightly higher than that before the photo-excitation.

1. Introduction

Samarium mono-sulfide (SmS) is a typical valence fluctuating compound that exhibits the pressure-induced valence transition from divalent with black insulating (BI) phase to trivalent with golden metallic (GM) phase, namely the black-golden (BG) transition at the pressure of about 0.65 GPa [1]. The BG transition also occurs by replacing Sm with yttrium (Y) owing to the smaller ionic radius of Y^{3+} than that of Sm^{2+} [2-4]. The Y substituted material by 17 %, $Sm_{0.83}Y_{0.17}S$, is located just before the border between black and golden phases [5]. This material exhibits a metallic character but black colour, namely black metallic (BM) phase, which is different from the BI and GM phases appearing in SmS [6,7]. In the BM phase, divalent (Sm^{2+}) is stable, but it is located very close to the GM phase with $Sm^{(3-\delta)+}$. Therefore, the valence transition from the BM phase to GM phase might occur due to a photo-excitation from the Sm^{2+} to $Sm^{(3-\delta)+}$ states.

In this study, to investigate the photo-induced valence transition and the electronic structure of the photo-induced phase, we performed an ultrafast pump-probe reflectivity measurement on $Sm_{0.83}Y_{0.17}S$. We found that the photo-excited state appears just after the pump-light irradiation and it decays to a new photo-induced phase, namely a metastable state with a double exponential curve. However, the metastable state was concluded not to be the GM phase, so the valence transition did not appear by the photo-excitation.

2. Experimental methods

Single crystals of $Sm_{0.83}Y_{0.17}S$ having the same NaCl-type crystal structure as SmS were grown by the Bridgman method. To perform the ultrafast pump-probe reflection spectroscopy, a Ti: sapphire regenerative amplifier (1.55 eV, 70 fs, 1 kHz) was used as the pump and probe light sources. The photon energy (wavelength) of the pump beam was fixed as 1.55 eV (800 nm). On the other hand, the photon energy (wavelength) of probe beam was tuned from 0.48 eV (2600 nm) to 2.07 eV (600 nm) by using an optical parametric amplifier. The spot size of the pump beam at the sample position is about 2 mm in diameter.

3. Results and Discussion

Figure 1(a) presents the time-evolution of the reflectivity change ($\Delta R/R$) at several photon energies. At higher energy side, the reflectivity greatly increases immediately after the photo-excitation. The time-evolution of $\Delta R/R$ after the great increase shows a double exponential decay with a constant background with a longer lifetime than 1 ns, but it will be recovered in the repetition time of 1 ms. This suggests that the photo-excited state is relaxed to a metastable state with the long lifetime.

Before the photo-excitation, the reflectivity spectrum increases with decreasing photon energy, suggesting the existence of carriers explained by the Drude model as shown by the red line in Fig. 1(b). Using this spectrum, we can obtain time-dependent reflectivity spectra. The spectra just after photo-excitation and in the metastable state are shown by green squares in Figs. 1(b) and 1(c), respectively.

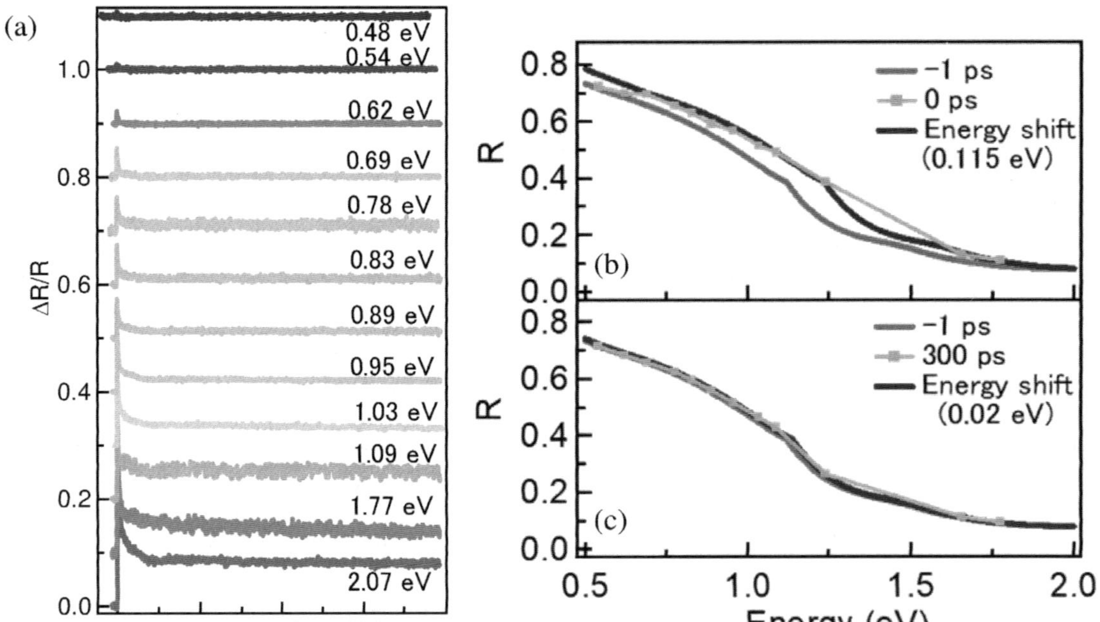

Figure 1 (a) Time-evolution of reflectivity change ($\Delta R/R$) after the photo-excitation at several probing photon energies. Each curve is shifted in the vertical direction by the step of 0.1 eV. (b) Reflectivity spectrum at 0 ps (just after the photo-excitation) indicated by green line. Red line is the spectrum at -1 ps (before the photo-excitation). Blue line is the same as the red line but the energy is shifted by 0.115eV. (c) Reflectivity spectrum in the metastable state at 300 ps after the photo-excitation indicated by green line. Red line is the spectrum before photo-excitation (same as the red line in (b)) and blue line is the same spectrum as the red line but the energy is shifted by 0.02 eV.

The spectrum after the photo-excitation (green line) agree with the energy-shifted spectrum (blue line) in Figs. 1(b) and 1(c), respectively.

In the Drude model, the energy shift of the slope as well as the plasma edge ($\hbar\omega_0$) corresponds to the change of the carrier density, i.e., the increase of the reflectivity intensity ($\Delta R/R$) after the photo-excitation can be explained by the increase of the Drude weight as well as the increase of carrier density. The increase of the carrier density corresponds to twice as much as the energy shift. As shown in Fig. 1(b), the spectrum after the photo-excitation (green line) is very similar to the spectrum before the photo-excitation ($\hbar\omega_0 \sim 1.7$ eV) with the energy shift by 0.115 eV (blue line), i.e., the carrier density increases by about 16 % due to the photo-excitation.

In the metastable state, on the other hand, the obtained spectrum (green line) is consistent with the spectrum with the energy shift by 0.02eV (blue line) as shown in Fig. 1(c). The increase of the carrier density by the photo-excitation is only about 3 %.

Figure 2(a) shows the excitation-power dependence of $\Delta R/R$ at 1.55 eV (probe beam). With increasing excitation power, the reflectivity increases. The time-evolution of $\Delta R/R$ can be expressed by the equation with a double decay process as shown by black lines in Fig. 2(a); $\Delta R/R = I_1 \exp(-t/\tau_1) + I_2 \exp(-t/\tau_2) + I_3$. The lifetimes ($\tau_1$ = ~0.35 ps and τ_2 = ~4.2 ps) do not change with the excitation-power shown in the inset of Fig. 2(a), but the constant value I_3 only changes. Figures 2(b) and 2(c) show the excitation-power dependences of the peak intensity of $\Delta R/R$, which is the intensity of the photo-excited state, and the constant term I_3, which is the intensity of the metastable state, respectively. As shown in Fig. 2(b), the peak intensity of $\Delta R/R$ monotonically increases as the excitation-power increases. This result suggests that the plasma frequency increases with increasing excitation-power, i.e. the number of the photo-excited carriers increases. In the same manner, as shown in Fig. 2(c), I_3 also monotonically increases with increasing excitation-power, suggesting that the photo-excitation only produces photo-carriers but does not cause a phase transition.

Even though the carrier density increases after the photo-excitation, the metastable state is not the same as the GM phase. The BG transition might need other effects such as lattice relaxation.

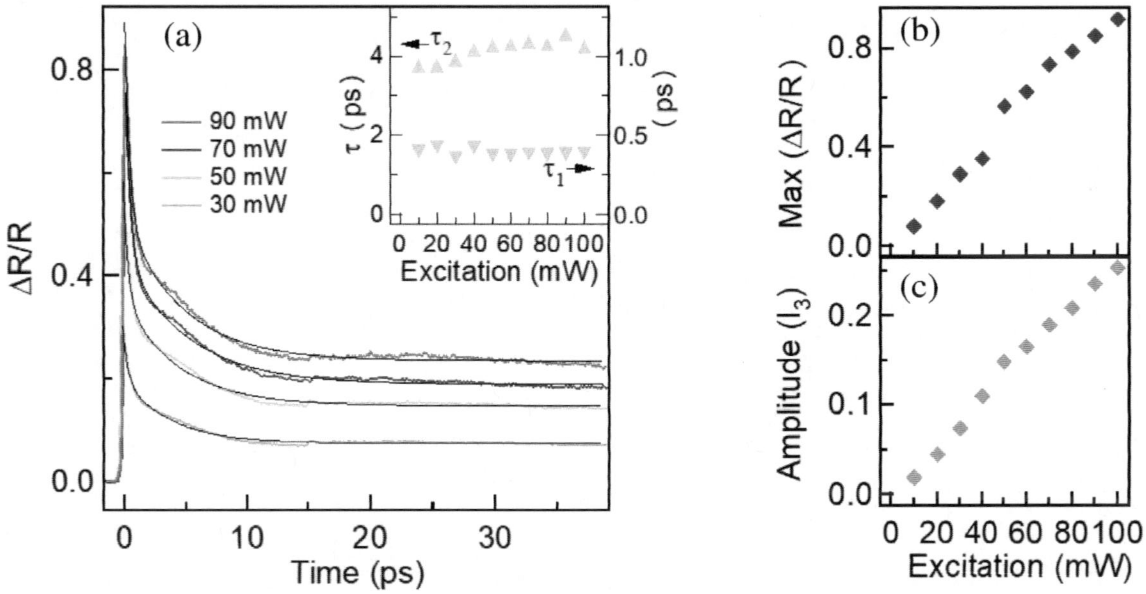

Figure 2 Excitation-power dependence of the time-evolutions of $\Delta R/R$ (colour lines; experimental results, black lines; fitting curves) (a), lifetimes τ_1 and τ_2 (inset of (a)), the maximum value of $\Delta R/R$ (b), and the constant value I_3 (the intensity of the metastable state) (c). The photon energy of the probe light was set as 1.55 eV.

IOP Conf. Series: Journal of Physics: Conf. Series **1220** (2019) 012005 doi:10.1088/1742-6596/1220/1/012005

4. Conclusion

We performed pump-probe measurements on $Sm_{0.83}Y_{0.17}S$ to investigate the ultrafast electronic relaxation dynamics of valence fluctuation and valence transition. We observed a huge increase ($\Delta R/R > 0.8$) of the reflectivity just after the photo-excitation and a double exponential relaxation to a metastable state. The reflectivity spectrum of the photo-excited state can be explained by the increase of the Drude weight, suggesting the carrier density increases after the photo-excitation. The metastable state has a long lifetime (>1 ns) and slightly high carrier density. However, the metastable state is not the GM phase but is still in the BM phase even with the maximum excitation-power at 100 mW.

Acknowledgments

This study was partly supported by JPSJ Grant-in-Aid for Scientific Research (C) (Grant Number 18K04836) and the Photon and Quantum Basic Research Coordinated Development Program by MEXT, Japan.

References

[1] A. Jayaraman et al., 1970 Phys. Rev. Lett. **25** 1430.
[2] K. Imura et al., 2015 J. Phys.: Conf. Ser. **592** 012028.
[3] M.Kaneko et al., 2014 JPS Conf. Proc. **3** 011080.
[4] L. J. Tao and F. Holtzberg, 1975 Phys. Rev. B **11** 3842.
[5] K. Imura et al., 2013 J. Kor. Phys. Soc. **62**, 2028.
[6] M. B. Maple and D. Wohlleben, 1971 Phys. Rev. Lett. **27** 511.
[7] S. D. Bader, N. E. Phillips, and D. B. McWhan, 1973 Phys. Rev. B **7** 4686.

IOP Conf. Series: Journal of Physics: Conf. Series **1220** (2019) 012006 doi:10.1088/1742-6596/1220/1/012006

Nonlinear optical dynamics of a 2D semiconductor quantum dot super-crystal: Emerging multistability, self-oscillations and chaos

V A Malyshev[1], **P Á Zapatero**[2], **A V Malyshev**[2,3], **R F Malikov**[4] **and I V Ryzhov**[5]

[1] Zernike Institute for Advanced Materials, University of Groningen, Nijenborgh 4, 9747 AG Groningen, The Netherlands

[2] GISC, Departamento de Física de Materiales, Universidad Complutense, E-28040 Madrid, Spain

[3] Ioffe Physical-Technical Institute, 26 Politechnicheskaya str., 194021 St. Petersburg, Russia

[4] Akmullah State Pedagogical University of Bashkortostan, 450000 Ufa, Russia

[5] Hertsen State Pedagogical University, 191186 St. Petersburg, Russia

E-mail: `v.malyshev@rug.nl`; `a.malyshev@fis.ucm.es`

Abstract. We conduct a theoretical study of the nonlinear optical dynamics of a 2D super-crystal comprising regularly spaced identical semiconductor quantum dots (SQDs), subjected to a resonant continuous wave excitation. A single SQD is considered as three-level ladder-like systems involving the ground, one-exciton and bi-exction states. We show that the super-crystal reveals a rich nonlinear dynamics, exhibiting multistability, self-oscillations and chaos. The behaviour is driven by the retarded SQD-SQD interactions and bi-exciton binding energy.

1. Introduction

In the last decade, the so-called *metamaterials*, a class of new materials not existing in nature, received a great deal of attention [1, 2]). Super-crystals comprising regularly spaced identical SQDs represent one of examples of metamaterials with tunable optical properties which can be controlled by the geometry and chemical composition of components, as well as the lattice arrangement. There exists a variety of methods to fabricate such systems [3, 4].

In this communication, we report on a theoretical study of the nonlinear optical dynamics of a super-crystal comprising a square lattice of spherical SQDs (preliminary results have been published in Ref. [5]). A single SQD is considered as a point-dipole system, that is accurate if the variation of the optical field over the particle size d is small ($kd \ll 1$, k being the wavenumber) and $a > (3/2)d$, where a is the lattice constant [6]. We assume that these conditions hold. Due to the high density of SQDs and high oscillator strengths of the SQD's transitions, the total (retarded) dipole-dipole SQD-SQD interactions have to be taken into account, which is elaborated in the paper within the mean-field approximation. The real part of the SQD-SQD interaction results in the dynamic shift of the SQD's energy levels, whereas the imaginary part describes the collective radiative decay of SQDs, both depending on the population differences between the levels. These two effects are crucial for the nonlinear dynamics of the SQD super-crystal. As a result, in addition to bistability, analogous to that for a thin layer of two-level

IOP Conf. Series: Journal of Physics: Conf. Series **1220** (2019) 012006 doi:10.1088/1742-6596/1220/1/012006

emitters [8], we found multistability, periodic and aperiodic self-oscillations and chaotic regimes in the super-crystal optical response.

2. Theoretical background

We model single SQD by a three-level ladder-like system with the ground, one-exciton, and bi-exciton states ($|1\rangle$, $|2\rangle$, and $|3\rangle$, respectively) with corresponding energies $\varepsilon_1 = 0$, $\varepsilon_2 = \hbar\omega_2$, and $\varepsilon_3 = \hbar\omega_3 = \hbar(2\omega_2 - \Delta_B)$, where $\hbar\Delta_B$ is bi-exciton binding energy (see, e.g., Refs. [7]). The dipole allowed transitions $|1\rangle \leftrightarrow |2\rangle$ and $|2\rangle \leftrightarrow |3\rangle$ are characterized by the transition dipole moments $\mathbf{d_{21}}$ and $\mathbf{d_{32}}$, the latter are assumed to be real, aligned, and $\mathbf{d_{32}} = \mu\mathbf{d_{21}} \equiv \mu\mathbf{d}$. The states $|3\rangle$ and $|2\rangle$ spontaneously decay with rates γ_{32} and γ_{21}, accordingly, and $\gamma_{32} = \mu^2\gamma_{21} \equiv \mu^2\gamma$. The bi-exciton state $|3\rangle$ is dipole forbidden and can be populated either via consecutive $|1\rangle \to |2\rangle \to |3\rangle$ transitions or via the simultaneous absorption of two photons of frequency $\omega_3/2$. In what follows, we will consider both options.

It is assumed that the system is subjected to a resonant external field $\boldsymbol{E}_0\cos(\omega t)$ incident normally to the super-crystal and polarized along the transition dipole moments. The optical dynamics of SQDs is described by means of the quantum master equation for the density operator ρ, which in the rotating (with the frequency of the external field ω) frame reads

$$\dot{\rho} = -\frac{i}{\hbar}\left[H^{RWA}, \rho\right] + \frac{\gamma_{21}}{2}\left([\sigma_{12}\rho, \sigma_{21}] + [\sigma_{12}, \rho\,\sigma_{21}]\right) + \frac{\gamma_{32}}{2}\left([\sigma_{23}\rho, \sigma_{32}] + [\sigma_{23}, \rho\,\sigma_{32}]\right), \tag{1a}$$

$$H^{RWA} = \hbar\left[\Delta_{21}\sigma_{22} + (\Delta_{21} + \Delta_{32})\sigma_{33}\right] - i\hbar\Omega\left(\sigma_{21} + \mu\sigma_{32}\right) + h.c., \tag{1b}$$

$$\Omega = \Omega_0 + (\gamma_R + i\Delta_L)(\rho_{21} + \mu\rho_{32}). \tag{1c}$$

In Eq. (1a), $\hbar$ is the reduced Plank constant, H^{RWA} is the SQD Hamiltonian in the rotating wave approximation (RWA), square brackets denote the commutator, and the two other terms represent the relaxation operator, where $\sigma_{ij} = |i\rangle\langle j|$ $(i, j = 1, 2, 3)$. In Eq. (1b), $\Delta_{21} = \omega_2 - \omega$ and $\Delta_{32} = \omega_3 - \omega_2 - \omega$ are detuning away from the resonance of the corresponding transitions. $\Omega_0 = dE_0/\hbar$ and $\Omega = dE_{SQD}/\hbar$ are the Rabi amplitudes of the external (E_0) and total field (E_{SQD}), acting on a SQD. The latter is the sum of the applied field E_0 and the field produced by all others SQDs in place of a given one [the second term in Eq. (1c)], taken in the mean-field approximation. The quantities γ_R and Δ_L are the amplitudes of the far-zone and near-zone fields, describing the collective SQD relaxation and energy level shifts, respectively. They are given by the following expressions [9]: $\gamma_R = 4.51\gamma/(ka)^2$ and $\Delta_L = -3.35\gamma/(ka)^3$.

3. Results

In calculations, we set $\mu = \sqrt{2/3}$, $\gamma_R = 100$, $\Delta_L = 1000$ [9] (here and in the following, all quantities are in units of γ). The results of calculations are presented in Figs. 1 and 2. Figure 1 displays the steady-state solutions to Eqs. (1a) - (1c) obtained by the method invented in Ref. [9]. Plots (a) and (b) show the dependence of the mean-field magnitude $|\Omega|$ (left column) on the external field magnitude $|\Omega_0|$ calculated for the one-photon ($\omega_0 = \omega_2$, $\Delta_{21} = 0$, $\Delta_{32} = -\Delta_B$) and two-photon ($\omega_0 = \omega_3/2$, $\Delta_{21} = -\Delta_{32} = \Delta_B/2$ resonance, respectively. As is seen from the figures, the mean-field magnitude $|\Omega|$ may have several solutions (up to five at $\Delta_B = -25$) for a given value of the external field magnitude $|\Omega_0|$, that signals emerging instabilities. We uncovered the stability of different branches computing the Lyapunov exponents λ [10]. For this purpose, we calculated the eigenvalues of the Jacoby matrix (of rank eight) of the linearized Eqs. (1a) - (1c) as a function of $|\Omega|$ and selected from those an exponent with the maximal real part, Max{Re[λ]}, which determines the character of evolution of a small deviation from the steady-state solution. At Max{Re[λ]} < 0, a given solution is stable (deviation decreases) and *vice versa.* Calculated in this way Max{Re[λ]} are plotted in the right panels of Fig. 1. The

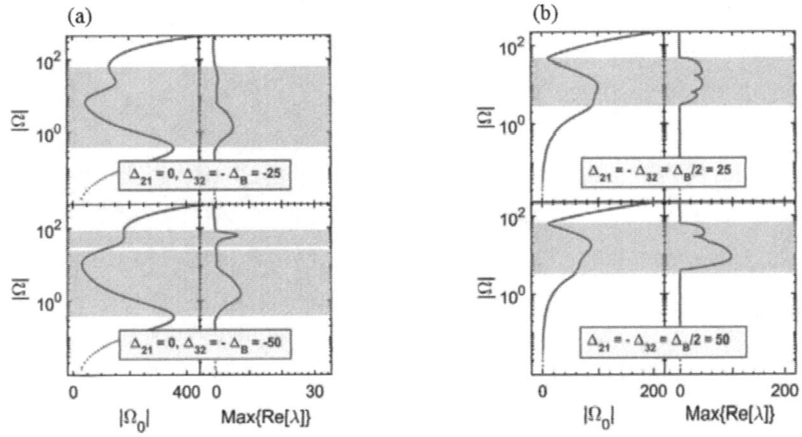

Figure 1. (a) - Steady-state solutions to Eqs. (1a) - (1c) for the case of one-photon resonance ($\Delta_{21} = 0, \Delta_{32} = -\Delta_B$) for different values of the bi-exciton binding energy Δ_B (shown in the plots). The left column displays the $|\Omega|$-vs-$|\Omega_0|$ dependence. The shaded regions show unstable parts of the stationary solutions (marked in red), obtained by analyzing the Lyapunov exponents λ, the maximum values of the real parts of which, Max$\{$Re$[\lambda]\}$, are depicted in the right column. (b) - Same as in (a), only for the case of two-photon resonance ($\Delta_{21} = -\Delta_{32} = -\Delta_B/2$.)

shaded regions show the unstable (Max$\{$Re$[\lambda]\} > 0$) parts of the steady-state solutions. We stress that not only branches with the negative slope are unstable, that is always the case, but those with the positive slope as well. This occurs for both the one- and two-photon resonance excitation, in the upper branch for the former and in the lower one for the latter.

To uncover the nature of instabilities of different branches of the steady-state solutions (shown in Fig. 1), we solved numerically Eqs. (1a) - (1c). Two types of initial conditions were used: (i) - the system starts from one of the steady-states solutions and (i) - the system initially is in the ground state [$\rho_{11}(0) = 1$]. Figure 2 shows the dynamics of the mean-field Ω calculated for the one-photon resonance ($\omega_0 = \omega_2$, $\Delta_{21} = 0$, $\Delta_{32} = -\Delta_B$) - plot (a) and two-photon resonance ($\omega_0 = \omega_3/2$, $\Delta_{21} = -\Delta_{32} = \Delta_B/2$) - plot (b).

Figure 2(a) displays the results obtained for $\Delta_B = 50$. Two unstable points of the the steady-state solution were considered: ($|\Omega_0| = 90, |\Omega| = 23.5655$) - upper row and ($|\Omega_0| = 140.0002$ - lower row (see the inserts). The left column in panel (a) shows the time evolution of the mean-field magnitude $|\Omega|$. As is seen, after some delay, the system reaches an attractor (sustained sgnal), the character of which significantly differs for considered points. The upper row demonstrates results for ($|\Omega_0| = 90, |\Omega| = 23.5655$); the dynamics looks like simple self-oscillations (see the upper-right inset for a blow up of the dynamics of $|\Omega|$). The Fourier spectrum of the sustained signal (middle column) contains a couple of well-defined harmonics. Accordingly, the phase map in the phase space (Re$[\Omega]$, Im$[\Omega]$) (right panel) represents a simple closed curve, commonly called limit cycle [10]. By contrast, for ($|\Omega_0| = 140.0002, |\Omega| = 32.3261$), the dynamics manifests signatures of aperiodic oscillations. This is reflected in the Fourier spectrum of the attractor which consists of incommensurated frequencies [10]. The phase space map of the attractor represents an open trajectory, almost filling a finite area in the phase space.

Figure 2(b) shows the evolution of the system, being initially in the ground state [$\rho_{11}(0) = 1$], calculated for the external field magnitudes $|\Omega_0| = 91$ (upper row) and $|\Omega_0| = 94$ (lower row) at $\Delta_B = 50$. Here, in both cases we observe a highly irregular behavior of the system. The Fourier spectrum of the signal represents a broad structureless quasi-continuum, suggesting

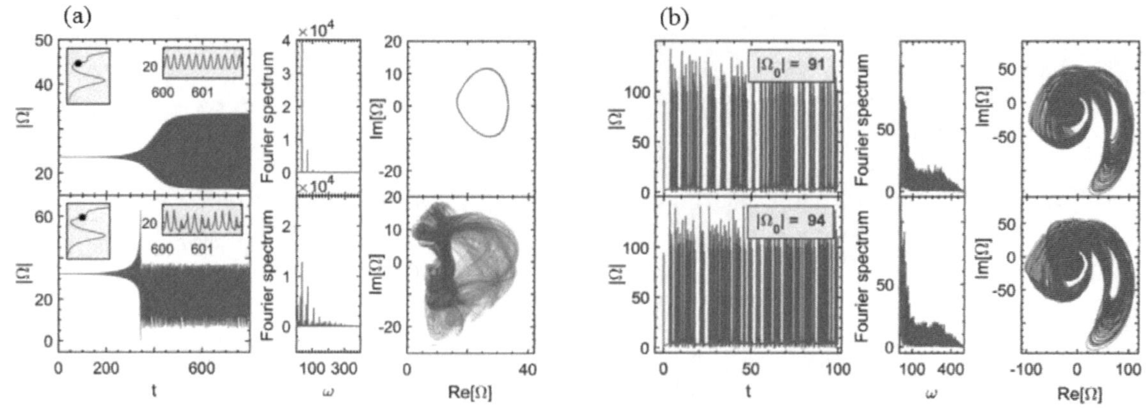

Figure 2. (a) - Time-domain dynamics of the mean-field magnitude $|\Omega|$ (left column), the Fourier spectrum (middle column) and phase space map of the attractor in the $(\mathrm{Re}[\Omega], \mathrm{Im}[\Omega])$ plane (right column) for the case of the one-photon resonance ($\Delta_{21} = 0$, $\Delta_{32} = -\Delta_B = -50$). The inserts show fragments of the dynamics on a short time interval (upper-right insets) and the steady-state solution (upper-left insets), where the thick dots indicate the points on the steady-state solutions: ($|\Omega_0| = 90, |\Omega| = 23.5655$) - upper row, ($|\Omega_0| = 140.0002, |\Omega| = 32.3261$) - lower row, for which the calculations were performed. (b) - Same as in (a), but now for the case of the two-photon resonance ($\Delta_{21} = -\Delta_{32} = -\Delta_B/2 =$ - 25) for the external field magnitudes $|\Omega_0| = 91$ (upper raw) and $|\Omega_0| = 94$ (lower raw) and the ground state as the initial condition.

that the oscillatory regime found is of chaotic nature [10]. The phase-space map shows the corresponding attractor. In the chaotic regime, the system behavior is very sensitive to the initial conditions, that may be used for the encription of information [11].

4. Summary

We conducted a theoretical study of the nonlinear optical dynamics of a 2D SQD super-crystal subject to a resonant continuous wave excitation, modeling an isolated SQD a three-level ladder-like system with the ground, one-exciton and biexciton states. The set of parameters used in our study is typical for SQDs emitting in the visible range, such as, for instance, CdSe and CdSe/ZnSe. It turned out that the super-crystal optical response might demonstrate multistability, self-oscillations and dynamic chaos. All this makes such systems promising for practical applications in all-optical information processing and optical computing.

References

[1] Zheludev N I 2010 *Science* **328** 582
[2] Soukoulis C M and Wegener M 2010 *Science* **330** 1633
[3] Evers W H *et al.* 2013 *Nano Lett.* **13** 2317
[4] Boneschanscher M P *et al.* 2014 *Science* **344** 1377
[5] Malikov R F, Ryzhov I V and Malyshev V A 2017 *EPJ Web of Conferences* **161** 02014
[6] S. Y. Park S Y and Stroud D 2004 *Phys. Rev. B* **69** 125418
[7] Stufler S *et al.* 2006 *Phys. Rev. B* **73** 125304
[8] Malikov R F and Malyshev V A 2017 *Opt. Spectrosc.* **122** 955
[9] Zapatero P Á *et al.* 2004 Nonlinear optical response of a two-dimensional quantum dot supercrystal: Emerging multistability, periodic/aperiodic self-oscillations, and hyperchaos *Preprint* ArXiv:1806.00387v1
[10] Eckmann J P and Ruelle D 1985 *Rev. Mod. Phys.* **57** 617
[11] Gao T and Chen Z 2008 *Phys. Lett. A* **372** 394

A New Class of All-Inorganic Perovskite Microplate for Lasing

Juan Du[1], Zhengzheng Liu[1], Yuxin Leng[1], Zhiping Hu[2], Xiaosheng Tang[2], Miao Zhou[2], Takayoshi Kobayashi[3]
[1]State Key Laboratory of High Field Laser Physics, Shanghai Institute of Optics and Fine Mechanics, Shanghai, China
[2]Key Laboratory of Optoelectronic Technology and Systems, Chongqing University, Chongqing 400044, China
[3]Advanced Ultrafast Laser Research Center and Brain Science Inspired Life Support Research Center, University of Electro-Communications, 1-5-1, Chofugaoka, Chofu, Tokyo 182-8585, Japan

Email: kobayashi@ils.uec.ac.jp, dujuan@mail.siom.ac.cn

Abstract. We report a stable low-threshold amplified spontaneous emission performance from a new type of all-inorganic perovskite $CsPb_2Br_5$ microplate with superior crystallization, enhanced stability, and tunability under both one- and two-photon excitation for the first time.

1. Introduction

Recently there are many reports about solution-processed metal halide perovskite nanocrystal structures presenting a promising optoelectronic applications [1-5]. Especially, $CsPbX_3$ quantum dots (QDs) have exhibited bright emission with photoluminescence (PL) quantum yields (QYs) of 90%. Comparing with zero- and one-dimensional nanocrystals, perovskites with two-dimensional crystals have also shown excellent optical properties as the reason of its sufficient gain toward whispering gallery mode lasing [6]. Therefore, they are expected to play an important role as its potential building blocks for the optical devices in

Fig. 1 (a) Scanning emission microscopy and (b) transmission electron microscope image of an individual microplate. (c) Scanning emission microscopy images of $CsPb_2Br_5$ microplate with 1.5 h. (d) side view of the relaxed $CsPb_2Br_5$ unit cell, with a schematic of the alternating Cs^+ and $(Pb_2Br_5)^-$ layers strongly coupled with each other.

IOP Conf. Series: Journal of Physics: Conf. Series **1220** (2019) 012007 doi:10.1088/1742-6596/1220/1/012007

future based on the emergent perovskite materials. Comparing with liner absorption, multi-photon absorption and emission have several unique advantages including large penetration depth, high spatial resolution and little damage to the targeted samples, rendering wide applications in biological and medical fields. Therefore, two-photon pumping lasers have been regarded as a promising strategy to achieve frequency up-conversion [7,8]. In the present study, we report all-inorganic perovskite $CsPb_2Br_5$ with novel physical properties by material synthesis, characterization, optical measurement together with first principles calculations, which show that $CsPb_2Br_5$ crystallizes in a tetragonal phase with a unique sandwiched structure with alternating Cs^+ and $(Pb_2Br_5)^-$ layers chemically bonded with each other. We demonstrated the excellent amplified spontaneous emission (ASE) performance from a new type of all-inorganic perovskite $CsPb_2Br_5$ microplate with superior crystallization, enhanced stability, and tunable optical properties. To the best of our knowledge, this is the first report on the $CsPb_2Br_5$ microplate.

2. Results and discussion

2.1 The morphology and composition of perovskite microplate

Using a facile solution-phase synthesis approach, we prepared semiconductor perovskite $CsPb_2Br_5$ microplates which exhibit high stability with several months and excellent crystal quality with a uniform size. Figure 1 a and b show scanning emission microscopy and transmission electron microscope image of $CsPb_2Br_5$ microplate with the size about micrometre scale. Being different from $CsPbX_3$, the growth process of $CsPb_2Br_5$ microplates displays a unique helical form crystal layered structure like bamboo, as shown in Fig 1c. To learn the crystallization of $CsPb_2Br_5$, we performed DFT-based first-principles calculations using the plane-wave-basis-set and the projector-augmented-wave method, as implemented in the Vienna ab initio simulation package (VASP) code to explore the atomic structure and electronic properties. It is found that $CsPb_2Br_5$ crystallizes in a tetragonal phase. It should be noted that $CsPb_2Br_5$ perovskite is characterized by a sandwiched structure with alternating Cs+ and $(Pb_2Br_5)^-$ layers chemically bonding with each other (Fig. 1d).

2.2 The ASE characterization for perovskite microplate

As shown in Fig 2a, when $CsPb_2Br_5$ microplate is excited by a Ti^{3+}doped sapphire pulse laser (700-1000 nm with peak at ~800nm, ~100fs, ~1W, 1kHz) stimulated emission could be obtained via coupling among the scattered light by individual $CsPb_2Br_5$ microplates. A bright ASE emission due to conduction-to-valence band transition located at the middle of the film was clearly observed in atmosphere condition when it was excited by 800 nm laser (Fig 2b). Figure 2c shows the power-dependent emission spectra of $CsPb_2Br_5$ microplate under the 800-nm fs-laser excitation. When the pumping density is under threshold, emission peak centre 528 nm and the intensity increases slowly with a full width at half maximum (FWHM) of 20 nm. However, above the threshold, the lasing peak red-shifted to ~540 nm with a FWHM of 6 nm appears. Figure 2d shows the corresponding lasing emission spectra under one-photon

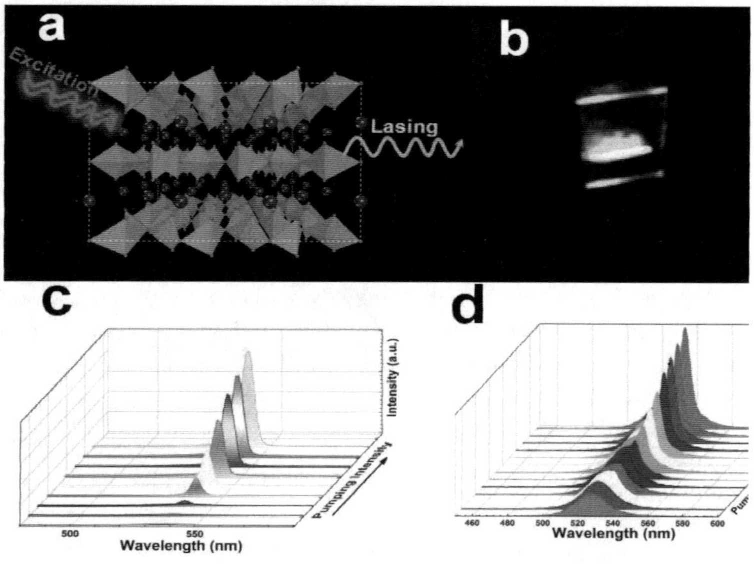

Fig.2. (a) The schematic representation of $CsPb_2Br_5$ microplates excited by 800 nm fs pulse laser focused linearly on the sample with a cylindrical lens. (b) The photograph of the glass covered with $CsPb_2Br_5$ microplates excited by Ti^{3+}doped sapphire pulse laser above lasing threshold excitation. (c) Pump-fluence dependence of the emission spectrum from $CsPb_2Br_5$ microplates under 800 nm and (d) 400 nm fs laser excitation.

IOP Conf. Series: Journal of Physics: Conf. Series **1220** (2019) 012007 doi:10.1088/1742-6596/1220/1/012007

excitation, which is consistent with the lasing excited by one-photon. Thus, $CsPb_2Br_5$ may have great potential for two-photon excited up-conversion devices.

2.3 Two-photon absorption crossection in perovskite

The pump intensity dependence of intensity and FWHM of the ASE signal with one- and two-photon absorption (OPA and TPA) were measured. The results depicted in Fig. 3 and Fig. 4, respectively, clearly show threshold intensities of ASE. They are respectively $I_{th}(OPA)$ = 1.94mJ/cm^2 and $I_{th}(TPA)$ = 21.5mJ/cm^2. The threshold of the ASE when the population inversion is achieved by the pump pulse, which is common for OPA and TPA. This is given by the relation between OPA and TPA cross sections and TPA threshold flux $\sigma_{OPA} = \Phi_{th}(TPA)\sigma_{TPA}$. From the TPA threshold intensity above mentioned and the relation, σ_{TPA} is estimated as $\sigma_{OPA} = \Phi_{th}(TPA)\sigma_{TPA}$ σ_{TPA} =6200GM (Goeppert Myer: 10^{50}cm^{4}s/photon). This crossection is very large in comparison with well know high TPA molecular crystal of anthracene, which is reported to have 4GM at 31.6 x 10^3 cm^{-1} (^1A_u absorption band) and 400GM at 33.4 x 10^3 cm^{-1} (X absorption band). The material may be useful for nonlinear optical devices such as three dimensional optical memory device utilizing deep reaching photoexcitation inducing photochemical hole-burning thanks to the large TPA crossection.

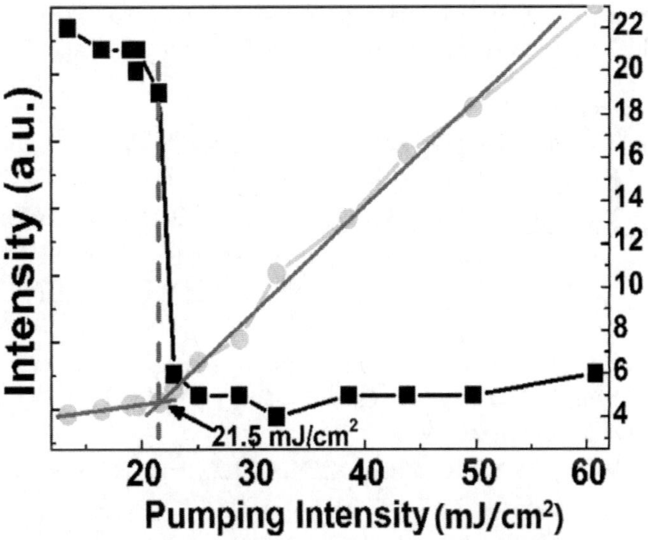

Fig. 3. Intensity dependence and FWHM of ASE spectrum plotted against two-photon pump intensity with the 800nm fs pulse.

3. Conclusion

In summary, we have studied $CsPb_2Br_5$ microplate as a typical material of all-inorganic perovskite, and demonstrated its large optical gain and corresponding emission. With a robust band gap of ~2.44 eV, $CsPb_2Br_5$ microplate exhibits stable low-threshold amplified spontaneous emission under both one- and two-photon excitation, which is related to its unique spatially distinguished valence/conduction band edge states originating from the intrinsic sandwiched structure. The material has extremely large two-photon absorption crossection. These results are expected to shed new light on future design and development of novel perovskite nanomaterials for optoelectronic devices including photoluminescence light sources and nonlinear material with gigantic two-photon absorbing medium.

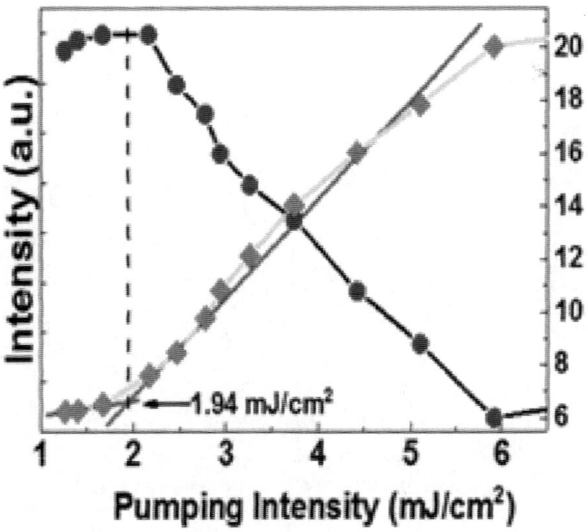

Fig. 4. Intensity dependence and FWHM (green curve with red squares) and intensity (black curve with blue circles) of ASE spectrum plotted against one-photon pump intensity with the 400nm fs pulse. Left ordinate is plotted linearly with equally separated notches.

Acknowledgment

This work was supported by 100 Talents Program of CAS, Strategic Priority Research Program of CAS [Grant No. XDB16], National Natural Science Foundation of China [Grant No. 61475169, 61521093, 11127901], and the initial funding of 100 Young Talents Plan at Chongqing University (0210001104430).

References

[1] Zhou H., Chen Q., Li G., 2014 *Science* **345** 542.

[2] Kim, Y. H., Cho, H., Heo, J. H., 2015 *Adv. Mater* **27** 1248.

[3] Yakunin, S., Protesescu, L., Krieg, F., 2015 *Nat. Commun.* **6** 8056.

[4] Jeon, N. J., Noh, J. H., Yang, W. S., 2015 *Nature* **517** 476.

[5] Xu, Y., Chen, Q., Zhang, C., Wang, R., Wu, H., Zhang, X., Xiao, M., 2016 *J. Am. Chem. Soc.* **138** 3761.

[6] Zhang, Q., Ha, S. T., Liu, X., Sum, T. C., Xiong, Q., 2014 *Nano Lett.* **14** 5995.

[7] Wang, Y., Ta, V. D., Gao, Y., He, T. C., Chen, R., Mutlugun, E. H., Demir, V., Sun, H. D., 2014 *Adv. Mater.* **26** 2954.

[8] Wang, Y., Li, X., Zhao, X., Xiao, L., Zeng, H., Sun, H., 2016 *Nano Lett.* **16** 448

IOP Conf. Series: Journal of Physics: Conf. Series **1220** (2019) 012008 doi:10.1088/1742-6596/1220/1/012008

Dynamics analysis of nanoparticles optically driven by a Laguerre-Gaussian beam with optical spin

Mamoru Tamura[1,2], **Takashige Omatsu**[3,4] **and Takuya Iida**[1,2,*]
[1]Graduate School of Science, Osaka Prefecture University, Sakai-shi, Osaka, 599-8570, Japan
[2]Research Institute for Light-induced Acceleration System, Osaka Prefecture University, Sakai-shi, Osaka, 599-8570, Japan
[3]Graduate School of Advanced Integration Science, Chiba University, Chiba-shi, Chiba, 263-8522, Japan
[4]Molecular Chirality Research Center, Chiba University, Chiba-shi, Chiba, 263-8522, Japan

E-mail: t-iida@p.s.osakafu-u.ac.jp

Abstract. We theoretically discover that the spin angular momentum (SAM) enables to modulate the orbital torque through the inter-particle light-induced force (IP-LIF). Laguerre-Gaussian beam with the orbital angular momentum (OAM) can induce the orbital motion for the optically trapped objects. In addition, the SAM can also accelerate or decelerate the orbital motion due to the IP-LIF. Our discovery provides a new physical aspect, i.e. the IP-LIF plays an important role in the many-body dynamics of nanoparticles via the SAM-OAM coupling.

1. Introduction

Laguerre-Gaussian (LG) beam with a helical wave-front carries an orbital angular momentum (OAM). The OAM enables the orbital motion of trapped objects, i.e. the OAM transfer effect [1], in optical tweezers, allowing a contactless manipulation of small objects by a focused laser [2]. The spinning motion can also be induced by a SAM associated with the circularly polarized light. However, the orbital and spinning motions are not independently determined by the OAM and SAM. For example, it was reported that the OAM forces the spinning motion of a trapped object with a sufficiently large size in comparison with the laser beam radius. Thus, a spinning velocity of the trapped object should be determined by the total angular momentum defined as the sum of OAM and SAM [3]. The circularly polarized LG beam with both positive (or negative) OAM and SAM deforms the irradiated surface of photo-sensitive organic material, such as azo-polymer, to establish helical structures, whereas the LG beam with either of OAM and SAM with a negative sign does not allow such helical mass-transport [4]. The above mentioned SAM-OAM coupling effects result from the mutual contribution of the both OAM and SAM.

In recent years, we discovered that the SAM contributes to the orbital motion of the trapped nanoparticles (NPs), in which, through the inter-particle light-induced force (IP-LIF) [5–7], the orbital motion of trapped NPs is accelerated or decelerated by SAM of the left- or right-circular polarization (LCP or RCP). Our model based on the IP-LIF is entirely different from the conventional SAM-OAM coupling in the tightly focused circularly polarized LG beams previously reported in [8], and it should provide a new physical insight concerning interaction between optical fields and matter. In this paper, we reveal that the SAM certainly contributes to the orbital torque through the IP-LIF.

IOP Conf. Series: Journal of Physics: Conf. Series **1220** (2019) 012008 doi:10.1088/1742-6596/1220/1/012008

2. Theoretical framework

As the target of this study, the single NP and paired NPs (gold, diameter 70 nm) are irradiated with the circularly polarized LG beam. Based on the manner of discrete dipole approximation, NPs are discretized by the cubic cells (edge length: 2nm) with the induced polarization $\mathbf{P}_i(\omega) = \chi(\omega)\mathbf{E}_i(\omega)$, where i is the index of cells. To consider the gold NPs dispersed in the water, the electric susceptibility is given by $\chi(\omega) = \varepsilon_{\rm p}(\omega) - \varepsilon_{\rm med}$ with the water dielectric constant $\varepsilon_{\rm med} = 1.33^2$ and the gold dielectric function by the Drude-critical points model [9] with fitting the experimental result [10]. $\mathbf{E}_i(\omega)$ is the response field expressed as,

$$\mathbf{E}_i(\omega) = \mathbf{E}^{\rm inc}(\mathbf{r}_i, \omega) + \sum_{j\neq i} \mathbf{G}(\mathbf{r}_i - \mathbf{r}_j, \omega) \cdot \mathbf{P}_j(\omega)V_j + \mathbf{S}_i(\omega) \cdot \mathbf{P}_i(\omega), \tag{1}$$

where $\mathbf{E}^{\rm inc}(\mathbf{r}, \omega)$ is the electric field of incident light, $\mathbf{G}(\mathbf{r}, \omega)$ is the Green's function of electromagnetic field in a homogeneous medium, V_i is the volume of i-th cell, and $\mathbf{S}_i(\omega) \cdot \mathbf{P}_i(\omega) = \int_{V_i} \mathbf{G}(\mathbf{r}, \omega) \cdot \mathbf{P}(\mathbf{r}, \omega) d\mathbf{r}$ is the analytical integration of $\mathbf{G}(\mathbf{r}, \omega)$ for $j = i$. For the circularly polarized LG beam propagating in $+z$-direction, $\mathbf{E}^{\rm inc}(\mathbf{r}, \omega)$ is expressed as $\mathbf{E}^{\rm inc}(\mathbf{r}, \omega) = [-\sqrt{2}{\rm i}w_0 Q(z,\omega)\rho]^{|\ell|} {\rm e}^{{\rm i}\ell\varphi} \mathbf{E}^{\rm G}(\mathbf{r}, \omega)$ for the arbitrary ℓ and $p = 0$, where ρ and φ are variables of cylindrical coordinates. $\mathbf{E}^{\rm G}(\mathbf{r}, \omega) = -\mathbf{E}_0 {\rm i} w_0^2 Q(z, \omega) {\rm exp}[{\rm i}Q(z, \omega)\rho^2 + {\rm i}kz]$ is the expression of Gaussian beam, $w_0 = 0.5\,\mu$m is the spot radius, k is the wavenumber in the medium, $Q(z, \omega)$ is written by $Q(z, \omega) = k/(2z - {\rm i}kw_0^2)$, and $\mathbf{E}_0 = \hat{\boldsymbol{\varepsilon}}\sqrt{2nI_0/c\epsilon_{\rm med}}$ ($I_0 = 2P/|\ell|!\pi w_0^2$, $P = 1$ W is power) gives the polarization and amplitude of electric field. The polarization vector is assumed as $\hat{\boldsymbol{\varepsilon}} = (1, \pm{\rm i}, 0)/\sqrt{2}$, where $+$ and $-$ correspond to the LCP and RCP, respectively. From $\sigma^-\hbar = \frac{{\rm i}\hbar}{2}(\hat{\boldsymbol{\varepsilon}} \times \hat{\boldsymbol{\varepsilon}}^*)_z = \pm\hbar$, LCP is $\sigma = 1$ and RCP is $\sigma = -1$ for the SAM. Under these conditions, $\mathbf{P}_i(\omega)$ and $\mathbf{E}_i(\omega)$ can be obtained by self-consistently solving the simultaneous equations.

Based on the general expression of time-averaged LIF derived from the Lorentz force [7], the LIF acting on i-th cell is described as,

$$\left\langle \mathbf{F}_i^{\rm cell}(\omega) \right\rangle = \frac{1}{2} {\rm Re}\left\{ [\nabla \mathbf{E}_i(\omega)]^* \cdot \mathbf{P}_i(\omega)V_i \right\}. \tag{2}$$

In addition, a sum of equation (2) yields the LIF acting on k-th NP,

$$\left\langle \mathbf{F}_k^{\rm NP}(\omega) \right\rangle = \sum_{i\in M_k} \left\langle \mathbf{F}_i^{\rm cell}(\omega) \right\rangle = \sum_{i\in M_k} \frac{1}{2} {\rm Re}\left\{ \left[\nabla \mathbf{E}^{\rm inc}(\mathbf{r}_i, \omega) + \sum_{j\notin M_k} \nabla \mathbf{G}(\mathbf{r}_i - \mathbf{r}_j, \omega) \cdot \mathbf{P}_j(\omega)V_j \right]^* \cdot \mathbf{P}_i(\omega)V_i \right\}, \tag{3}$$

where M_k is a set of the cell indices composing the k-th NP, and equation (1) is substituted for $\mathbf{E}_i(\omega)$, while the differential of third term in equation (1) is negligible. In [] of equation (3), the first term is caused by the incident light, whereas the second term is caused by the scattered light via the other cells, where the sum for $i, j \in M_k$ is approximately cancelled because of the action-reaction forces between dipoles. Therefore, the second term originated only from the cells composing other NPs, is namely IP-LIF, which is important for considering the orbital torque from the SAM. Then, the total torque around z-axis acting on NPs is simply expressed as,

$$\langle \tau_z(\omega) \rangle = \sum_i \left[\mathbf{r}_i \times \left\langle \mathbf{F}_i^{\rm cell}(\omega) \right\rangle \right]_z. \tag{4}$$

However, to consider the mechanical motion of NPs, they should be regarded as a rigid body whose translational motion is driven by the force acting on its center of mass (CM): $\langle \mathbf{F}_k^{\rm NP}(\omega) \rangle$. Thus, this consideration yields another defination of orbital torque based on $\langle \mathbf{F}_k^{\rm NP}(\omega) \rangle$,

$$\langle T_z(\omega) \rangle = \sum_k \left[\mathbf{r}_k^{\rm CM} \times \left\langle \mathbf{F}_k^{\rm NP}(\omega) \right\rangle \right]_z = \sum_k \left[\mathbf{r}_k^{\rm CM} \times \sum_{i\in M_k} \left\langle \mathbf{F}_i^{\rm cell}(\omega) \right\rangle \right]_z; \tag{5}$$

where $\mathbf{r}_k^{\rm CM} = (x_k^{\rm CM}, y_k^{\rm CM}, 0)$ is the CM position of k-th NP.

3. Orbital torque acting on the nanoparticles

Based on the configuration of figure 1(a), we estimated numerically the optical torque for the single NP by employing equation (4), and we found that the LCP-induced optical torque is relatively higher than the RCP-induced one at any wavelengths as shown in figure 1(b). In contrast, the equation (5) provided us that LCP-induced optical torque is fully identical with the RCP-induced one as shown in figure 1(c). The origin of this difference is the LIF caused by the second term of right-hand side in equation (1), which contributes to the spinning of NPs around the CM of themselves under the circular polarization. In the equation (3) for the single NP, this term is cancelled because of the sum for $i, j \in M_k$ (of course, the torque for spinning is not cancelled). Therefore, equation (4) using the LIF before the cancellation yielded the different torque for LCP and RCP, whereas equation (5) after the cancellation yielded the identical torque. This indicates the orbital motion of CM of single NP is not modulated by the SAM of circular polarization, where the IP-LIF is absent.

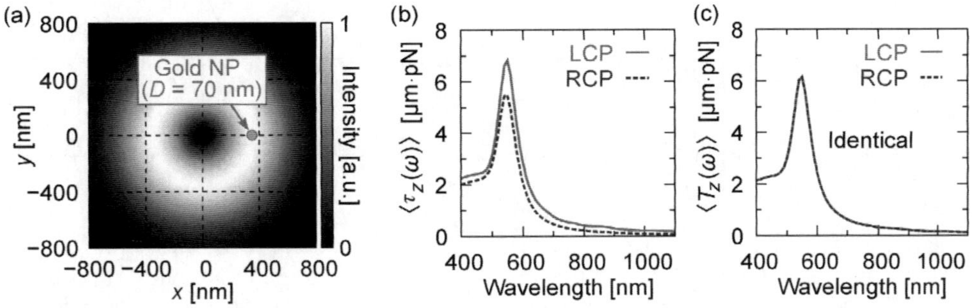

Figure 1. (a) Intensity distribution of $\mathbf{E}^{\rm inc}(\mathbf{r}, \omega)$ and the configuration of single NP (diameter $D = 70\,\rm nm$) located at the intensity maximum ($\mathbf{r}_1^{\rm CM} = (w_0/\sqrt{2}, 0, 0)$). (b)(c) Torque calculated by equation (4) and (5), respectively. Red solid and blue dashed lines were for LCP and RCP.

Next, as shown in figure 2, based on the configuration (a), the orbital torque acting on paired NPs was calculated by the equation (5). Although the torque was calculated for the LIF acting on the CM of NPs, there was a clear difference between LCP and RCP. This indicates the faster orbital motion can be expected for LCP, if the NPs keep the same relative position in the cylindrical coordinates during the motion.

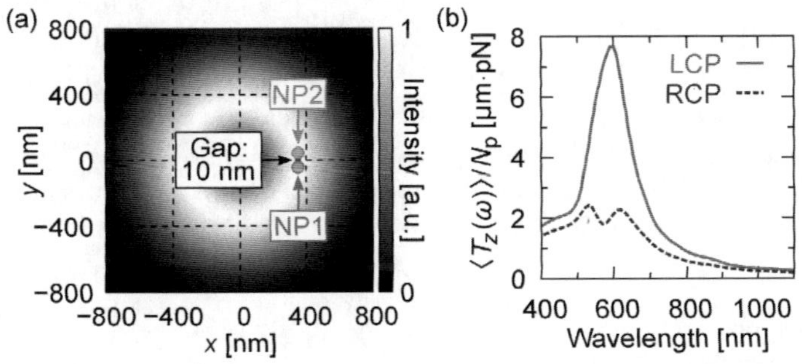

Figure 2. (a) Configuration of NPs, where they were located at the same x, z position with figure 1(a), but different y position: $y_k^{\rm CM} = \mp(D + g)/2$ ($-$ for NP1 and $+$ for NP2, gap length $g = 10\,\rm nm$). (b) Torque calculated by equation (5) (averaged over the NPs by $N_{\rm p} = 2$).

The reason why the torque for LCP and RCP differed is due to the presence of IP-LIF given by the second term in [] of equation (3). Figure 3 shows the LIF acting on the paired NPs, including IP-LIF. As shown in figure 3(c) and (d), the y-component of LIF $\langle F_{y,k}^{\rm NP}(\omega)\rangle$ was identical for LCP and RCP. In both cases, the NPs were attracted each other due to the IP-LIF. In addition, the IP-LIF was cancelled in the sum because of the action and reaction. The remained positive value of sum was caused by the OAM and would transport NPs for $+y$-direction in this configuration.

On the other hand, the x-component of LIF $\langle F_{x,k}^{\rm NP}(\omega)\rangle$ was inverted for LCP and RCP, where the zero of summed LIF indicated that $\langle F_{x,k}^{\rm NP}(\omega)\rangle$ was originated only from the IP-LIF. Under the circularly polarized light with SAM, the NPs are rotated around not only each CM, but also the CM of paired NPs as the two-body problem. Thus, the IP-LIF rotates NPs anti-clockwisely and clockwisely for the LCP and RCP. In particular, since the NPs were not located on the x-axis, $\langle F_{x,k}^{\rm NP}(\omega)\rangle$ also remained in equation (5) as $\mathbf{r}_k^{\rm CM}\times\langle\mathbf{F}_k^{\rm NP}(\omega)\rangle = x_k^{\rm CM}\langle F_{y,k}^{\rm NP}(\omega)\rangle - y_k^{\rm CM}\langle F_{x,k}^{\rm NP}(\omega)\rangle$, and modulated the orbital torque. Given $y_k^{\rm CM} = \mp(D+g)/2 = \mp 0.04\,\mu{\rm m}$ and $\langle F_{x,k}^{\rm NP}(\omega)\rangle = \pm 70\,{\rm pN}$ for the peak wavelength 600 nm in figure 3(a) LCP, $-y_k^{\rm CM}\langle F_{x,k}^{\rm NP}(\omega)\rangle$ was about $2.8\,\mu{\rm m}\cdot{\rm pN}$. Its twice value $5.6\,\mu{\rm m}\cdot{\rm pN}$ was almost corresponding to the difference between LCP and RCP at 600 nm in figure 2(b), where the factor of 2 came from the difference of SAM: $+1-(-1)=2$, while the factor of 2 from $N_{\rm p}=2$ was not considered because figure 2(b) showed $\langle T_z(\omega)\rangle/N_{\rm p}$. As above, since $\langle F_{x,k}^{\rm NP}(\omega)\rangle$ was originated only form IP-LIF, therefore, we can conclude that the difference in the orbital torque in figure 2(b) is originated from the SAM through the IP-LIF.

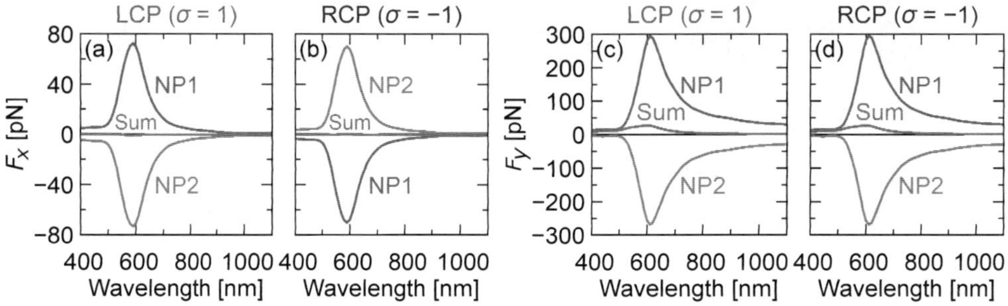

Figure 3. LIFs acting on each NP and their sum. (a)(b) x-component and (c)(d) y-component of LIF for (a)(c) LCP ($\sigma=1$) and (b)(d) RCP ($\sigma=-1$) are shown.

4. Conculusion

We have discovered that the SAM enables the modulation of the orbital torque through the IP-LIF. In particular, the IP-LIF contributes significantly to the CM motion of NPs. These results, manifesting the mutual induction of mechanical motion by the SAM and OAM, contribute to the further development of laser manipulation and laser materials processing by using LG beams.

Acknowledgments

The authors thank Prof. H. Ishihara and Prof. S. Tokonami for their support and encouragement. This work was supported by Scientific Research on Innovative Areas (No. JP16H06507), Grant-in-Aid for Scientific Research (A) (No. JP18H03884, No. JP17H00856), Grant-in-Aid for Scientific Research (B) (No. JP15H03010, No. JP18H03522), Challenging Research (Exploratory) (No. JP17K19070) from JSPS KAKENHI, JST-Mirai Program (No. JPMJMI18GA), the Key Project Grant Program of Osaka Prefecture University, and the Cooperative Research Program of "Network Joint Research Center for Materials and Devices".

References

[1] Allen L, Beijersbergen M W, Spreeuw R J C and Woerdman J P 1992 *Phys. Rev. A* **45** 8185
[2] Ashkin A 1997 *Proc. Natl. Acad. Sci. USA* **94** 4853
[3] Simpson N B, Dholakia K, Allen L and Padgett M J 1997 *Opt. Lett.* **22** 52
[4] Barada D, Juman G, Yoshida I, Miyamoto K, Kawata S, Ohno S and Omatsu T 2016 *Appl. Phys. Lett.* **108** 051108
[5] Burns M M, Fournier J-M and Golovchenko J A 1989 *Phys. Rev. Lett.* **63** 1233
[6] Iida T and Ishihara H 2006 *Phys. Rev. Lett.* **97** 117402
[7] Iida T and Ishihara H 2008 *Phys. Rev. B* **77** 245319
[8] Zhao Y, Edgar J S, Jeffries G D M, McGloin D and Chiu D T 2007 *Phys. Rev. Lett.* **99** 073901
[9] Vial A and Laroche T 2008 *Appl. Phys. B* **93** 139
[10] Johnson P B and Christy R W 1972 *Phys. Rev. B* **6** 4370

IOP Conf. Series: Journal of Physics: Conf. Series **1220** (2019) 012009 doi:10.1088/1742-6596/1220/1/012009

Bayesian Spectroscopy with a Replica Exchange Monte Carlo Method on an Excitonic Absorption Spectrum of a Cu_2O Thin Crystal

Kazunori Iwamitsu[1], **Masato Okada**[2,3], **Ichiro Akai**[4]

[1]Faculty of Science, Kumamoto University, 2-39-1 Kurokami Chuo-ku, Kumamoto 860-8555, Japan.

[2]Department of Complexity Science and Engineering, The University of Tokyo, 5-1-5 Kashiwanoha Kashiwa, Chiba 277-8561, Japan.

[3]Research and Service Division of Materials Data and Integrated System, National Institute for Material Science, 1-2-1 Sengen, Tsukuba, Ibaraki, 305-0047, Japan.

[4]Institute of Pulsed Power Science, Kumamoto University, 2-39-1 Kurokami Chuo-ku, Kumamoto 860-8555, Japan.

E-mail: `iwamitsu@kumamoto-u.ac.jp`

Abstract. To study a bi-axial stress effect in a Cu_2O thin crystal sandwiched by paired MgO substrates, we have applied Bayesian spectroscopy to decompose an absorption spectrum with employing a replica exchange Monte Carlo (RXMC) method. The absorption spectrum includes broad bands of an inter-band transition and an exciton continuous state simultaneously with excitonic resonance transitions. However, it had been difficult to decompose correctly because of a bottleneck in optimal solution search efficiency. To relieve the bottleneck, we introduced the RXMC method and succeeded in decomposing all spectral components. Posterior probabilities of the band-gap energy and the exciton binding energy distribute at lower energy side than those in stress-free bulk crystals, which provides statistical evidence to show that the bi-axial stress remains in the Cu_2O thin crystal sandwiched by MgO substrates.

1. Introduction

The exciton in Cu_2O is an important system for studying excitonic Bose-Einstein condensation. In an experiment of Cu_2O bulk crystals performed by Yoshioka *et al.* [1], a uniaxial stress was imposed to form a trapping potential for saving cold excitons spatially. On the other hand, we prepared Cu_2O thin crystals, which were recrystallized epitaxially in the small gap between paired MgO substrates [2]. In such thin crystals, although a bi-axial stress will be imposed at their interfaces due to a lattice mismatch between Cu_2O and MgO, the energy shift of the yellow exciton state would be small because the lattice mismatch is sufficiently small. So that, we applied Bayesian inference [3] to evaluate the small energy shift of the yellow exciton state [4]. This absorption spectrum includes not only the absorption lines X_n of the yellow-exciton series but also the absorption bands of an inter-band transition B and an exciton continuous state X_C. As a result, search efficiency of an optimal solution through the Markov chain Monte Carlo (MCMC) samplings deteriorates, so we had been forced to fix the spectral parameters for B and X_C [4].

In this paper, in order to improve on the analysis performed in Ref. [4], here we employ a replica exchange Monte Carlo (RXMC) method [3, 5] to improve the search efficiency of the optimal solution and to obtain posterior probability distributions of all spectral parameters also including B and X_C bands.

2. Bayesian Spectroscopy and Spectral Functions

To study the bi-axial stress effect in Cu_2O thin crystals sandwiched by paired MgO substrates, the small changes of the band-gap energy E_g and the exciton binding energy R_y will be discussed, so that we have to decompose a complicated absorption spectrum. Bayesian spectroscopy [6], in which Bayesian inference is utilized to perform spectral decomposition, can be used to evaluate probability distributions of all spectral parameters and provide us a statistical evidence of their small changes ΔE_g and ΔR_y. Let's consider an analysis of a measured spectrum $D := \{(x_1, y_1), \cdots, (x_N, y_N)\}$ explained by a physical model $\{G(x_i; \Theta)\}$, where Θ is a parameter set for all spectral components. Since random noises are superposed in the measured spectrum y_i, an error function is defined as $E(\Theta) := \sum_{i=1}^{N} [y_i - G(x_i; \Theta)]^2 / (2N)$. When the superposed noises are Gaussian random noises distributed with a standard deviation $\sigma_\mathrm{noise} := b^{-0.5}$, the conditional probability of D under Θ determined can be written as $P(D|\Theta, b) \propto \exp[-b\, N\, E(\Theta)]$ by using b and $E(\Theta)$ [4]. The Bayesian spectroscopy enables us to obtain posterior probability distributions $P(\Theta|D)$ of the spectral parameters Θ. However, when we use a Metropolis method [8, 4] for MCMC sampling, it is necessary to know the noise intensity beforehand because σ_noise determines the distribution width of $P(\Theta|D)$. On the other hand, the RXMC method makes it possible to evaluate $P(\Theta|D)$ as well as the noise intensity only from the measured data D [7].

Here, we consider a joint probability $P(\Theta, D, b)$ among Θ, D and b. According to the causality, $P(\Theta, D, b)$ is expressed as $P(D|\Theta, b)P(\Theta|b)P(b)$. By using a Bayesian theorem [9], $P(\Theta, D, b)$ can be reversely factorized as $P(\Theta, D, b) = P(\Theta|D, b)P(D|b)P(b)$. Consequently, the conditional probability $P(\Theta|D, b)$ of Θ under D and b given can be written as Eq. (1).

$$P(\Theta|D, b) = \frac{P(D|\Theta, b)P(\Theta|b)}{P(D|b)} \propto \exp[-b\, N\, E(\Theta)]\, P(\Theta|b), \tag{1}$$

where the denominator for normalization is given by marginalization of the numerator and it is defined as a partition function $Z(b)$ $[= \int d\Theta\, P(D|\Theta, b)P(\Theta|b)]$ on the analogy of statistical mechanics. Subsequently, a b-dependent Bayes free energy is defined as $F(b) := -\ln Z(b)$ [7, 10].

To determine the noise intensity ($b = \sigma_\mathrm{noise}^{-2}$) without any assumptions, we have to estimate a $\hat{b}$ that maximizes $P(b|D)$. Although the conditional probability $P(b|D)$ of b under D given is expressed as $P(D|b)P(b)/P(D)$, $P(b|D)$ is proportional to $P(D|b)$, where, since the measurement has been already completed and any assumptions are not introduced for b, the prior probabilities $P(D)$ and $P(b)$ become constant. Consequently, the $\hat{b}$ can be estimated by minimization of $F(b)$ as $\hat{b} := \arg\max_b P(b|D) = \arg\min_b F(b)$ [7, 10]. By using $\hat{b}$, the conditional probability of Θ can be sampled by the RXMC method on the basis of $P(\Theta|D, \hat{b}) \propto \exp\left[-\hat{b}\, N\, E(\Theta)\right] P(\Theta|\hat{b})$.

Open circles in Fig. 1 show the measured absorption spectrum in a Cu_2O thin crystal, and it includes spectral components of excitonic series X_n, an excitonic continuous band X_C and an inter-band transition B. For X_C and B, we used spectral functions for direct-gap and dipole-forbidden transition described in [11, 4] with considering the respective inhomogeneous broadening factors γ_{X_C} and γ_B. For X_n, Voigt functions based on an asymmetric Lorentzian function [12, 4] were employed, in which the same inhomogeneous broadening factor γ_X was introduced. Asymmetric factors A_2 and A_3 were considered only for X_2 and X_3, respectively. Although a common homogeneous broadening factor Γ_X was used for X_n, a different homogeneous broadening factor Γ_2 was required for X_2. For X_n, the transition energy

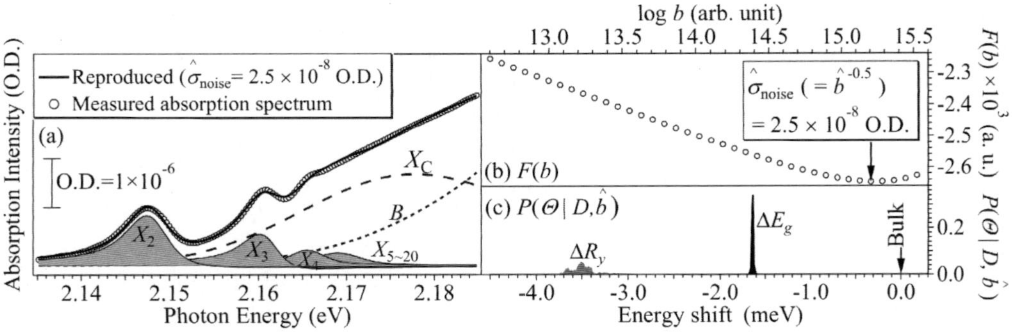

Figure 1. (a) Absorption spectra of the measured, reproduced and the respective spectral components. (b) b-dependent Bayes free energy. (c) Posterior probability distributions of energy shifts for E_g and R_y.

E_n and absorption intensity f_n were parameterized as Eq. (2) with E_g, R_y, f_0 and the quantum number n [13].

$$E_n = E_g - \frac{R_y}{n^2} \quad (\text{for} \quad n = 2, \cdots, 20), \quad f_n = f_0 \left(\frac{n^2 - 1}{n^5} \right) \quad (\text{for} \quad n = 2, 5, \cdots, 20). \qquad (2)$$

As exceptions for f_n in Eq. (2), independent absorption intensities f_3 and f_4 were required as described in [4]. As the result, the parameter set Θ for $G(x_i; \Theta)$ has fourteen spectral parameters: $\Theta = \{E_g, R_y, f_0, f_3, f_4, \Gamma_2, \Gamma_X, A_2, A_3, \gamma_X, f_{X_C}, \gamma_{X_C}, f_B, \gamma_B\}$.

3. Results and Discussion

Figure 1(b) shows the variation of Bayes free energy as a function of b. $F(b)$ minimizes at $\hat{b}$ as marked by a downward arrow, and this $\hat{b}$ value corresponds to the standard deviation of random noise distribution ($\hat{\sigma}_{\text{noise}} = \hat{b}^{-0.5}$) being $\hat{\sigma}_{\text{noise}} = 2.5 \times 10^{-8}$ in optical density (O.D.).

Table 1. $\overline{\Theta}$: mean values of Θ weighted by $P(\Theta|D, \hat{b})$, σ_Θ: distribution widths of $P(\Theta|D, \hat{b})$ shown by the standard deviation. †: f_{X_C}, γ_{X_C}, f_B and γ_B were fixed in [4].

$\overline{\Theta} \pm \sigma_\Theta$	$\overline{E_g} \pm \sigma_{E_g}$ (meV)	$\overline{R_y} \pm \sigma_{R_y}$ (meV)	$\overline{f_0} \pm \sigma_{f_0}$ $\times 10^{-8}$	$\overline{f_3} \pm \sigma_{f_3}$ $\times 10^{-8}$	$\overline{f_4} \pm \sigma_{f_4}$ $\times 10^{-8}$
RXMC	2171.36 ± 0.01	93.3 ± 0.1	8.9 ± 0.1	0.385 ± 0.009	0.21 ± 0.01
Ref. [4]	2171.392 ± 0.003	93.6 ± 0.1	8.8 ± 0.2	0.36 ± 0.01	0.18 ± 0.01
$\overline{\Theta} \pm \sigma_\Theta$	$\overline{\Gamma_2} \pm \sigma_{\Gamma_2}$ (meV)	$\overline{\Gamma_X} \pm \sigma_{\Gamma_X}$ (meV)	$\overline{A_2} \pm \sigma_{A_2}$	$\overline{A_3} \pm \sigma_{A_3}$	$\overline{\gamma_X} \pm \sigma_{\gamma_X}$ (meV)
RXMC	2.35 ± 0.07	1.69 ± 0.08	-0.197 ± 0.008	-0.34 ± 0.02	1.2 ± 0.1
Ref. [4]	2.4 ± 0.1	1.6 ± 0.2	-0.201 ± 0.008	-0.34 ± 0.02	1.2 ± 0.2
$\overline{\Theta} \pm \sigma_\Theta$	$\overline{f_{X_C}} \pm \sigma_{f_{X_C}}$ $\times 10^{-8}$	$\overline{\gamma_{X_C}} \pm \sigma_{\gamma_{X_C}}$ (meV)	$\overline{f_B} \pm \sigma_{f_B}$ $\times 10^{-4}$	$\overline{\gamma_B} \pm \sigma_{\gamma_B}$ (meV)	
RXMC	0.97 ± 0.01	13.9 ± 0.1	8.79 ± 0.09	15.06 ± 0.05	
Ref. [4]	$0.95^\dagger$	$14^\dagger$	$9.0^\dagger$	$15^\dagger$	

By using $\hat{b}$, posterior probability distributions $P(\Theta|D,\hat{b})$ for all spectral parameters were obtained through RXMC samplings. In Table 1, mean values $\overline{\Theta}$ and posterior probability distribution widths σ_Θ of Θ are summarized, where $\overline{\Theta}$ are weighted means by $P(\Theta|D,\hat{b})$ and σ_Θ are indicated in standard deviations of $P(\Theta|D,\hat{b})$ distributions. The first lines of the respective groups are the results by the RXMC method. The second lines in upper two groups are the results by the Metropolis method [4], and the second lines in the bottom group is the fixed values used in [4]. The optimal solutions $\overline{\Theta}$ obtained by the RXMC method coincide approximately with the previous result in [4], in which the spectral parameters for X_C and B were fixed to be assumed ones. In contrast, it became possible for the first time to obtain optimal solutions $\overline{\Theta}$ of fourteen spectral parameters by using RXMC method, although it was difficult with the Metropolis method [4]. By using $\overline{\Theta}$, a reproduced spectrum and the respective spectral components are displayed in Fig. 1(a), and the reproduced one can explain well the measured spectrum.

In Cu_2O thin crystals sandwiched by paired MgO substrates, we have offered a statistical evidence of that E_g and R_y become smaller than those in stress-free Cu_2O crystals in [4]. This result is completely confirmed as seen in Fig. 1(c), which shows posterior probability distributions of their energy shifts ΔE_g and ΔR_y from bulk crystals [14]. These $P(\Theta|D,\hat{b})$ have narrow distribution widths and appear in the negative shift region, and have no significant distribution at the point where the shift amount is zero.

4. Summary

By using RXMC method, Bayesian spectroscopy was applied to decompose the absorption spectrum for study of the bi-axial stress effect in Cu_2O thin crystals sandwiched by paired MgO substrates. Since the spectrum includes the absorption bands B and X_C as well as the exciton resonances X_n, there are fourteen spectral parameters by considering the inhomogeneous broadening factors. Although such spectral decomposition is difficult for the Metropolis method, we succeeded by the RXMC method, and obtained the posterior probability distributions of all spectral parameters. The measured spectrum was reproduced well by the optimal solution. On the basis of these probability distributions, we confirmed that E_g and R_y in our Cu_2O thin crystals decreases evidently from those in the stress-free crystals.

Acknowledgments

This work was supported by KAKENHI Grants-in-Aid (Nos. 16H01552, 16H04002, and 16K13824) and Grants-in-Aid for Scientific Research on Innovative Areas (JP25120009) from the Japan Society for the Promotion of Science (JSPS), and CREST (JPMJCR1761) from the Japan Science and Technology Agency (JST).

References

[1] Yoshioka K, Chae E and Kuwata-Gonokami M 2011 *Nature Commun.* **2** 328
[2] Iwamitsu K, Aihara S, Ota A, Ichikawa F, Shimamoto T and Akai I 2014 *J. Phys. Soc. Jpn.* **83** 124714
[3] Nagata K, Sugita S and Okada M 2012 *Neural Networks* **28** 82
[4] Iwamitsu K, Aihara S, Okada M and Akai I 2016 *J. Phys. Soc. Jpn.* **85** 094716
[5] Hukushima K and Nemoto K, 1996 *J. Phys. Soc. Jpn.* **65** 1604
[6] Akai I, Iwamitsu K and Okada M, 2018 *J. Phys.: Conf. Ser.* **1036** 012022
[7] Tokuda S, Nagata K and Okada M 2017 *J. Phys. Soc. Jpn.* **86** 024001
[8] Metropolis N, Rosenbluth A W, Rosenbluth M N, Teller A H and Teller E 1953 *J. Chem. Phys.* **21** 1087
[9] Bayes T and Price R 1763 *Phil. Trans. Roy. Soc.* **53** 370
[10] Iwamitsu K, Furukawa Y, Nakayama M, Okada M and Akai I 2018 *J. Lumin.* **197** 18
[11] Sturge M D 1987 in *Excitons* ed. Rashba E I and Sturge M D (Elsevier, Amsterdam)
[12] Toyozawa Y 1964 *J. Phys. Chem. Solids* **25** 59
[13] Elliott R J 1957 *Phys. Rev.* **108** 1384
[14] Gross E F 1956 *Nuovo Cimento Suppl.* **3** 672

Photocarrier injection by two-photon excitation in rubrene single crystal

K Goto, T Takayama, K Ohata, K Matsushita, and I Akimoto

Department of Applied Physics, Wakayama Univ., Sakaedani 930, Wakayama 640-8510, Japan

akimoto@sys.wakayama-u.ac.jp

Abstract. We performed photocarrier injection by two-photon excitation in rubrene crystal using laser pulses of nanosecond-duration at various wavelengths under an external electric field. Based on the excitation spectra, it is revealed that photocarriers are injected at excitation energies corresponding to the intrinsic absorption by two-photon process, in contrast to the midgap-states-mediated injection by one-photon process. This result means that the two-photon excitation method is useful to inject photocarriers free from the surface traps.

1. Introduction

Organic semiconductors are promising materials for the low cost and flexible electronics devices. A rubrene crystal, which exhibits an intrinsic mobility of 20 cm^2/Vs [1] higher than that of amorphous silicon, attracts a lot of attentions as a high mobility organic semiconductor. Therefore, this material has been considered to be a good model to demonstrate the band transport mechanism in a molecular crystal. However, precise measurements of transport parameters, such as effective masses and scattering time, have been challenging so far, because of the difficulty in carrier injection at low temperature and unstable quality of specimen during the measurement period. Optical injection of free carriers will be a reliable method to further investigations, because carriers can be injected at a controlled density without doping or interface at the electrode. In our previous study [2], we found that photocarriers are injected by dissociation of neutral Frenkel excitons under an electric field higher than 1 kV/cm at room temperature. However, decrease of carriers by trapping was a severe problem for measuring transport parameters at lower temperatures.

In this study, we investigate the way to inject carriers free from traps. We perform the two-photon excitation method under an electric field with pulsed light at wavelengths of transparent under a linear response. It has been reported that the mid-gap states, which are generated on the surface by the oxidation of a rubrene crystal [3], are intermediate states for optical carrier generation [4] and trap carriers and excitons under one-photon excitation [2]. Such states potentially disturb carrier transport by trapping and scattering. Based on the excitation spectra of transient photocarrier generation, influence of such surface states for the carrier injection is examined in comparison between the two-photon excitation and one-photon excitation at room temperature.

2. Experiment

Rubrene single crystal was prepared by the physical vapor transport method in gas phase after sublimated purification of a commercial rubrene powder (Sigma-Aldrich, R2206). The typical size of the grown crystal was $1 \times 2 \times 0.2$ mm^3, where the largest plane was the crystalline *ab*-plane. The

IOP Conf. Series: Journal of Physics: Conf. Series **1220** (2019) 012010 doi:10.1088/1742-6596/1220/1/012010

crystal was sandwiched with two transparent electrodes of indium-tin-oxide (ITO) glasses. An electric field was applied along the crystalline *c*-axis using a DC voltage source meter (Keithley, 6487J). The sample was mounted on a cold finger of a cryostat and irradiated through the ITO substrate by a nanosecond laser pulses from a tunable pulsed laser (Spectra-physics, MOPO, 5 ns, 10 Hz). One-photon excitation was performed in the wavelength ranges from 520 to 680 nm with typical incident pulse energy of 5 μJ, and two-photon excitation was performed in the wavelength range from 800 to 1200 nm with typical incident pulse energy of 100 μJ with loose focusing. Transient photocurrent was measured as a voltage drop across a 1 kΩ load resistance by an oscilloscope (Tektronix, TDS 3032) at room temperature as described elsewhere [2]. The detected voltage value was converted to that of current. Furthermore, we measured microwave response to the photocarriers generaed by the two-photon excitation inside a cylindrical microwave cavity at frequency of 9.6 GHz (Bruker, MD5W1).

3. Results and discussion

Figure 1 shows temporal profiles of transient photocurrent measured at various excitation wavelengths by (a) one-photon and (b) two-photon processes under an electric field of 71.4 kV/cm. The peak intensities of the transient response by one-photon excitation were approximately 20 times larger than those by two-photon excitation. However, we consider that the intensity of the displacement current is related to the generated carrier number but not proportional to it, because somewhat cancellation effect might appear in the displacement current for carriers generated deeply inside a crystal. The signal decayed three times faster by one-photon excitation than by two-photon excitation probably due to traps at the surface.

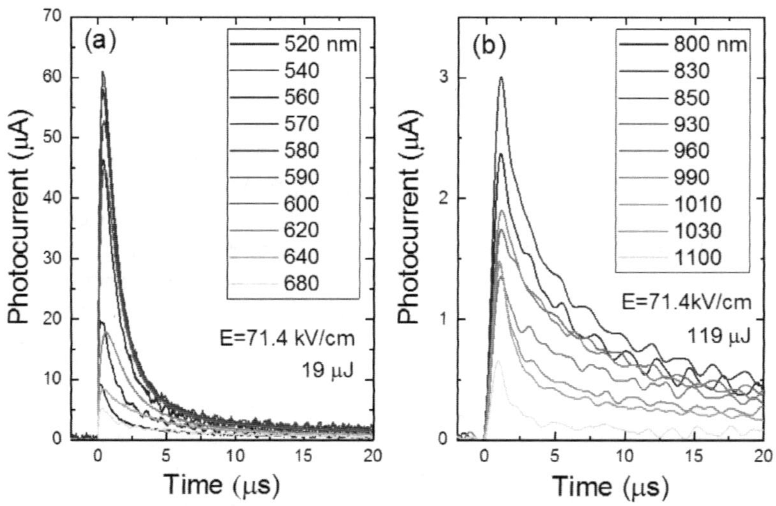

Figure 1. Time profiles of transient photocurrent at room temperature excited by excitation wavelengths (a) in the range from 520 nm to 680 nm for one-photon process and (b) in the range from 800 nm to 1200 nm for two-photon process.

Figure 2 shows the excitation pulse energy dependence of the transient photocurrent peak intensity excited at wavelengths of 530 nm and 830 nm for one- and two-photon excitations, respectively. Photocurrent injected at 530 nm increased linearly to the excitation pulse energy ensuring a one-photon process, whereas photocurrent injected at 830 nm increased as square of the excitation pulse energy ensuring a two-photon process. Therefore, we confirmed that the photocurrents can be injected by both one- and two-photon excitation processes under an external electric field. Note that the crystal was damaged by the excitation pulse energy over 285 μJ with a loose focus into a few hundred μm.

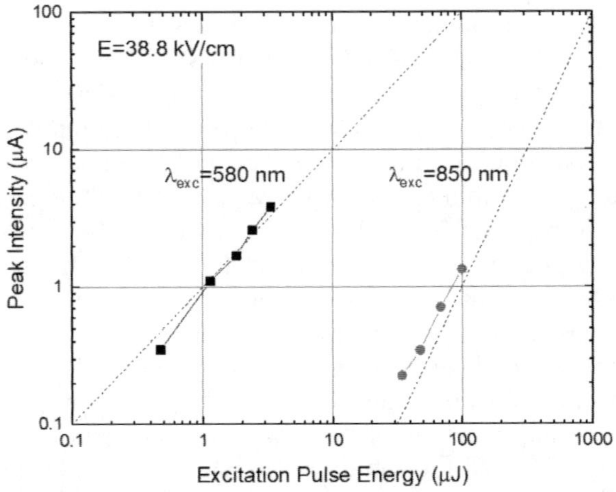

Figure 2. Excitation pulse energy dependence of the transient photocurrent peak intensity excited at 530 nm (black squares) and 830 nm (red circles). The broken lines guide the linear (left) and square (right) dependence of excitation pulse energy.

Figure 3 shows the excitation spectra of the time-integrated photocurrent, which was converted to the value in the unit of charge, measured by one- and two- photon excitations. By the one-photon excitation, a maximum was apparent around the photon energy of 2.2 eV (open circles) below the intrinsic absorption edge of rubrene crystal. This broad peak coincides with the mid-gap level which is attributed to the oxidized state on the crystal surface [3] and an intermediate state of the photocarrier generation [4]. On the other hand, by the two-photon excitation, the photocurrent appeared above the photon energy of 2.4 eV (solid squares) which is corresponding to the absorption edge for polarized light in the crystalline *ab*-plane. The spectral shape roughly traced the absorption spectrum [5] up to 3.1 eV. From these results, we emphasize that the two-photon excitation process is appropriate to generate carriers deeply inside a crystal where carriers are not affected by the surface conditions.

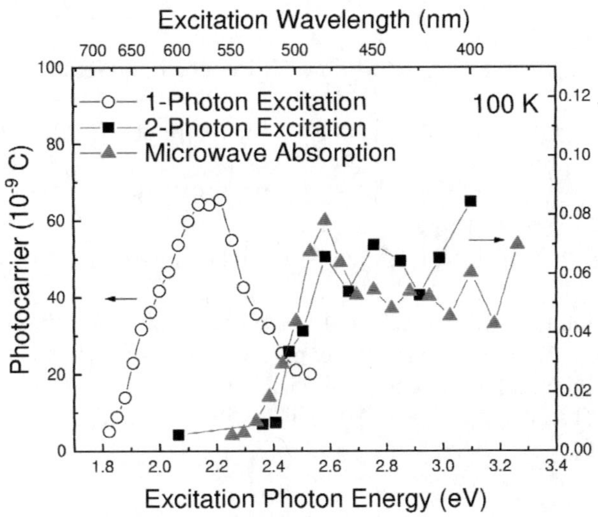

Figure 3. Excitation spectra of photocurrent injected by one- (open circles) and two- (solid squares) photon excitations. Excitation spectrum of microwave response to photocurrent injected by the two-photon excitation (pale triangles) was also shown. Data points obtained by the two-photon excitation were plotted at twice the photon energy.

In a rubrene crystal, the lowest energy edge of absorption is attributed to an electric dipolar transition from the highest occupied molecular orbital (HOMO) to the lowest unoccupied molecular orbital (LUMO). The symmetries of HOMO and LUMO are belonging to the irreducible representations of A_g and A_u, respectively, in the point group C_{2h} of a rubrene molecule. The electric dipole transition between HOMO and LUMO for one-photon excitation polarized in the crystalline *ab*-plane is allowed only by the assistance of molecular vibration of b_g-mode, so-called vibronic transition [6]. Observed absorption and photoluminescence spectra polarized in the *ab*-plane is explained by the vibronic transitions with a Raman-active b_g mode of 1520 cm^{-1} (188 meV) [7] under the Franck-Condon model. For the two-photon excitation polarized in the crystalline *ab*-plane, the HOMO-LUMO transition is allowed only by the assistance of molecular vibration of b_u-mode. Referring to an infrared absorption spectrum of rubrene [8], IR-active mode at 1494 cm^{-1} (185 meV) may allow the transition by two-photon excitation. Then, the excitation spectrum by two-photon process is roughly similar to the absorption spectrum for the polarized light in the crystalline *ab*-plane.

Finally, we add an experimental fact that a part of microwave response to the photocarriers injected by the two-photon excitation exhibited the similar excitation spectrum to that of photocarriers by the two-photon excitation (pale triangles in Fig.3). This fact indicated that the two-photon excitation method matches to the spectroscopy monitoring by microwave, because carriers injected deeply inside a crystal can be probed. It is a hint to future study by the cyclotron resonance of the free carriers in an organic crystal in order to measure the transport parameters precisely.

4. Conclusion

We measured photocurrent in a rubrene crystal with polarized light in *ab*-crystal plane by one- and two-photon excitations under an electric field. We found that photocarriers are injected by two-photon excitation free from the mid-gap levels which strongly affect the carrier injection under one-photon excitation. The excitation spectrum of photocurrent injected by two-photon excitation roughly coincided with the intrinsic absorption of rubrene crystal, which was achieved by HOMO-LUMO transition with the assistance of molecular vibrations. Therefore, Frenkel excitons generated deeply inside a crystal dissociate into charge carriers under an external electric field. The two-photon excitation method is effective to inject carriers which can move free from the surface traps for future investigation of carrier transport.

References

[1] Podzorov V, Menard E, Borissov A, Kiryukhin V, Rogers J A, Gershenson M E, 2004 *Phys. Rev. Lett.* **93** 086602

[2] Goto K, Takayama T, Ohata K, Matsushita Y, and Akimoto I, 2019 *J. Lumin.* **205** 46 (DOI: 10.1016/j.jlumin.2018.08.068)

[3] Mitrofanov O, Lang D V, Kloc C, Wikberg J M, Siegrist T, So W-Y, Sergent M A, and Ramirez A P, 2006 *Phys. Rev. Lett.* **97** 166601

[4] Najafov H, Biaggio I, Podzorov V, Calhoun M F, and Gershenson M E, 2006 *Phys. Rev. Lett.* **96** 056604

[5] Najafov H, Lee B, Zhou Q, Feldman L C and Podzorov V, 2010 *Nature Materials* **9** 938-943

[6] Irkhin P, Ryasnyanskiy A, Koehler M, and Biaggio I, 2012 *Phys. Rev.* B **86** 085143

[7] Weinberg-Wolf J R, McNeil L E, Liu S and Kloc C, 2007 *J. Phys.: Condens. Matter* **19** 276204

[8] SDBSWeb: https://sdbs.db.aist.go.jp (National Institute of Advanced Industrial Science and Technology, date of access), SDBS No. 2488.

IOP Conf. Series: Journal of Physics: Conf. Series **1220** (2019) 012011 doi:10.1088/1742-6596/1220/1/012011

Excitation energy dependence for electron traps in $CaTiO_3$:Pr, Al single crystals

Y Nanai[1], A Igarashi[2] and H Kamioka[2]

[1]Department of Electrical Engineering and Electronics, Aoyama Gakuin University, Sagamihara, Kanagawa, 252-5258, Japan

[2]Department of Physics, Nihon University, Setagaya, Tokyo, 156-8550, Japan

E-mail: nanai@ee.aoyama.ac.jp

Abstract. We investigated the red afterglow process of $CaTiO_3$:Pr, Al through photoluminescence and thermoluminescence measurements by using fabricated single crystals. The crystals with several cubic millimeter in size were obtained through a flux method. We found that the afterglow properties varied with the excitation energy. At room temperature, the red afterglow was observed for the charge transfer from Pr^{3+} to Ti^{4+} excitation and the direct excitation of Pr^{3+}, while not observed for the interband excitation of $CaTiO_3$ due to the relaxation process going through $4d^1 5d^1$ state of Pr^{3+}. For the direct excitation of Pr^{3+} at 80 K, the excited electrons did not reached to trap states for afterglow due to a potential barrier, which can overcome with thermal assistance at room temperature. For the charge transfer excitation at 80 K, the excited electrons could reach the traps with excess energy.

1. Introduction

Persistent luminescence phosphors have attracted attention for applications in warning sign, soft illumination and medical imaging. In particular, such red phosphors having long emission time are demanded to realize white afterglow. $CaTiO_3$(CTO) doped with Pr^{3+} ions is a promising candidate. Several reports of the red afterglow on $CaTiO_3$:Pr (CTO:Pr) and $CaTiO_3$:Pr, Al (CTO:Pr, Al) have been made so far [1–7]. Their red emission originates from $^1D_2 \rightarrow {}^3H_4$ transition of Pr^{3+} ions. There are three major excitation bands for this emission. That is, interband transition in CTO (Interband), metal to metal charge transfer from Pr^{3+} to Ti^{4+} (MMCT) and direct excitation in Pr^{3+} ions ($^3H_4 \rightarrow {}^3P_J$). The afterglow characteristics are largely dependent on the correlation of the energy levels of these excitation bands. However, this has not been discussed much yet. In the past reports, furthermore, the afterglow process has been discussed for powder samples. In order to discuss the excited states in more detail, it is necessary to perform spectroscopic measurements on single crystal samples.

In this work, we fabricated single crystals of CTO:Pr and CTO:Pr, Al by a flux method and investigated the relaxation processes of the excited states through photoluminescence (PL) decay and thermoluminescence (TL) glow curves with their excitation energy dependence. The temperature dependence of the relaxation processes at each excitation energy were discussed in detail based on configuration coordination diagrams.

IOP Conf. Series: Journal of Physics: Conf. Series **1220** (2019) 012011 doi:10.1088/1742-6596/1220/1/012011

2. Experimental Procedures

The single crystals of CTO:Pr and CTO:Pr, Al were grown by a flux method. A mixtures of $CaCO_3$, TiO_2, $Pr_2(CO_3)_3{\cdot}8H_2O$, Al_2O_3 powders were prepared in the molar ratio of $CaCO_3 : TiO_2 : Pr_2(CO_3)_3{\cdot}8H_2O : Al_2O_3 = 0.999 : 1-2x : 0.0005 : x$ ($x = 0$ and 0.0005). The mixture and KF flux were heated at 1100°C for 12 h in a platinum crucible. Then, they were cooled slowly back to 800°C for 40 h. Finally, we obtained several crystals from the crucible through washing the flux by hot water at room temperature. The crystal structures were identified as CTO by powder X-ray diffraction measurements. The existence of Al ions in the CTO:Pr, Al samples was confirmed through energy dispersive X-ray spectrometry measurements.

A solid-state UV laser (266 nm) and a charge-coupled device camera were used for normal PL measurements. A monochromatic output from a xenon arc lamp and a photomultiplier with a band-pass filter were used for photoluminescence excitation (PLE), PL decay curve and TL glow curve measurements, where the red emission of Pr^{3+} was only detected. In the PL decay curve measurements, the samples were exposed to an excitation light for 10 min. And then, the PL intensities were monitored for 20 min after the excitation light was blocked by a mechanical shutter. In the TL glow curve measurement, the samples were cooled to 80 K and then exposed to excitation light for 60 min. After the exposure, the TL intensity was monitored while the sample was gradually heated with a rate of 5 K/min.

3. Results and Discussions

Figure 1 shows the photographs of typical single crystal samples of (*a*) CTO:Pr and (*b*) CTO:Pr, Al. The former was light brown and $2 \times 2 \times 2$ mm in size, and the latter was light greenish yellow and $2 \times 2 \times 1$ mm in size. It was found that their body colors correlated with the density of impurities and/or vacancies were changed by Al-doping. These crystals show red emission under 365 nm excitation. Figure 2 shows the PL and PLE spectra of CTO:Pr and CTO:Pr, Al crystals. The $^1D_2 \to {}^3H_4$ emission of Pr^{3+} was commonly observed in the PL spectra. In the PLE spectra, we confirmed the three excitation bands of "Interband", MMCT and "$^3H_4 \to {}^3P_J$", as mentioned above. The band shape for MMCT was clearly affected by Al-doping. This indicates that the Al-doping changes the ligand field around Pr^{3+} ions through structural relaxation or charge compensation in CTO.

Figure 3 shows the PL decay curves of CTO:Pr and CTO:Pr, Al crystals, where they were

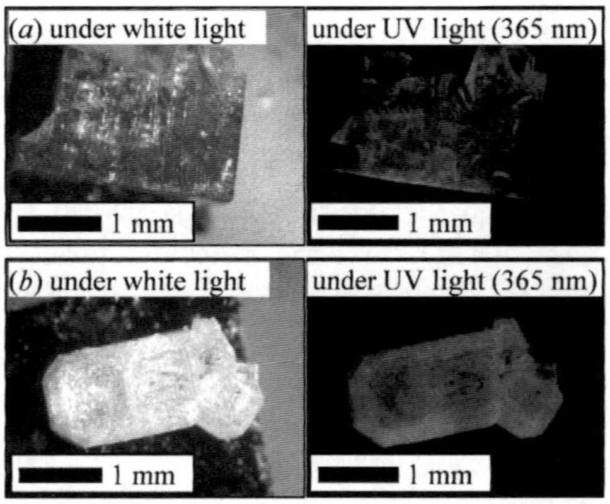

Figure 1. Photographs of (*a*) $CaTiO_3$:Pr and (*b*) $CaTiO_3$:Pr, Al single crystals.

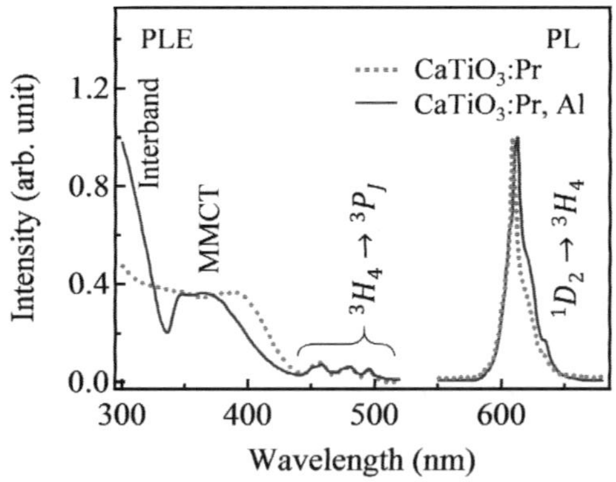

Figure 2. PL and PLE spectra of $CaTiO_3$:Pr and $CaTiO_3$:Pr, Al. Excitation wavelength for PL spectra is 266 nm. Detection wavelength for PLE spectra is 610 nm.

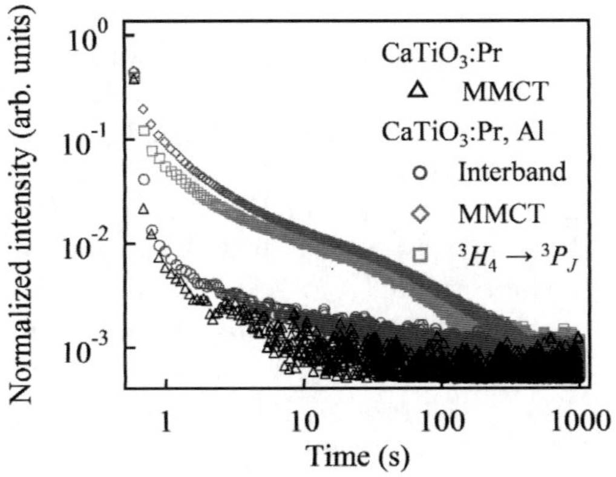

Figure 3. PL decay curves of CaTiO$_3$:Pr and CaTiO$_3$:Pr, Al for several excitation bands. These initial intensities are normalized to be 1 at time zero.

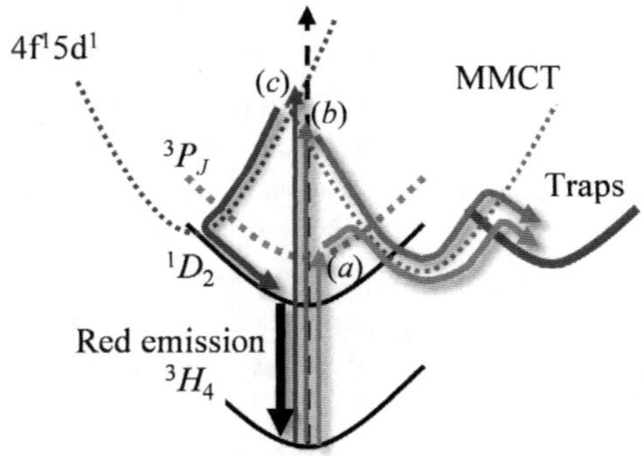

Figure 4. Configuration coordinate diagram for the (a) "$^3H_4 \rightarrow {}^3P_J$", (b) MMCT and (c) "Interband" excitation and relaxation processes in CaTiO$_3$:Pr, Al at room temperature.

excited at the excitation bands of "Interband", MMCT and "$^3H_4 \rightarrow {}^3P_J$". The corresponding excitation wavelengths were 300, 370 and 455 nm, respectively. A promising long afterglow was confirmed in CTO:Pr, Al rather than in CTO:Pr. This indicates that Al-doping contributes the formation of deep electron traps. The excited electrons can smoothly transfer to some electron traps induced by Al doping through the MMCT state. However, the holes generated complementally to the Pr ground level are thought to stay without moving because the ground level is far from the valence band maximum as 0.4 eV [4]. Therefore, the observed afterglow phenomena are considered to be mainly based on electron traps. The afterglow time τ was defined as the time which the initial PL intensity became 1/1000. CTO:Pr, Al had quite long afterglow time as $\tau = 375$ s and 221 s for the excitation bands of the MMCT and "$^3H_4 \rightarrow {}^3P_J$", respectively. On the other hand, the "Interband" excitation resulted in quite short τ of 24 s. The tendency that the relaxation time for the "Interband" excitation is much faster than the

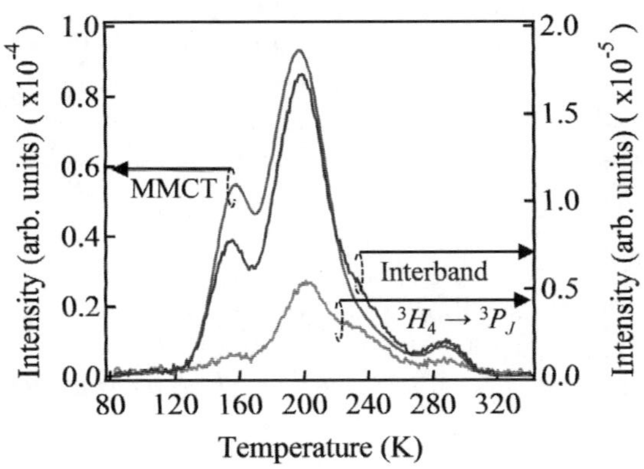

Figure 5. TL glow curves of CaTiO$_3$:Pr, Al for "Interband", MMCT and "$^3H_4 \rightarrow {}^3P_J$" excitation bands.

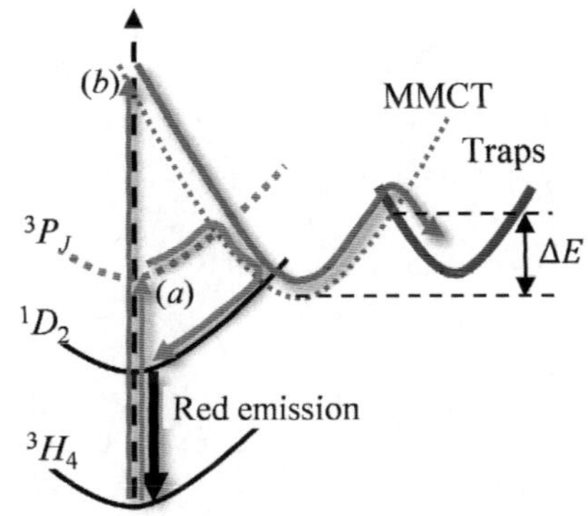

Figure 6. Configuration coordinate diagram for the (a) "$^3H_4 \rightarrow {}^3P_J$" and (b) MMCT excitation and relaxation processes in CaTiO$_3$:Pr, Al at 80 K.

other was similarly observed also when the "Interband" and MMCT were excited by a 266 nm and 355 nm laser, respectively. This suggests that the relaxation process for the "Interband" excitation is quite different from that for other excitations regardless of excitation intensity.

The excitation energy dependence of the relaxation processes can be interpreted by using a configuration coordinate diagram of energy levels for Pr^{3+} in CTO:Pr, Al. The relaxation processes after the three excitation bands were summarized in figure 4. For the "$^3H_4 \rightarrow {}^3P_J$" and MMCT excitations, excited electrons lead to some electron traps contributing to afterglow, as indicated by the processes (*a*) and (*b*) in the diagram, respectively. According to the report of Zhang et al., $4f^15d^1$ excitation band of Pr^{3+} locates around 300 nm in the PLE spectrum of CTO:Pr [8]. Therefore, the "Interband" excitation leads Pr^{3+} ion to the $4f^15d^1$ state and then to the $^1D_2$ state without reaching to the traps, as indicated by the process (*c*).

Figure 5 shows TL glow curves of CTO:Pr, Al for the three excitation bands. Each glow curve includes several intrinsic peaks. They correspond to electron traps, and the peak near 290 K contributes to the afterglow in particular. The TL intensity for the MMCT excitation was much greater than that of the other. It should be noted that the TL intensity for the "$^3H_4 \rightarrow {}^3P_J$" excitation was smaller than that for the MMCT excitation, while the afterglow time was about the same at room temperature. Regarding the peak around 290 K, we also measured the intensity by changing the starting temperature of the glow measurement. Then, the peak intensity gradually increased as the starting temperature was changed from 80 K to 180 K. This suggests that an activation energy is necessary for the state transition from the MMCT to the electron trap state. The excitation energy dependence of the relaxation processes at 80 K can be also interpreted by using the energy diagram for Pr^{3+} in CTO:Pr, Al. The relaxation processes from the two excitation bands were summarized in figure 6. For the "$^3H_4 \rightarrow {}^3P_J$" excitation, excited electrons transfer to traps through the MMCT state hardly occurs due to the potential barrier (ΔE), as indicated by the process (*a*). The barrier can be overcome by thermal assistance at room temperature. For the MMCT excitation, on the other hand, electrons can reach to traps even at 80 K with the excess energy, as indicated by the process (*b*).

4. Conclusion

The single crystals of the red afterglow phosphor CTO:Pr and CTO:Pr, Al were grown by a flux method. They showed commonly the PL for $^1D_2 \rightarrow {}^3H_4$ transition of Pr^{3+} ions. The PLE spectra had three excitation bands: "Interband", MMCT and "$^3H_4 \rightarrow {}^3P_J$". The red afterglow processes was investigated for the each excitation band, mainly focusing on CTO:Pr, Al with long afterglow time. For the "Interband" excitation, the afterglow time was much shorter than that for the MMCT and "$^3H_4 \rightarrow {}^3P_J$" excitations. The time reduction was caused by the direct relaxation from $4f^15d^1$ to $^1D_2$ state of Pr^{3+}. In the TL glow curve measurement, it was found that excited electrons for the "$^3H_4 \rightarrow {}^3P_J$" excitation hardly move to traps at 80 K due to a potential barrier. For the MMCT excitation, the electrons overcome the barrier using excess energy even at low temperature.

References

[1] Pan Y, Su Q, Xu H, Chen T, Ge W, Yang C and Wu M 2003 *J. Solid State Chem.* **174** 69
[2] Jia W, Jia D, Rodriguez T, Evans D R, Meltzer R S and Yen W M 2006 *J. Lumin.* **13** 119
[3] Boutinaud P, Sarakha L, Cavalli E, Bettinelli M, Dorenbos P and Mahiou R 2009 *J. Phys. D: Appl. Phys.* **42** 045106
[4] Katayama Y, Ueda J and Tanabe S 2014 *J. Lumin.* **148** 290
[5] Yin S, Chen D and Tang W 2007 *J. Alloys Compd.* **441** 327
[6] Yin S, Chen D and Tang W 2007 *J. Mater. Sci.* **42** 2886
[7] Li S and Liang X 2008 *J. Mater. Sci.: Mater. Electron* **19** 1147
[8] Zhang X, Cao C, Zhang C, Xie S, Xu G, Zhang J and Wang X-J 2010 *Mater. Res. Bull.* **45** 1832

IOP Conf. Series: Journal of Physics: Conf. Series **1220** (2019) 012012 doi:10.1088/1742-6596/1220/1/012012

Ultrafast photocontrol of proton-mediated organic ferroelectric cocrystal

T Umanodan[1], T Ishikawa[1], S Koshihara[1,2], S Horiuchi[3] and Y Okimoto[1]

[1]Department of Chemistry, Tokyo Institute of Technology, Meguro, Tokyo 152-8551, Japan
[2]Institute of Innovative Research (IIR), Tokyo Institute of Technology, Meguro, Tokyo 152-8551, Japan
[3]National Institute of Advanced Industrial Science and Technology (AIST), Tsukuba, Ibaraki 305-8565, Japan

E-mail: umanodan.t.aa@m.titech.ac.jp

Abstract. Photoirradiation effect in a ferroelectric cocrystal Hdppz-Hca is investigated in terms of time-resolved second harmonic (SH) generation measurements. By the photoexcitation with a visible pulse (530 nm) causing the intramolecular transition of Hca molecule, the SH intensity was suddenly suppressed by ≈30% on the time scale of picosecond, while the reflectivity scarcely changed. This result is discussed based on reversal of the polarized domain by the photoexcitation.

1. Introduction

Ferroelectricity is one of the most important topics in condensed matter physics on account of several applications, e.g., mechanical actuator, capacitor, memory etc [1]. In addition to conventional oxide systems, some organic ferroelectrics have been developed. The organic system possesses several advantages in the light of various practical applications, such as mechanical flexibility, printability and cost-effectiveness. Among them, proton-mediated supramolecular cocrystals composed of proton-donor and -acceptor (DA) molecules have been developed as novel organic ferroelectric materials. Among them, the ionic cocrystal Hdppz-Hca (dppz = 2,3-di(2pyridinyl)pyrazine, Hca = chloranilate, as depicted in Figure 1(a)) shows excellent ferroelectricity with high Curie temperature (T_c = 402 K) and a large spontaneous polarization [2].

In Hdppz-Hca cocrystal, protons mediating hydrogen bonds exhibit ordering and disordering transition similar to that observed in KH_2PO_4 (KDP) [3]. Below T_c, the proton takes one of the stable positions in the double welled potential showing long range ordering, which realizes the ferroelectricity breaking the inversion symmetry. Above T_c, by contrast, protons can thermally cross the potential barrier, which restores the inversion symmetry of the crystal. Thus, proton dynamics plays an essential role in the ferroelectric nature of Hdppz-Hca cocrystal.

In addition to the chemical approach, there have been reported some studies on photo-control of ferroelectric materials using femtosecond pulsed laser system [4–6]. In the previous study, we reported the photoinduced dynamics in terms of the time-resolved second harmonic (SH) spectroscopy which can detect the temporal changes in macroscopic ferroelectricity [7,8]. The photoexcitaion with a visible light (530 nm) causing intra-molecular transition (π-π^*) of Hca molecule [9] suppressed the SH intensity on the time scale of nanosecond, suggesting the macroscopic disordering of protons through the π electron-proton coupling by light [7]. However, the initial state of this observed photoexcited dynamics has not been studied. Here, we investigate the femtosecond dynamics of Hdppz-Hca cocrystal and discuss the origin of the ultrafast photonic change of the ferroelectricity.

2. Method

A rectangular single crystal of Hdppz-Hca was grown by the diffusion method in acetone solution as reported in [2]. The orientation of the crystallographic axes was determined by X-ray diffraction and polarized reflection spectroscopy.

The transient relative changes in SH intensity ($\Delta I_{SH}/I_{SH}$) and in reflectivity ($\Delta R/R$) were obtained by a conventional pump-probe technique using a regenerated mode-locked amplified Ti:sapphire laser system with about 120 fs pulse width, a 1 kHz repetition rate, and approximately 1.55 eV photon energy. The amplified light was separated into two beams and we used one beam for pumping the crystal and the other as a probe pulse to detect the reflectance and SH light. To photoexcite the π-π^* transition of Hca molecule, we converted the fundamental pulse into signal and idler pulses using an optical parametric amplifier and then obtained 530 nm pulse by sum-frequency generation from the signal and the fundamental pulse. After irradiation with the visible pulse, we monitored the change in reflectivity at around 1.55 eV and the SH light of the 1.55 eV (i.e., about 3.1 eV) generated from the sample.

The SH signal was detected by a photomultiplier after the SH light went through high-pass filters and a grating-type monochromator to cut out the fundamental pulses, while the reflectance light was directly detected by a Si photodiode. The polarizations of the pump and probe pulse were along the *c*-axis (the polarization axis). All the measurements were done at room temperature.

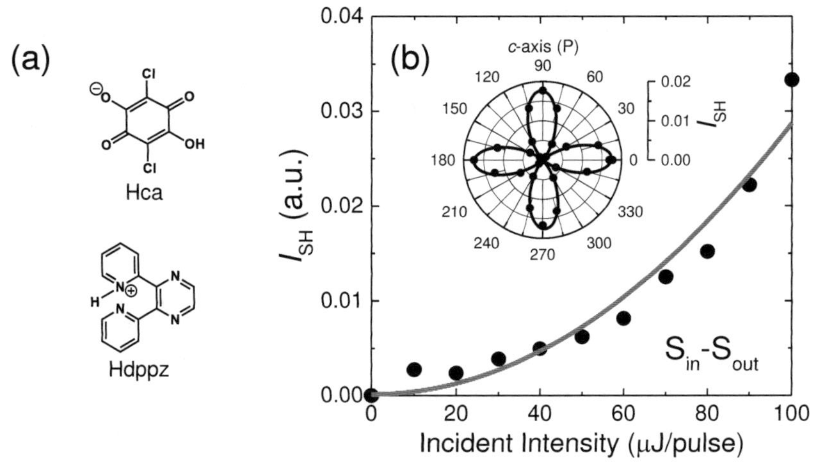

Figure 1. (a) Structural formula of Hca and dppz molecule. (b) Power dependence of SH intensity (I_{SH}) of Hdppz-Hca. The inset shows the polar plot of the observed I_{SH}.

3. Results and discussion

3.1. Bulk SH signal generated from Hdppz-Hca cocrystal

We first investigated the bulk SH signal generated from Hdppz-Hca cocrystal to evaluate the degree of $\chi^{(2)}$ of the sample. In Figure 1(b), we show the incident pulse intensity dependence of the SH signal (I_{SH}) in Hdppz-Hca by filled circles. The polarization of the incident light is along the polarization axis (*c*-axis) and the SH measurements were done in the geometry of S-in and S-out configuration. The value of I_{SH} nonlinearly increases as the incident intensity increases, and is well fitted by the parabolic curve as shown by the solid line, directly indicating SH generation from the cocrystal.

For reference, in the inset of Figure 1(b), we show the polar plot of the intensity of the *c*-axis polarized SH pulse as a function of the incident angle. The angles of 90° and 270° in the inset denote that the polarization of the incident pulse corresponds to the *c*-axis (i.e., S-in configuration). The SH intensity is plotted along the radial direction and its relative scale is written in the inset. The SH shows remarkable angle dependence and the pattern is a four-leaf shape in the directions of the parallel and perpendicular to the *c*-axis. To estimate $\chi^{(2)}$ of the sample, we analyze the obtained angle dependence considering the crystal symmetry of Hdppz-Hca (monoclinic, space group *Cc*, point group *m* [2]). Under these circumstances, the observed angle dependence can be described as

$$I_{SH} \propto (d_{31}\cos^2\theta + d_{33}\sin^2\theta + 2d_{35}\cos\theta\sin\theta)^2, \quad (1)$$

where d_{31}, d_{33} and d_{35} are components of the nonlinear optical coefficient tensor [10]. We performed a fitting analysis based on the equation (1) and the result is shown by the solid line in the inset of Figure 1(b). The values of the obtained components are $d_{31}/d_{33} \approx -0.99$, and $d_{35} \approx 0$.

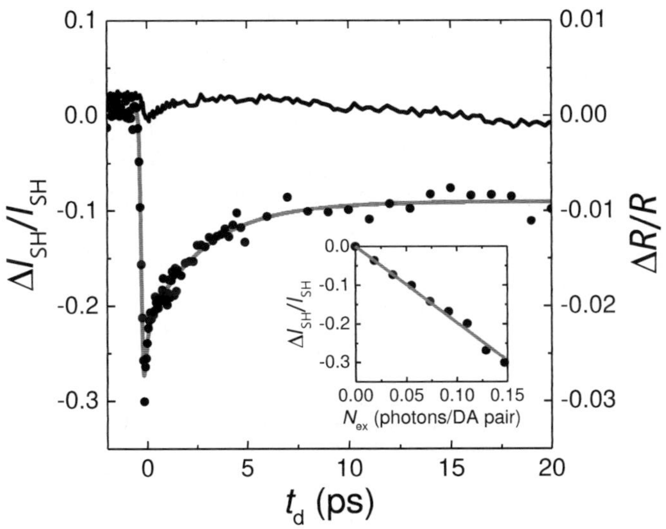

Figure 2. Time profile of $\Delta I_{SH}/I_{SH}$ (filled circles) and $\Delta R/R$ (black line). The red line shows the fitting curve of $\Delta I_{SH}/I_{SH}$. The inset shows the number of photons for the excitation per DA pair (N_{ex}) dependence of $\Delta I_{SH}/I_{SH}$ at ≈0 ps.

3.2. Time resolved SHG spectroscopy

Next, we discuss the photoexcited state in Hdppz-Hca cocrystal with irradiation of the visible pulse. Figure 2 shows the delay time (t_d) dependence of the relative change in SH intensity ($\Delta I_{SH}/I_{SH}$). As depicted by the filled circles, the SH intensity suddenly decreases by ≈30% just after photoexcitation and shows a fast recovery. To evaluate the relaxation dynamics of $\Delta I_{SH}/I_{SH}$, we fitted the profile by the following function:

$$f(t) = I_1\exp\left(-\frac{t}{\tau_1}\right) + I_2\exp\left(-\frac{t}{\tau_2}\right) + I_3. \quad (2)$$

The first and second terms denote exponential decays of the photoexcited state, and I_1, I_2, τ_1 and τ_2 amplitude of the change and the decay time of the first and second terms, respectively. The third term is a constant component which is constant when t_d is positive ($t_d > 0$). We performed the fitting analysis considering the convolution between $f(t)$ and the response function of the present fs laser system. We plotted the fitting result based on equation (2) by the red line, which well fit the observed time profile. The obtained relaxation times are $\tau_1 \approx 0.15$ ps and $\tau_2 \approx 3.0$ ps.

For comparison, we display the time dependence of the relative change in reflectivity ($\Delta R/R$) by the solid line in Figure 2. Like the time profiles of $\Delta I_{SH}/I_{SH}$, $\Delta R/R$ also shows decaying by the relaxation of the π^* electrons after the photoexcitation. However, the amplitude of the reduction was even smaller than that of $\Delta I_{SH}/I_{SH}$, which is quite similar to the case of the inorganic polar material, $BiCoO_3$ [6]. This suggests that the π–π* excitation causes some photoirradiation effects that largely change SH intensity but hardly modulates reflectivity in this system.

Here, let us discuss microscopic origins of the observed $\Delta I_{SH}/I_{SH}$. Very lately, Iwano *et al.* [11] noticed a croconic acid crystal that also shows the proton mediated ferroelectricity and theoretically revealed that the photoexcited π^* electrons adiabatically act on the protons and cause resultant reversal of polarized domain. This domain reversal driven by the cooperative proton movement is an origin of the observed photonic suppression of the SH signal. It is worth noting that this is consistent with the observed small reflectance change by light, because reflectivity itself is not affected by polarized domain orientation of a crystal. Under the scenario, the two relaxation times can be assigned as time for the domain reversal (τ_1) and relaxation from the depolarized domain (τ_2).

The inset of Figure 2 shows excitation power dependence of $\Delta I_{SH}/I_{SH}$ just after the photoexcitation ($t_d \approx 0$ ps). The horizontal axis (N_{ex}) denotes number of photons for the excitation per DA pair. (The data of $\Delta I_{SH}/I_{SH}$ displayed in Figure 2(a) has been obtained with $N_{ex} \approx 0.15$). $\Delta I_{SH}/I_{SH}$ linearly increases with the photons, indicating that $\Delta I_{SH}/I_{SH}$ is almost proportional to the number of the injected photons. The maximum value of $|\Delta I_{SH}/I_{SH}|$ is about ≈30% just after the photoexcitation and, in that case, the efficiency of the photonic change is estimated as ≈3.5 DA pairs per one photon show the polarization reversals.

4. Summary

We investigated ultrafast nonlinear optical properties and the photoexcited state in a proton-mediated organic ferroelectric cocrystal, Hdppz-Hca. We irradiated the sample with 530 nm fs laser pulse to excite the intramolecular π-π^* transition of Hca molecule. After the irradiation, the SH intensity abruptly decreased by ≈30%, implying reversal of the ferroelectric domains driven by the photoinduced proton transfer. This observed results can be viewed as a novel technique of controlling the ferroelectricity by light at room temperature.

Acknowledgments
The authors thank S. Naruse and Y. Iita for their technical assistance. This research was supported by JSPS KAKENHI Grant No. JP16K05397.

References

[1] Lines M E and Glass A M 2001 *Principles and Applications of Ferroelectrics and Related Materials* (New York: Oxford University Press)
[2] Horiuchi S, Kumai R and Tokura Y 2013 *J. Am. Chem. Soc.* **135** 4492
[3] Blinc R 1960 *J. Phys. Chem. Solids.* **13** 204
[4] Yamamoto K, Iwai S, Boyko S, Kashiwazaki A, Hiramatsu F, Okabe C, Nishi N and Yakushi K 2008 *J. Phys. Soc. Jpn.* **77** 074709
[5] Miyamoto T, Yada H, Yamakawa H and Okamoto H 2013 *Nat. Commun.* **4** 2586
[6] Okimoto Y, Naruse S, Fukaya R, Ishikawa T, Koshihara S, Oka K, Azuma M, Tanaka K and Hirori H 2017 *Phys. Rev. Applied* **7** 064016
[7] Umanodan T, Tanaka S, Naruse S, Ishikawa T, Onda K, Koshihara S, Horiuchi S and Okimoto Y 2015 *J. Phys. Soc. Jpn.* **84** 073707
[8] Umanodan T, Kaneshima K, Takeuchi K, Ishii N, Itatani J, Hirori H, Sanari Y, Tanaka K, Kanemitsu Y, Ishikawa T, Koshihara S, Horiuchi S and Okimoto Y 2019 *J. Phys. Soc. Jpn.* **88**, 013705
[9] Kumai R, Horiuchi S, Okimoto Y and Tokura Y 2006 *J. Chem. Phys.* **125** 084715
[10] Authier A 2013 *International Tables for Crystallography* vol D 2nd online ed, ed A Authier (Chester: International Union of Crystallography) part 1 pp 18
[11] Iwano K, Shimoi Y, Miyamoto T, Hata D, Sotome M, Kida N, Horiuchi S and Okamoto H 2017 *Phys. Rev. Lett.* **118** 107404

IOP Conf. Series: Journal of Physics: Conf. Series **1220** (2019) 012013 doi:10.1088/1742-6596/1220/1/012013

Effect of modulation of ultrafast transient carrier dynamics by interface on terahertz signal

Shintaro Yamamoto, Osamu Kojima, Takashi Kita

Department of Electrical and Electronic Engineering, Graduate School of Engineering, Kobe University, 1-1 Rokkodai, Nada, Kobe, 657-8501, Japan

E-mail: 175t266t@stu.kobe-u.ac.jp

Abstract. The carrier dynamics in the nanostructured semiconductors related to the drift and diffusion currents strongly affects the device performance, such as the efficiency of the carrier injection into the active layer. We report on the effects of the photocurrent direction on the THz signals emitted from the GaAs crystal including an interface. The polarity inversion of the signal is caused by the change in photocurrent direction from diffusion to drift. The inversion is not affected by electron lifetimes. These results suggest that measurement of the THz wave is useful to consider the photocurrent direction.

1. Introduction

The light emitting devices based on the semiconductor quantum dots (QDs), such as the QD lasers in the ultrafast optical telecommunication systems [1-3] and the single photon sources in the information communication systems [4-6], have been studied. In general, the self-assembled QDs are grown in the Stranski-Krastanov mode, so that the potential is modulated by the lattice-mismatched strain [7, 8]. The potential modulation due to the strain, which changes the current injection direction, is important when the efficiency of QD based devices is considered. The carrier dynamics, in particular, injection process to InAs QDs from GaAs has been discussed from the results measured by a pump-probe technique [8, 9] and the time-resolved photoluminescence spectrum [10, 11]. On the other hand, it is difficult to directly define the direction of carrier diffusion using these techniques.

On the other hand, the observation of THz time-domain signals emitted from the semiconductor surfaces provides the diffusion direction of carrier excited by the ultrashort pulse lasers [12-14]. For instance, in the n-type (p-type) semiconductors the THz signal shows the initial increase (decrease), which originates from difference in the diffusion direction; from the surface to the substrate (from the substrate to the surface). Therefore, the effects of the potential modulation caused by the lattice-mismatched strain on the carrier diffusion direction can be revealed by measuring the THz time domain signal [15]. In this work, we demonstrate the calculation results of the THz signal emitted from the GaAs crystal including an interface, which causes the potential modulation, near the surface. The origins of THz signal are divided into two parts; from the surface to the interface and from the interface to the surface. When the direction of photocurrent in the deeper region is changed, the polarity of the THz signal is inverted. This inversion is discussed from the aspects of the balance between the diffusion and drift, and carrier lifetimes.

2. Calculation method

In the calculation of the THz signals, the potential structure shown in figure 1 was assumed for simplicity. In the GaAs crystal with the total thickness of 650 nm, the interface modulating the potential locates at 150 nm from the surface, which is similar potential structure including the δ-doping layer. In the calculation, the weak electric field was considered. Hence, the overshoot is not caused. The region from the surface to the interface and that from the interface to the bottom are referred as L_1 and L_2, respectively. In the L_1 region, the electrons move toward the interface due to the diffusion and drift. On the other hand, in the L_2 region, while the electrons move to the substrate due to diffusion, those move toward the interface due to drift. The photocurrent generated by ultrashort pulse can be expressed as follows [10]. The signal rise was considered by Gaussian profile.

$$J(t) \propto \frac{e^{-\frac{\gamma t}{2}}}{t}\left\{\sin\left(\frac{\omega_0+\omega_{L1}}{2}\right)\sin\left(\frac{\omega_0-\omega_{L1}}{2}\right) - B\sin\left(\frac{\omega_{L1}+\omega_{L2}}{2}\right)\sin\left(\frac{\omega_{L1}-\omega_{L2}}{2}\right)\right\}, \quad (1)$$

where $\omega_0 = (\omega_{exc}^2 - \gamma^2)^{\frac{1}{2}}$, $\omega_{L1} = \left(\omega_{exc}^2\exp(-\alpha L_1) - \frac{\gamma^2}{4}\right)^{\frac{1}{2}}$ and $\omega_{L2} = \left(\omega_{exc}^2\exp(-\alpha L_2) - \frac{\gamma^2}{4}\right)^{\frac{1}{2}}$.

And $\omega_{exc} = \left(\frac{e^2 n_{exc}}{\kappa\kappa_0 m^*}\right)^{\frac{1}{2}}$ is the plasma frequency of photoexcited carriers at the surface. α is the absorption coefficient in the GaAs crystal. κ and κ_0 are the relative permittivity and permittivity of vacuum, respectively. m^* is the reduced mass of an electron-hole pair in GaAs crystal, and e is the elementary charge. n_{exc} is the density of photoexcited electron-hole pairs at the surface. In the calculation, parameters $n_{exc}=1\times10^{17}$ cm^{-3}, $\gamma=5$ ps^{-1}, $\alpha=1.3$ μm^{-1}, $\kappa=12.9$ and $\kappa_0=8.8\times10^{-11}$ m^{-3}kg^{-1}s^{4}A^2 were used [16-18]. Here, the parameter B expresses the ratio of drift and diffusion in the L_2 region in figure 1. The THz signals were obtained by numerical differentiation of the equation (1).

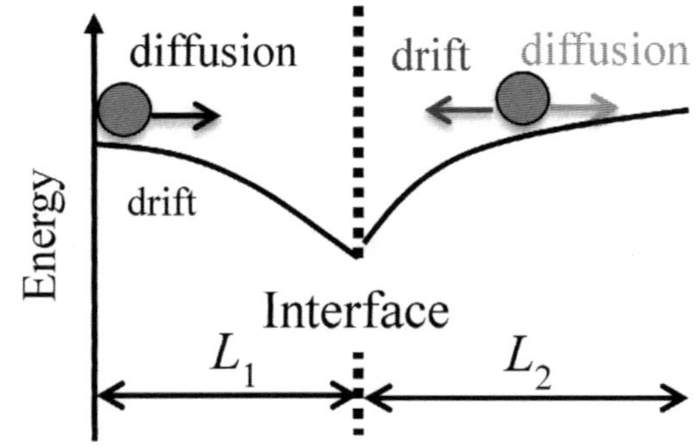

Figure 1. The potential structure for calculations. The dotted line locates on the interface. In the L_2, the direction of carrier movement by drift and diffusion is opposite.

3. Results and discussion

Figure 2(a) indicates the dependence of the THz signals on the value of B. At B=0 there is no contribution of the carriers in the L_2 region; the signals results from the photocurrent only in the L_1 region. Decreasing B, which corresponds to an increase in the contribution due to diffusion current from the L_2 region, the amplitude of the signal gradually increases. On the other hand, increasing B, which increases the drift current from the substrate to the surface, the amplitude decreases until B=0.5 and then increases.

Focusing on the amplitude at t=0 in figure 2(a), the polarity of the signal is inverted from positive to negative. To clarify the polarity inversion, the amplitude at t=0 of the THz signals was plotted in figure 2(b) by closed circles. Increasing the value of B is the amplitude monotonically decreases and the polarity inversion occurs at around B=0.6. Furthermore, the frequency changes at B=0.6 where polarity inversion, as shown in figure 2(b) by open circles. Here, the frequency is much larger than the reported experimental values. The frequency in equation (1) is sensitive to n_{exc}. Therefore, the calculated frequency may not be correct. However, the dependence on B is considered to be worth discussing. The polarity inversion does not originate from the initial carrier distribution due to the light absorption. As indicated in figure 2(c), the initial carrier distribution in the L_1 and L_2 regions are hardly changed. The vertical line shows the location of the interface. The light intensities with the photon energies of 1.38 eV, 1.46 eV and 1.55 eV in a GaAs crystal calculated using the absorption coefficient [18] were indicated by the gray, dotted and solid curves, respectively. Considering the number of carriers generated by ultrashort pulse depending on the light intensity, there is no remarkable difference in the photon energies. Therefore, as the physical mechanism to invert the polarity of THz signals, the effect of the potential modulation caused by the interface was considered.

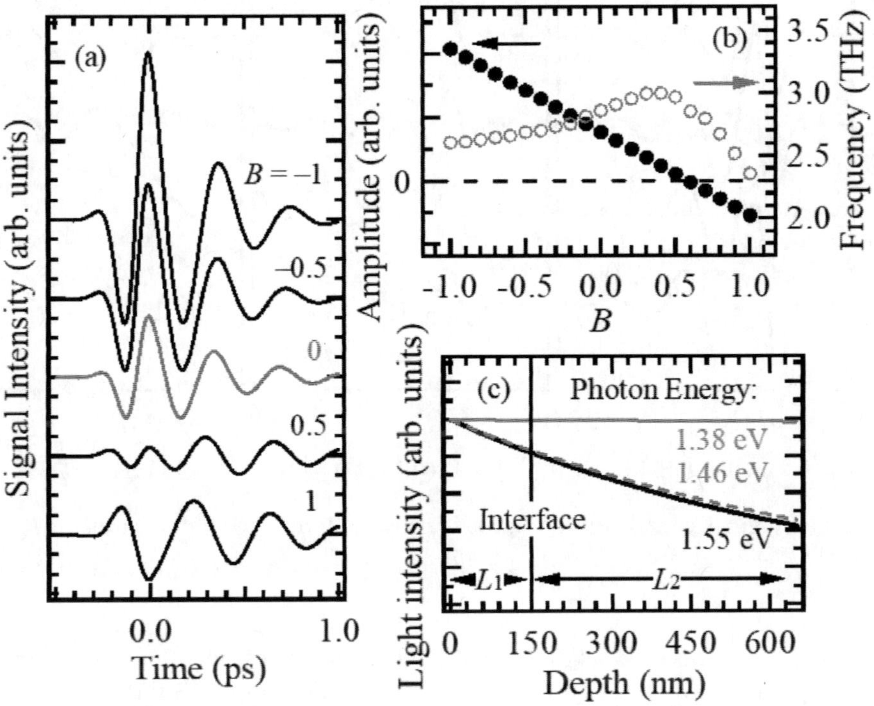

Figure 2. (a) The THz signal calculated for various B values. (b) The amplitude at t=0 and the frequency were plotted by the closed and open circles, respectively. (c) The light intensity profile in the GaAs crystal calculated for the light with the energies of 1.38 eV, 1.46 eV and 1.55 eV. The vertical line indicates the position of the interface.

The polarity inversion is not changed by the carrier lifetimes. The terahertz signals depending on the electron lifetime calculated for the various B values are shown in figure 3(a), 3(b) and 3(c). Clearly, the signal does not depend on the electron lifetimes. In the case of γ_{L2}=2.0 γ_{L1}, although electron in the L_2 region disappears much faster than that in the L_1 region. That signal inverted in the case of B=1. Furthermore, in all conditions, the variation of the electron lifetimes in the L_2 region hardly changes the signal amplitude. Our results suggest that the measurement of time-domain THz signal can give important information about the current direction in the deeper area.

4. Conclusion

We have investigated the effects of the change in the photocurrent direction on the THz signal emitted from the semiconductor surface, which caused by the interface. When the drift current increases in the region underneath the interface, the polarity of the THz signal was inverted. On the other hand, the signal, including polarity inversion, hardly depends on the electron lifetime. These results suggest that measurement of the THz wave is useful to consider the photocurrent direction and contributes to improve the efficiency of the devices based on QDs.

Acknowledgments

This work was supported by JSPS KAKENHI Grant No. 16KK0129 and the Research Foundation for the Electrotechnology of Chubu.

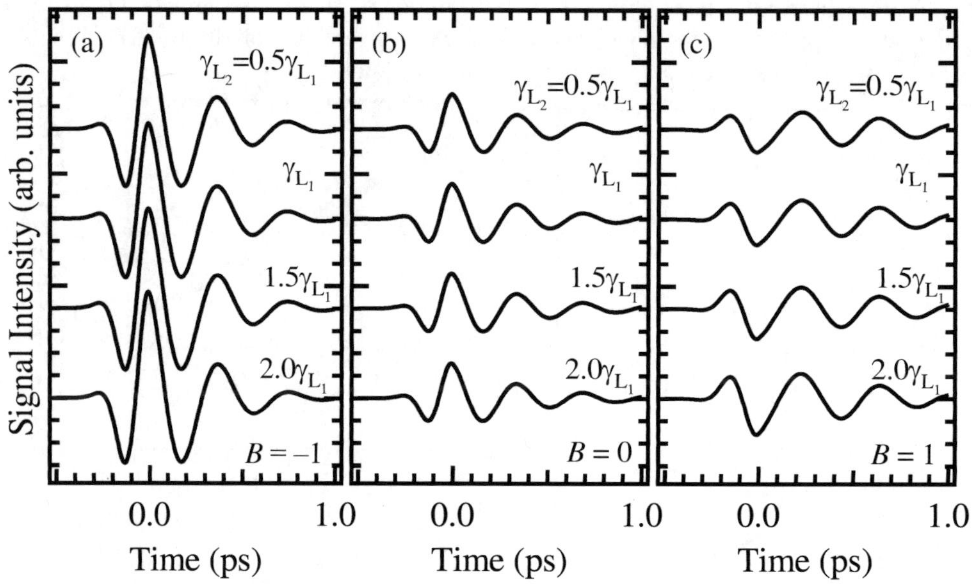

Figure 3. The THz signals calculated for various ratios of γ_{L2} and γ_{L2}. In the cases (a) $B=-1$, (b) B=0 and (c) B=1.

References

[1] Huffaker D L, Park G, Zou Z, Shchekin O B and Deppe D G 1998 *Appl. Phys. Lett.* **73** 2564
[2] Salhi A, Fortunato L, Martiradonna L, Cingolani R, De Vittorio M and Passaseo A 2006 *J. Appl. Phys.* **100** 123111
[3] Chen S, Li W, Wu J, JiangQ, Tang M, Shutts S, Elliott S N, Sobiesierski A, Seeds A J, Ross I, Smowton P M and Liu H 2016 *Nat. Photon.* **10** 307
[4] Chang W H, Chen W Y, Chang H S, Hsieh T P, Chyi J I and Hsu T M 2006 *Phys. Rev. Lett.* **96** 117401
[5] Stevenson R M, Young R J, Atkinson P, Cooper K, Ritchie D A and Shields A J 2006 *Nature* **466** 217
[6] Senellart P, Solomon G and White A 2017 *Nat. Nanotechnol.* **12** 1026
[7] Grundmann M, Stier O and Bimberg D 1995 *Phys. Rev.* B **52** 11969
[8] Jaskólski W, Zieliński M, Bryant G W and Aizpurua J 2006 *Phys. Rev.* B **74** 195339
[9] Feldmann J, Cundiff S T, Arzberger M, Böhm G and Abstreiter G 2001 *J. Appl. Phys.* **89** 1180
[10] Liu H-Y, Meng Z-M, Dai Q-F, Wu L-J, Guo Q, Hu W, Liu S-H, Lan S and Yang T 2008 *J. Appl. Phys.* **103** 083121
[11] Pulizzi F, Kent A J, Patanè A, Eaves L and Henini M 2004 *Appl. Phys. Lett.* **84** 3046
[12] Karachinsky L Y, Pellegrini S and Buller G S 2004 *Appl. Phys. Lett.* **84** 7
[13] Birkedal D, Hansen O, Sørensen C B, Jarasiunas K, Brorson S D and Keiding S R 1994 *Appl. Phys. Lett.* **65** 79
[14] Nakajima M, Hangyo M, Ohta M and Miyazaki H 2003 *Phys. Rev.* B **67** 195308
[15] Kojima O, Izumi R, Kita T 2018 *J. Phys. D: Appl. Phys.* **51**, 305102
[16] Shi Y, Yang Y, Xu X, Ma S, Yan W and Wang L 2006 *Appl. Phys. Lett.* **88** 161109
[17] Reklaitis A 2011 *J. Appl. Phys.* **109** 083108
[18] Aspnes D E and Studna A A, 1983 *Phys. Rev.* B **27** 985
[19] Reklaitis A and Reggiani L, 2000 *Phys. Rev.* B **62** 16773
[20] Casey H C Jr, Sell D D and Wecht K W 1975 *J. Appl. Phys.* **46** 250

Electronic state of nitrogen in doped titanium dioxide

H Funabiki[1], K Ozawa[1], and T Sekiya[1]

[1]Faculty of Engineering, Yokohama National University, 79-5 Tokiwadai, Hodogaya, Yokohama, 240-8501, Japan

funabiki-haruka-sx@ynu.jp

Abstract. A yellow N-doped TiO_2 powder was prepared by the sol-gel method using titanium tetra-isopropoxide and hydrazine monohydrate. An EPR signal was induced by the irradiation of light with $h\nu \geq 2.2$ eV to the yellow TiO_2 powder and the signal intensity gradually decreased in the dark. The EPR signal was due to thermally trapped carriers interacting with the doped N. The thermal activation energy of the trapped carriers was estimated to be 0.22 eV. XPS measurement was performed on the yellow powder as well as colorless and blue N-doped powders obtained by oxidation and reduction, respectively. The N1s bands at 399.5 eV can be assigned to nitrogen adsorbed on the surface. The atomic valence of nitrogen in the colorless, yellow and blue powders was positive, neutral and negative, respectively. The atomic valence of nitrogen in doped TiO_2 depends on the oxidation-reduction procedure.

1. Introduction

Titanium dioxide is a UV-driven photocatalytic material in practical use, whose valence and conduction bands are mainly formed by oxygen 2p and titanium 3d orbitals, respectively. The irradiation of UV light excites the valence electrons to the conduction band, resulting in the generation of holes in the valence band. The photogenerated holes and electrons contribute to the oxidation and reduction reactions at the surface, respectively. A first- principles calculation indicated that N-doped anatase titanium dioxide has an optical absorption band slightly below the band gap and is yellow [1]. Therefore, it is expected that such N-doped titanium dioxide can be used as a visible-light-sensitive photocatalyst.

We have investigated the properties of anatase titanium dioxide, particularly its defects and impurities [2, 3]. As-grown anatase single crystals grown by chemical vapor transport with NH_4Cl as a transport agent generally exhibited a pale blue color. Yellow or colorless single crystals were obtained by heat treatment of the pale blue crystals in oxygen. Reduction of the colorless crystals in a hydrogen gas flow resulted in a change in their color to pale blue. This reversible change in the crystalline color depending on the oxidation-reduction procedure appears to be due to the effect of impurities and/or a variety of defects. Optical absorption measurement shows that an absorption band at 2.9 eV is a characteristic of the yellow crystal but cannot be observed in colorless and pale blue crystals [2]. Electron paramagnetic resonance (EPR) measurement of the yellow crystals revealed the presence of a pair of EPR triplets and the inclusion of nitrogen as an impurity [3]. The nitrogen impurity does not detach from the crystal during the heat treatments for oxidation and reduction. Thus, the colorless and pale blue crystals must also include nitrogen. Therefore, the dependence of the change in the electronic state of nitrogen on the heat treatment is of interest. To investigate the electronic state of doped nitrogen in crystals with different colors, we prepared a titanium dioxide powder rich in nitrogen by the sol-gel method and measured its EPR and X-ray photoelectron spectroscopy (XPS) spectra.

2. Experimental

N-doped anatase titanium dioxide powders were obtained by the sol-gel method. Titanium tetra-isopropoxide and hydrazine monohydrate were dissolved in ethanol. The solution was stirred and refluxed at 80 °C for 3 h to obtain a precipitate. The precipitate was dried at room temperature and heated in an ammonia gas flow. Finally, a yellow powder was obtained by calcination at 500 °C. A colorless powder was also obtained by heat treatment of the yellow powder at 650 °C for 1 h in air. Reduction of the colorless powder at 650 °C for 1 h in a hydrogen gas flow resulted in a change in its color to blue. X-ray diffraction (XRD) spectra of the powders were obtained using a SmartLab diffractometer (Rigaku). The diffusion reflectance was measured by a UV365 spectrometer (Shimadzu). EPR measurement with visible-light irradiation was performed on the yellow powder using a JES-FA200 spectrometer (JEOL). Resonance field and microwave frequency were calibrated using a R5372 microwave counter (Advantest) and an ES-FC5 NMR field meter (JEOL). The temperature of sample was provided by an ES-DVT4 thermal controller (JEOL). A tungsten lamp was used as the light source. XPS measurement was performed on the yellow, colorless and blue powders using a Quantera SXM-II spectrometer (ULVAC-PHI).

3. Results and Discussion

3.1. Heat treatment of sol-gel-derived powders

XRD measurement revealed that the sol-gel-derived yellow powder is composed of anatase TiO_2 single phase. After the heat treatment, a partial phase transition to rutile was observed. The fraction of the anatase phase was estimated to be 0.821 and 0.599 in the colorless and blue powders, respectively. In the diffusion reflectance spectra, a decrease in reflectance corresponding to the absorption band at 2.9 eV was observed in the yellow powder. The band disappeared after the heat treatment for oxidation or reduction. The diffusion reflectance spectra indicated that the colorless and blue powders were partly composed of rutile phase TiO_2 because of the shift of the absorption edge to 3.0 eV.

3.2. EPR measurement

Figure 1 shows the temperature dependence of the EPR signals of the N-doped yellow anatase powder measured under the irradiation with a W-lamp of whole wavelengths. The lineshape of the observed EPR signal is different from that of the single crystal because, in the powder, the principal axes of the crystallites have random orientations relative to the direction of the external magnetic field. The observed EPR signal appears to be derived from the paramagnetic species of $S = 1/2$ interacting with a nuclear spin of $I = 1$. The spectrum simulated using the parameter sets (g_{xx}, g_{yy}, g_{zz}) = (2.0070, 2.0060, 2.0047) and (A_{xx}, A_{yy}, A_{zz}) = (2.26, 3.60, 31.8) in the unit of Gauss by EasySpin software [4] shows

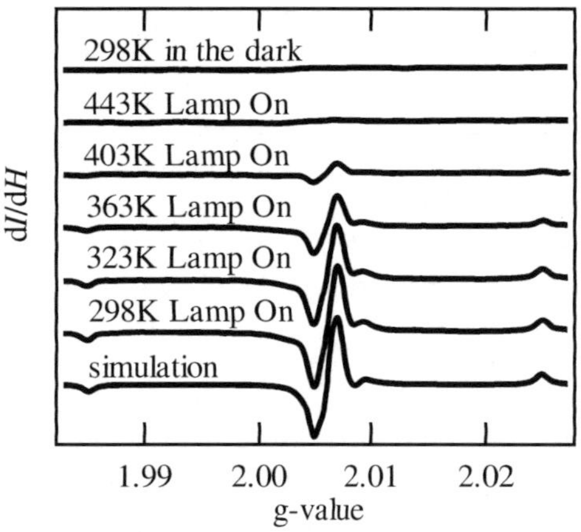

Figure 1. Temperature dependence of EPR signal and powder pattern simulation.

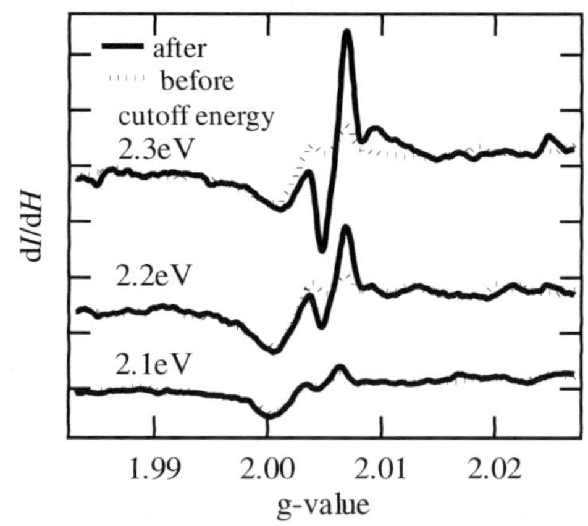

Figure 2. EPR signals of yellow powder with light irradiation of different photon energy.

good agreement with the observed spectrum of the N-doped powder. The parameter set used for spectral simulation of the powder is similar to that of yellow single crystal [3]. This indicates that the spin structure in the yellow powder is the same as that in the single crystal. Figure 1 shows that the intensity of the EPR signal decreases with increasing temperature, and the signal cannot be observed above 443 K. In addition, the signal was not observed after cooling to room temperature in the dark.

The EPR signal of the yellow powder was induced by the irradiation of visible-light whose energy was selected using short-wavelength cutoff filters. It was initially confirmed that the yellow powder kept in the dark for 2 weeks had weak EPR signals derived from surface defects [5]. Figure 2 shows the change in the EPR spectra of the yellow anatase powder irradiated by cutoff light with different energy for 5 min. The powder pattern of EPR signals was induced by the irradiation of light with $h\nu \geq 2.2$ eV, in agreement with a previous report [6]. The EPR signal gradually decayed in the dark. Figure 3 shows temporal changes in the intensity of EPR signals shortly after stopping light irradiation for different temperatures, where each signal intensity is normalized to the intensity under light irradiation. To estimate the thermal stability of the trapped carriers, the decay curves were analyzed using the stretched exponential function $I(t) \propto \exp\left[-(t/\tau)^{\beta}\right]$, where τ and β represent the average decay time and decay exponent, respectively [7]. The average decay time decreases with increasing temperature. The values of β were estimated to be 0.3 which was independent from temperature. The decay of the intensity of the EPR signal is due to a relaxation process caused by thermal activation of the trapped carriers. We assume that the lifetime has an activated behavior and can be described as $\tau(T) = \tau_0 \exp(-\Delta E/T)$, where ΔE and k are the thermal activation energy and Boltzmann constant, respectively. In Figure 4, the activation energy was estimated to be 0.22 eV from the good line fitting based on $\tau(T)$.

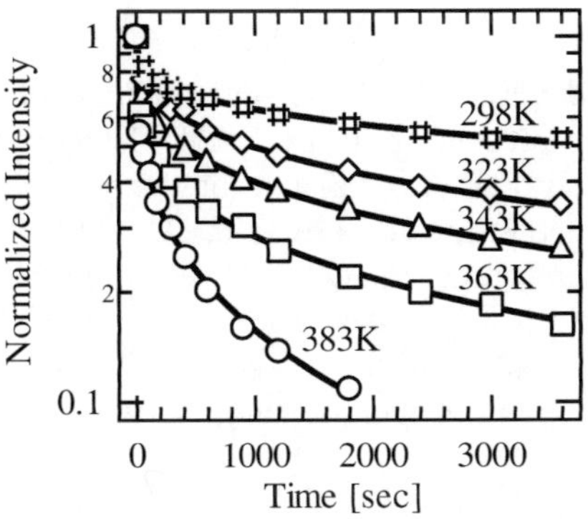

Figure 3. Temperature dependence of the decay curve of the EPR signal.

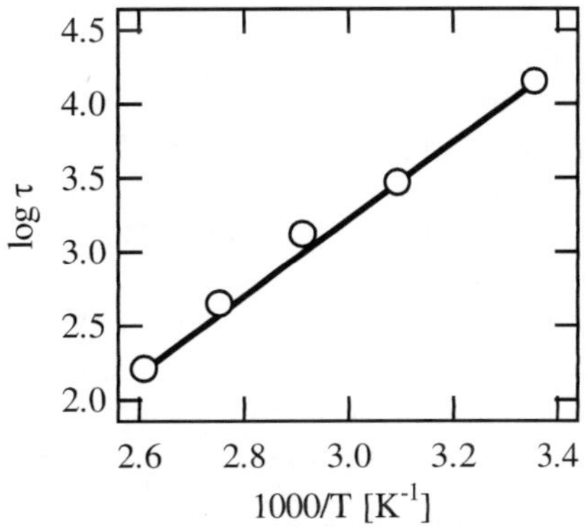

Figure 4. Temperature dependence of the average decay time.

3.3. XPS measurement

The XPS measurement revealed that the binding energies of the O1s and Ti2p bands were independent of the color of the powder. This indicates the atomic valences of O and Ti in the powders were independent of the crystalline phase because of the structural similarity between rutile and anatase. Figure 5 shows XPS N1s spectra measured for the yellow, colorless and blue powders. The N1s band at 399.5 eV was observed for all powders. To remove adsorbed N_2 from the surface of the powder particles, we performed sputtering using argon ions. Because of the reduced intensity of the N1s band at 399.5 eV due to the sputtering, the band was assigned to nitrogen adsorbed on the surface of the powders. The lack of change in the band intensity after the sputtering suggests that the bands at 402.9 and 396.5 eV can be assigned to doped nitrogen in the bulk of the colorless and blue powders, respectively. The binding energy of the bulk nitrogen in the yellow powder appears to be about 399.5 eV. These sussested that the change in the color of the powders mainly dependends mainly on the atomic valence of nitrogen. Therefore, the difference among these binding energies of N1s reveals that the atomic valence of doped nitrogen depends on the oxidation-reduction procedure. Many experimental results [8] have indicated a correlation between the binding energy and atomic valence; nitrogen with positive or negative atomic valence has a binding energy larger or smaller than 400 eV, respectively. The atomic valence of the doped nitrogen in the yellow powder resembles to be almost neutral, based on the adsorbed N_2 has a neutral atomic valence.

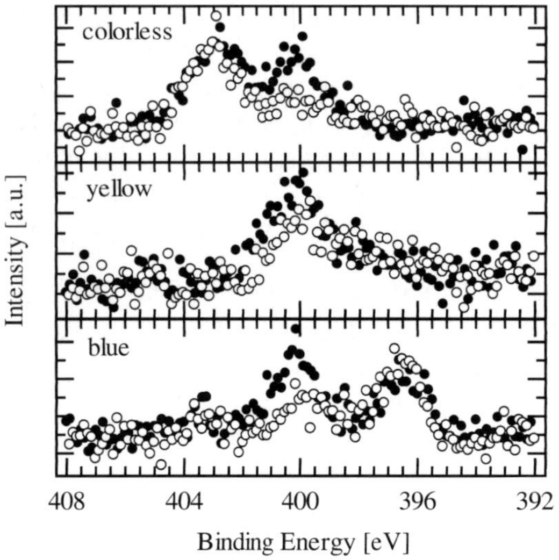

Figure 5. N1s XPS spectra of the N-doped TiO_2 powders. The black and white markers mean before and after sputtering, respectively.

4. Conclusion

The irradiation of light with $h\nu \geq 2.2$ eV to a sol-gel-derived yellow TiO_2 powder resulted in the appearance of a powder pattern of EPR signals, which gradually decayed in the dark. The decay of the intensity of the EPR signals is due to a relaxation process caused by thermal activation of the trapped carriers. The thermal activation energy of the trapped carriers was estimated to be 0.22 eV. In the XPS spectra of the colorless, yellow and blue powders, N1s bands were observed at 402.9, 399.5, and 396.5 eV, respectively. The atomic valence of the nitrogen in the yellow powder is almost neutral. The colorless and blue N-doped titanium dioxide powders also contained nitrogen with a positive and negative atomic valence, respectively. The atomic valence of doped nitrogen depends on whether the heat treatment was performed oxidation or reduction.

Acknowledgement

This work was partly supported by a Grant-in-Aid for Scientific Research from Japan Society for the Promotion of Science (No. 15K06421).

References

[1] Asahi R, Morikawa T, Ohwaki T, Aoki K and Taga Y 2001 *Science* **293** 269.
[2] Sekiya T, Yagisawa T, Kamiya N, Mulmi D, Kurita S, Murakami Y and Kodaira T 2004 *J. Phys. Soc. Jpn.* **73** 703.
[3] Sekiya T, Kamiya N, Ohya S, Kurita S and Kodaira T 2009 *J. Phys. Soc. Jpn.* **78** 114701.
[4] Stoll S and Schweiger A 2006 *J. Magnetic Resonance* **178** 42.
[5] Anpo M, Shima T and Kubokawa Y 1985 *Chem. Lett.* **14** 1799.
[6] Kokorin A, Konstantinova E and Kisch H 2013 *Kinetics and Catalysis* **54** 373.
[7] Redfield D and Bube R H *Photoinduced Defects in Semiconductors* (Cambridge University Press, Cambridge, 1996) .
[8] NIST X-ray Photoelectron Spectroscopy Database https://srdata.nist.gov/xps/Default.aspx.

Anisotropic dynamics of nanoparticles in clusters at a solid-liquid interface by laser trapping

Itsuo Hanasaki[1], **Chie Hosokawa**[2,3]

[1]Institute of Engineering, Tokyo University of Agriculture and Technology, Naka-cho 2-24-16, Koganei, Tokyo 184-8588, Japan
[2]Biomedical Research Institute, National Institute of Advanced Industrial Science and Technology (AIST), 1-8-31 Midorigaoka, Ikeda, Osaka 563-8577, Japan,
[3]Advanced Photonics and Biosensing Open Innovation Laboratory, AIST, Yamadaoka 2-1, Suita, Osaka 565-0871, Japan

E-mail: `hanasaki@cc.tuat.ac.jp`

Abstract. It is well-recognized that the Brownian motion of particles in fluids is random. Nevertheless, there can be characteristics depending on the specific physical conditions. We analyze the system of nanoparticle clusters formed by the laser trapping force field at the solid-liquid interface, based on the microscopy movie data. Since the laser trapping force field is basically a function of radial distance from the focal point in the two dimension at the liquid-solid interface, we examine the difference of displacement distributions in the radial and circumferential directions. The results show that the basic characteristics in this system depends on the laser power, and there is an anisotropy in the stochastic motion of the nanoparticles.

1. Introduction

When laser-trapping force field is applied in colloidal dispersion in the vicinity of the solid wall, a cluster of trapped nanoparticles can be obtained. In particular, depending on the details of the laser power etc., ring structures of nanoparticles are formed. The Brownian motion of particles in fluid can exhibit characteristics that originate from this specific configuration. This might affect the initiation events of crystallization by laser-induced perturbation [1]. There are rooms for the analysis beyond the overall diffusion coefficient, depending on the specific physical conditions on either particles themselves [2, 3] or the surrounding environments [4] and available amount as well as the precision of the data.We analyze the movie data obtained from the optical microscopy and camera system. We track the particles in the microscopy movie and evaluate the displacement distribution. There can be effective difference in the nature of Brownian motion in the radial direction and its perpendicular direction of the observation domain, where the origin of the radial direction is the focal point of the laser. We have obtained a displacement distributions that are different among the radial and the circumferential components, which we will discuss in this article.

2. Methodological details

2.1. Experimental setup for observation

The sample consists of 500-nm fluorescent dye-doped polystyrene particles in water at the concentration of 3.8×10^9 particles/mL contained in the 86-μm thick space as shown in Fig. 1.

IOP Conf. Series: Journal of Physics: Conf. Series **1220** (2019) 012015 doi:10.1088/1742-6596/1220/1/012015

The nanoparticles coated with fluorescent dye and they are observed by the inverted microscope with objective lens of ×60 with N.A.0.7 mounted with EMCCD camera, which captures the dynamics of the nanoparticles at a frame rate of 38.1 fps and an exposure time of 25 ms. The trapping optical force originates from the 1064-nm near-infrared laser beam of Nd:YVO_4. The laser power was adjusted by the angle of a half wave plate in front of the polarizing beam splitter.

2.2. Analysis of the movie data

After extraction of trajectory data set from the microscopy movie data in the steady state after the transient period of particles attraction based on the algorithm of Ref. [5], the displacements per each time interval of the camera frames are transformed from xy components to radial Δr_{r} and its tangential, i.e., circumferential components Δr_{t} with an origin defined as the laser spot:

$$\Delta r_{\mathrm{r}} = \Delta r_x \cos\theta + \Delta r_y \sin\theta, \quad (1)$$

$$\Delta r_{\mathrm{t}} = -\Delta r_x \sin\theta + \Delta r_y \cos\theta, \quad (2)$$

where Δr_x and Δr_y are the displacements in the x and y direction of the camera system, and θ is the angle comprised of the particle position and the x axis:

$$\theta = \sin^{-1}\left(\frac{y_{\mathrm{p}} - y_{\mathrm{o}}}{\sqrt{(x_{\mathrm{p}} - x_{\mathrm{o}})^2 + (y_{\mathrm{p}} - y_{\mathrm{o}})^2}}\right), \quad (3)$$

where $(x_{\mathrm{p}}, y_{\mathrm{p}})$ and $(x_{\mathrm{o}}, y_{\mathrm{o}})$ are the positions of the particle and laser spot, respectively.

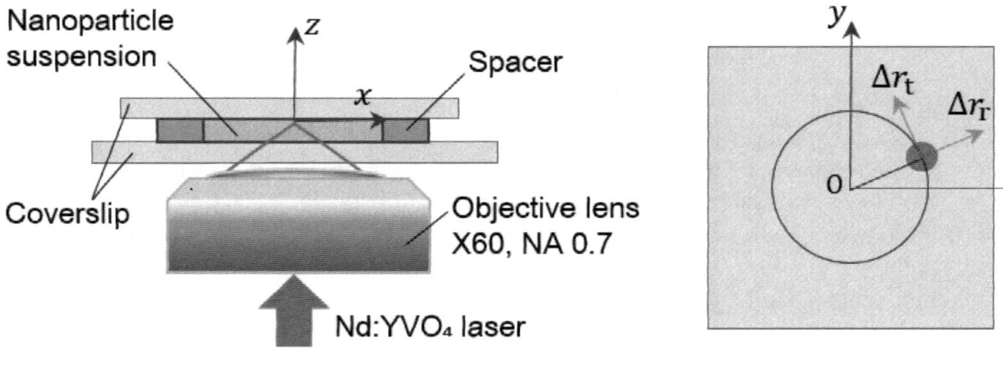

Figure 1. Schematic diagram of the measurement configuration of the sample dispersion. The observation is in the xy plane. The displacement distribution is evaluated in terms of $(\Delta r_{\mathrm{r}}, \Delta r_{\mathrm{t}})$ to take into account the effect of laser on the nanoparticle dynamics.

3. Results and Discussion

The laser induces a force field that is basically symmetric in the circumferential direction. There is a trapping effect depending on the distance from the focal point. In other words, the direct influence of the optical force is basically limited in the radial direction. Nevertheless, there is a confinement effect of the cluster in the circumferential direction. The displacement distribution

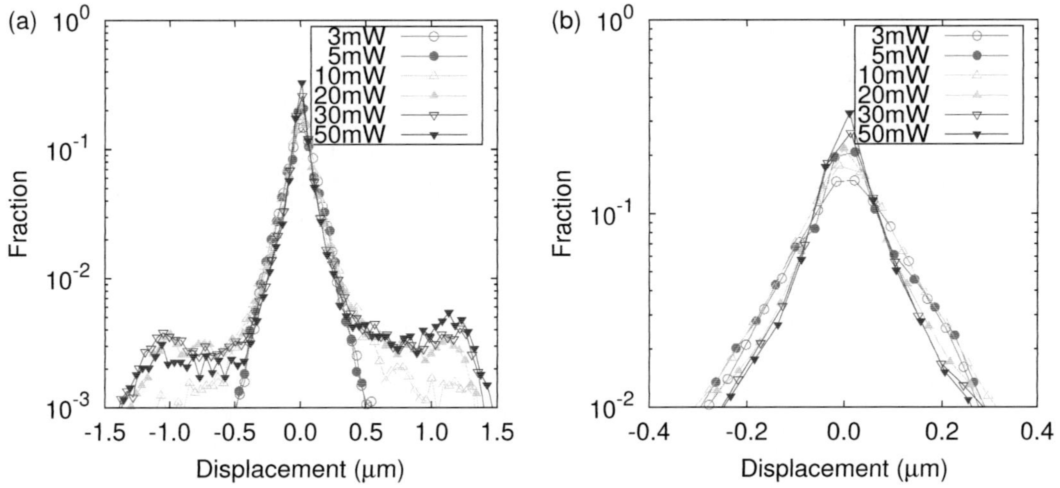

Figure 2. Radial component of the displacement distribution, where the time span for the displacements is 0.026 s: (a) overall distributions including the rare events, and (b) distributions around small displacements. The legends show the laser power to trap the nanoparticles.

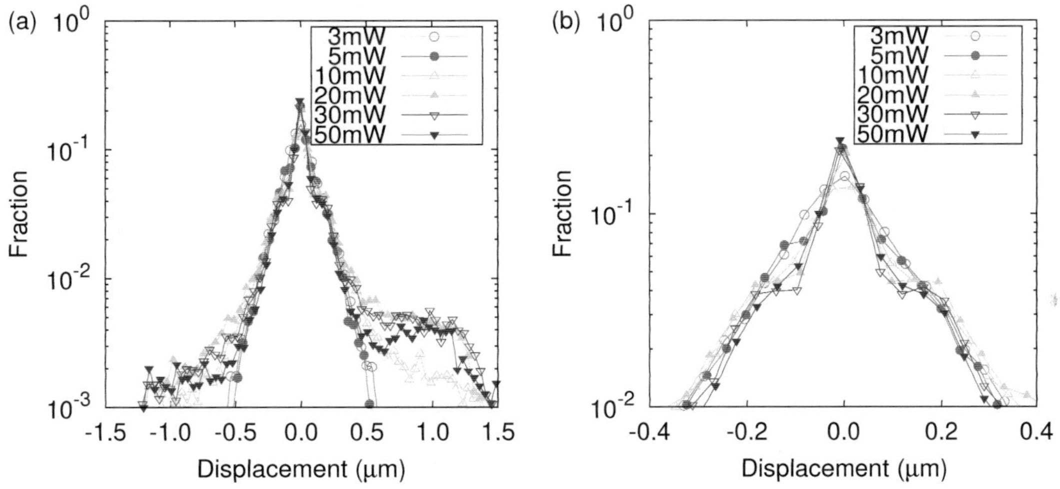

Figure 3. Circumferential component of the displacement distribution, where the time span for the displacements is 0.026 s: (a) overall distributions including the rare events, and (b) distributions around small displacements. The legends show the laser power to trap the nanoparticles.

is sharply peaked around zero regardless of the radial (cf. Figs. 2) and circumferential (cf. Figs. 3) directions or laser power.

However, there is a difference in the displacement distribution for the radial and circumferential directions as shown in Figs. 2(b) and 3(b). Whereas the radial displacements less than 0.3 μm is simply more sharply peaked at zero for larger laser powers, the distributions of circumferential displacements exhibit non-monotonicity for the larger laser powers. If the particle position distribution is skewed from the circular symmetry to be oval or prolate shapes, it would affect both of the radial and circumferential components based on Eqs. 2 and 3. Therefore, this is attributed to the difference in the radial and circumferential confinement effects.

Furthermore, substantially large displacements appear beyond certain value of the laser power as shown in Figs. 2(a) and 3(a). The clear dependence of the laser power is at least partly attributed to the effect of the scattering force. The scattering force is non-conservative type of the force acting on the nanoparticles. In other words, the nanoparticles steadily receives momentum from the laser in the incident direction (cf. Fig. 1). This qualitative difference from the pure conservative force field can induce the triggered stochastic motion of the nanoparticles. Although the detailed quantitative determination of the force field itself is beyond the scope of this article, it is worth noting that the NA of the lens is not so high nor the particle size is not so small to suppress the scattering force in this experiment. Even though the scattering force from the laser is independent of time, combination with the incessant Brownian motion with direct and indirect (i.e., hydrodynamic) interactions between the nearby nanoparticles results in the marked stochastic motion.

4. Concluding remarks

The nanoparticles trapped by the laser field forms clusters at the solid-liquid interfaces. The constituent nanoparticles experience different confinement effects in the radial and circumferential direction defined by the origin of laser spot and substrate plane. We analyzed the microscopy movie data to evaluate the anisotropy in the Brownian dynamics, and determined the existence of anisotropy in the dynamics. We have obtained counter-intuitive results that the non-monotonicity of the displacement distribution is more pronounced for the circumferential direction compared to the radial direction. This kind of anisotropy might affect the collective structure formation in the crystallization process depending on the combination with other physical conditions and materials.

Acknowledgement

This work was partly supported by JSPS KAKENHI Grant Numbers 17H05463 and JP16H06504 in Scientific Research on Innovative Areas "Nano-Material Optical-Manipulation", and JP17H01820.

References

[1] Sugiyama T, Yuyama K and Masuhara H 2012 *Accounts of Chemical Research* **45** 1946
[2] Hanasaki I and Isono Y 2012 *Phys. Rev. E* **85** 051134
[3] Matsuda Y, Hanasaki I, Iwao R, Yamaguchi H and Niimi T 2016 *Anal. Chem.* **88** 4502–4507
[4] Hanasaki I, Uehara S, Arai Y, Nagai T and Kawano S 2015 *Jpn. J. Appl. Phys.* **54** 125601
[5] Sbalzarini I and Koumoutsakos P 2005 *J. Struct. Biol.* **151** 182–195

IOP Conf. Series: Journal of Physics: Conf. Series **1220** (2019) 012016 doi:10.1088/1742-6596/1220/1/012016

Modulation of Graphene/Au(111) Interaction by Electrocatalytic Hydrogen Evolution Reaction

Tomohiro Fukushima, Kensuke Yasuda, Hidetaka Hasebe, Ruifeng Zhou, Hiro Minamimoto, and Kei Murakoshi

Department of Chemistry, Faculty of Science, Hokkaido University, Sapporo, Hokkaido 060-0810, Japan

Corrpesonding : kei@sci.hokudai.ac.jp

Abstract. Electronic structure of monolayer graphene with metal contacts can be modified by control of interfacial interaction. Here, we show that the strength of local interfacial interaction between graphene and Au(111) surface can be electrochemically controlled by the hydrogen evolution reaction process. Graphene/Au(111) electrodes were prepared by the chemical vapour deposition. Raman spectroscopy indicated existence of monolayer of graphene over Au surface. Local interaction of graphene/Au(111) interaction can be differentiated by intercalation of H_2 molecules at interface. This study shed lights on the possibility of proton penetration through the graphene to modify the strength of interaction between graphene and Au(111) surface.

1. Introduction

Two-dimensional materials have attracted attentions due to their multi-functionality and designability through the interfacial interaction [1]. One of the representative two-dimensional materials is graphene. Graphene shows excellent electronic properties [2], and their electronic structure can be modified by metal contacts [3]. On the other hand, one of the feasible properties of graphene is permeability of proton through basal plane of graphene [4,5], which is even not observed for He gases [6]. Recently it has been shown that electrocatalytic hydrogen evolution reaction process induces the formation of "nanobubble" between graphene sheets [7]. We envisioned that graphene/metal interaction can be modulated by intercalation of hydrogen through the nanobubble formation process, which can be also regarded as a novel method for controlling the interaction between metal and graphene contacts.

Here, we show that interaction between graphene and Au (111) can be modified by electrochemical hydrogen evolution reaction (HER). Graphene / Au (111) electrode was fabricated by chemical vapor deposition (CVD). Electrocatalytic HER process differentiates the graphene/Au(111) interaction evidenced by observation of Raman G band and Raman 2D band.

2. Experimental Section

2.1. Materials

Ag/AgCl electrodes were prepared by anodization of Ag wire in saturated KCl aqueous solution. Au beads were prepared by frame-annealing method. Ar gas (99.999%) and CH_4 gas (99.999%) was employed for CVD. Graphene/Au (111) electrodes were prepared with the according to literature procedure [8]. Monolayer graphene was synthesized through the CVD [9]. Quartz furnace was cleaned by oxidation of contaminates at 900 °C in air, and subsequent H_2/Ar treatment at 900 °C. Monolayer graphene was grown on the Au (111) surface by annealing at 1035 °C for 15 min under 250 sccm (standard cubic centimeter per minute) of Ar, and thereafter 1 sccm of CH_4 and 250 sccm of Ar flow for 3 min. This graphene-covered Au electrode was denoted as G/Au. The quality of crystallinity was confirmed from Raman spectroscopy.

2.2. Methods

Three-electrode cell was employed for electrochemical measurements. Ag/AgCl (saturated KCl aqueous solution) and Pt electrodes were used as reference electrode and counter electrode, respectively. All the electrochemical potentials were referenced to Ag/AgCl (saturated KCl aqueous solution). All solutions were bubble with Ar gas to deoxygenate for 20 min, and then electrochemical experiments were conducted. Microscopic Raman spectroscopy measurements were conducted by using Nanofinder under 514 nm laser excitation (3.0 mW, 60 sec).

3. Results and Discussion

Figure 1 shows cyclic voltammograms for Au and G/Au electrodes in 0.1 M H_2SO_4 aqueous solution. Decrease in electrochemical potential leads to the increase in cathodic current, which can be attributed to the electrocatalytic HER current. Furthermore, the current from G/Au electrodes is lower than one from Au electrodes. These observations suggest that graphene coating results in decrease of electrocatalytic activity for HER.

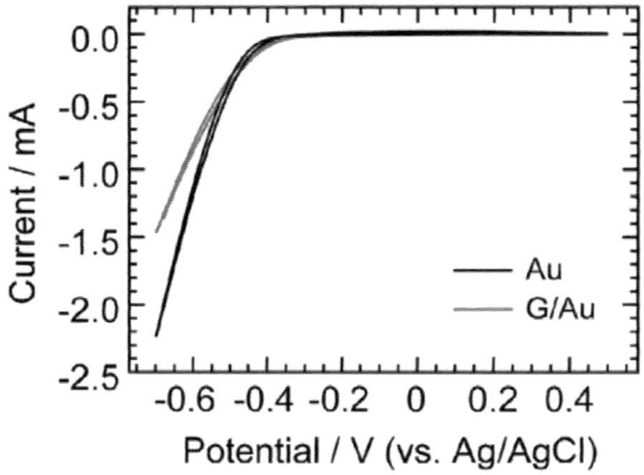

Figure 1. Cyclic voltammograms for Au (black line) and G/Au (red line) electrodes in 0.1 M H_2SO_4 aqueous solution. Scan rate 50 mV/sec. Immersed electrode surface area was 0.05 cm^2 approximately.

The structure of graphene was characterized from Raman spectroscopy. The G band (1560 cm^{-1}) is ascribed to the in-plane bond stretching motion between neighbouring carbon-carbon bonding, which confirms graphene formation over the Au(111) electrode. The 2D band (2700 cm^{-1}) is a double resonance phonon process of graphene and serves as a fingerprint for the number of graphene layers [8,10]. We confirmed the formation of monolayer graphene over Au(111) [8], and subsequently exposed G/Au electrodes under electrocatalytic HER condition.

Figure 2 shows the Raman spectra of G/Au after electrolysis at −0.4 V in 0.1 M H_2SO_4 aqueous solution for 1 min. G band and 2D band wavenumbers are variable for different spot sites, and G band wavenumbers are increased as 2D band wavenumbers increased. This is due to the thermal fluctuation of graphene lattice [11,12]. Since Au surface can interact with graphene via charge-doping as well as

IOP Conf. Series: Journal of Physics: Conf. Series **1220** (2019) 012016 doi:10.1088/1742-6596/1220/1/012016

dielectric screening [12,13], intercalation between graphene and Au is expected to decrease 2D band wavenumber. Figure 3 shows the Raman spectra of G/Au after electrolysis at −0.5 V in 0.1 M H_2SO_4 aqueous solution for 1 min. Positive correlation between G band wavenumber and 2D band wavenumber was not observed. In addition, decrease in 2D band wavenumbers were generally observed. We are assuming that this breakdown of positive correlation between G band and 2D band wavenumbers was induced by a decrease in the interfacial interaction via graphene nanobubble formation. Proton penetration through tunneling might trigger the switching of graphene/Au(111) interaction.

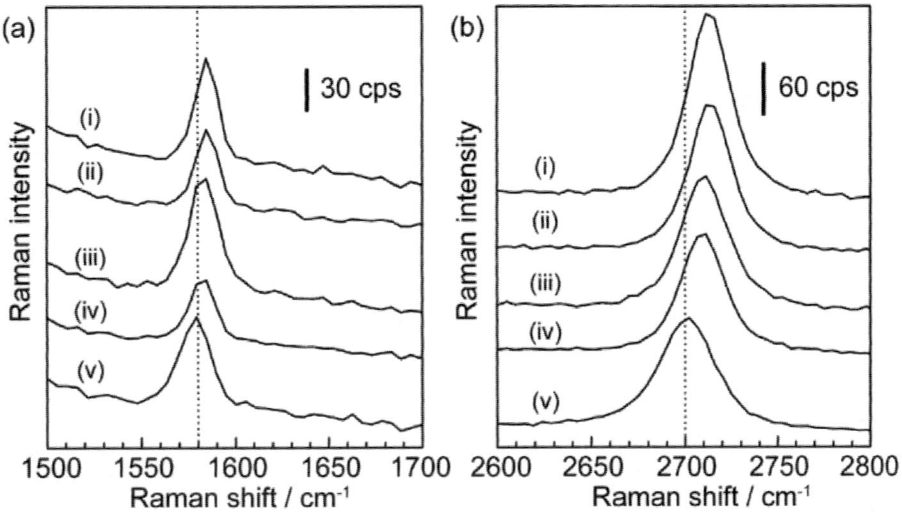

Figure 2. Raman spectra of G/Au after electrolysis at −0.4 V in 0.1 M H_2SO_4 aqueous solution for 1 min. Different spots (i)-(v) are measured for (a) G band region and (b) 2D band region. Dotted lines are shown for eye guide.

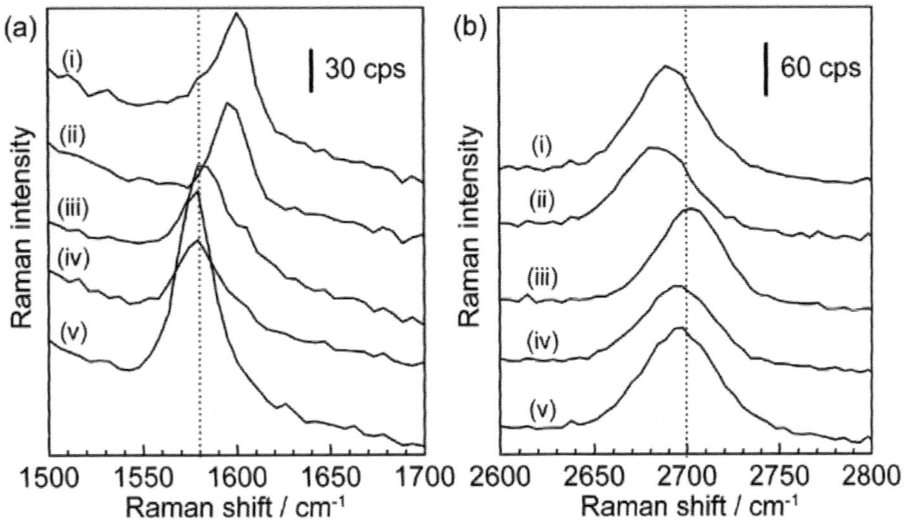

Figure 3. Raman spectra of G/Au after electrolysis at −0.5 V in 0.1 M H_2SO_4 aqueous solution for 1 min. Different spots (i)-(v) are measured for (a) G band region and (b) 2D band region. Dotted lines are shown for eye guide.

4. Conclusions

We demonstrated that graphene/Au(111) interfacial interaction can be modified by electrocatalytic hydrogen evolution reaction. Difference in interaction was observed for high overpotential in 0.1 M H_2SO_4. Intercalation of H_2 molecules into graphene/Au(111) interface during HER process might trigger the switching of graphene/Au(111) interaction. Detailed proton penetration mechanism and the effect of chemical modification of graphene surfaces are under investigation.

References

[1] Geim A K and Grigorieva I V 2013 Van der Waals heterostructures *Nature* **499** 419.
[2] Geim A K and Novoselov K S 2007 The rise of graphene *Nat. Mater.* **6** 183.
[3] Schulman D S, Arnold A J and Das S 2018 Contact engineering for 2D materials and devices *Chem. Soc. Rev.* **47** 3037.
[4] Hu S, Lozada-Hidalgo M, Wang F C, Mishchenko A, Schedin F, Nair R R, Hill E W, Boukhvalov D W, Katsnelson M I, Dryfe R A W, Grigorieva I V, Wu H A and Geim A K 2014 Proton transport through one-atom-thick crystals *Nature* **516** 227.
[5] Mishchenko A, Grigorenko A N, Dryfe R A W, Radha B, Grigorieva I V and Geim A K 2016 Sieving hydrogen isotopes through two-dimensional crystals *Science* **351** 68.
[6] Bunch J S, Verbridge S S, Alden J S, Van Der Zande A M, Parpia J M, Craighead H G and McEuen P L 2008 Impermeable atomic membranes from graphene sheets *Nano Lett.* **8** 2458.
[7] An H, Tan B H, Moo J G S, Liu S, Pumera M and Ohl C D 2017 Graphene Nanobubbles Produced by Water Splitting *Nano Lett.* **17** 2833.
[8] Yasuda S, Kumagai R, Nakashima K and Murakoshi K 2015 Electrochemical potential stabilization of reconstructed Au(111) structure by monolayer coverage with graphene *J. Phys. Chem. Lett.* **6** 3403.
[9] Oznuluer T, Pince E, Polat E O, Balci O, Salihoglu O and Kocabas C 2011 Synthesis of graphene on gold *Appl. Phys. Lett.* **98** 183101.
[10] Ferrari A C, Meyer J C, Scardaci V, Casiraghi C, Lazzeri M, Mauri F, Piscanec S, Jiang D, Novoselov K S, Roth S and Geim A K 2006 Raman spectrum of graphene and graphene layers *Phys. Rev. Lett.* **97** 187401.
[11] Lee J E, Ahn G, Shim J, Lee Y S and Ryu S 2012 Optical separation of mechanical strain from charge doping in graphene *Nat. Commun.* **3** 1024.
[12] Zhou R, Yasuda S, Minamimoto H and Murakoshi K 2018 Sensitive Raman probe of electronic interactions between monolayer graphene and substrate under electrochemical potential control *ACS Omega* **3** 2322.
[13] Varykhalov A, Scholz M R, Kim T K and Rader O 2010 Effect of noble-metal contacts on doping and band gap of graphene *Phys. Rev. B - Condens. Matter Mater. Phys.* **82** 121101.

IOP Conf. Series: Journal of Physics: Conf. Series **1220** (2019) 012017 doi:10.1088/1742-6596/1220/1/012017

Dynamics of the electric-field induced magnetization in antiferromagnetic chromium oxide observed by Faraday rotation

R Hikita[1], H Taniguchi[1], T Shinkai[1] and T Kohmoto[1,2]

[1] Graduate School of Science, Kobe University, Kobe 657-8501, Japan
[2] Molecular Photoscience Research Center, Kobe University, Kobe 657-8501, Japan

E-mail: kohomoto@kobe-u.ac.jp

Abstract. We observed the dynamics of the electric-field induced magnetization in antiferromagnetic chromium oxide (Cr_2O_3) by the Faraday-rotation measurement using a continuous-wave probe light in the millisecond region and a pulse probe light in the nanosecond region. It was found that the Faraday-rotation amplitude linearly depends on the electric field, decreases with increasing temperature, and disappears above the Néel temperature. In the nanosecond region, nanosecond rise of the electric-field induced Faraday-rotation signal was observed.

1. Introduction

In recent years, various types of multiferroic materials, in which ferroelectric and magnetic orders coexist, have been found. Many of them are antiferromagnets which have spin structures of spiral type, and their giant magnetoelectric effect has been attracting attention.

Antiferromagnetic chromium oxide (Cr_2O_3) is not multiferroic, but is known to show the linear magnetoelectric effect, in which the electric polarization is induced in proportion to the applied magnetic field and the magnetization is induced in proportion to the applied electric field. Cr_2O_3 is a material in which magnetoelectric effect was first observed a half century ago [1]. Since the linear magnetoelectric effect was discovered, Cr_2O_3 has attracted interest in the magnetoelectric effect and the magneto-optical effect in the experiments of the nonreciprocal rotation of the polarization plane [2] and the second harmonic generation [3]. Recently, the details of the room-temperature linear magnetoelectric effect was reported [4].

The electric-field induced magnetization has been observed as the response to the applied alternating electric field. The electromotive force in a pickup coil [1,4-7] or Faraday rotation of a probe light [8-10] was detected as the magnetization signal, where the alternating frequency of 10^3-10^6 Hz was used. However, the dynamics of electric-field induced magnetization in the time regions shorter than microseconds has not been reported so far. In the present study, we pay attention to the magnetization induced by the electric field in the magnetoelectric effect. We studied the dynamics of the electric-field induced magnetization in Cr_2O_3 using a Faraday-rotation measurement in the millisecond and nanosecond regions.

2. Sample and Experiment

We use a single crystal of Cr_2O_3 as the sample, whose thickness is 0.25 mm. Cr_2O_3, which has a corundum structure with a threefold symmetry axis (*c*-axis), is an antiferromagnet with the Néel temperature T_N = 307 K. Cr^{3+} ions align along the *c*-axis and are surrounded by distorted octahedra of O^{2-} ions. Below T_N, the spins of Cr^{3+} ions align antiferromagnetically along the *c*-axis, and Cr_2O_3 shows the linear magnetoelectric effect.

The linearly-polarized probe light enters the sample parallel to the *c*-axis, and the Faraday-rotation signal of the transmitted light is detected by a polarimeter. The sample is sandwiched between a transparent electrode (ITO) and a metal mirror, and an electric field is applied between them. Since the Faraday-rotation signal is proportional to the magnetization, the electric-field induced magnetization can be observed by Faraday rotation. In this study, the Faraday-rotation signal was observed in the millisecond and nanosecond regions. In the millisecond region, a continuous-wave (cw) probe light (1064 nm) and a square-wave electric field are used, and the waveform of the Faraday-rotation signal from the polarimeter is averaged on an oscilloscope. In the nanosecond region, a pulse probe light (800 nm) and a pulsed electric field, whose rise time is 4 ns and decay time is ~50 ns, are used. The delay between the probe pulse and the rise of the electric field is swept, and the change in the Faraday-rotation signal for the on/off of the electric-field pulse is lock-in detected.

3. Results and Discussion

3.1 Measurement in the millisecond region by cw probe

The electric-field induced Faraday-rotation signals observed in the millisecond region in the square-wave electric field of +4.0, +1.6, -1.6, and -4.0 MV/m at 290 K is shown in figure 1. The electric field is switched on/off in a period of 2 ms. The rise and fall time of the Faraday-rotation signal is determined by the electric-field pulse. Figure 2 shows the electric-field dependence of the Faraday-rotation amplitude obtained from the millisecond signal at 290 K. The Faraday-rotation amplitude linearly depends on the electric field. This result suggests the linear magnetoelectric effect.

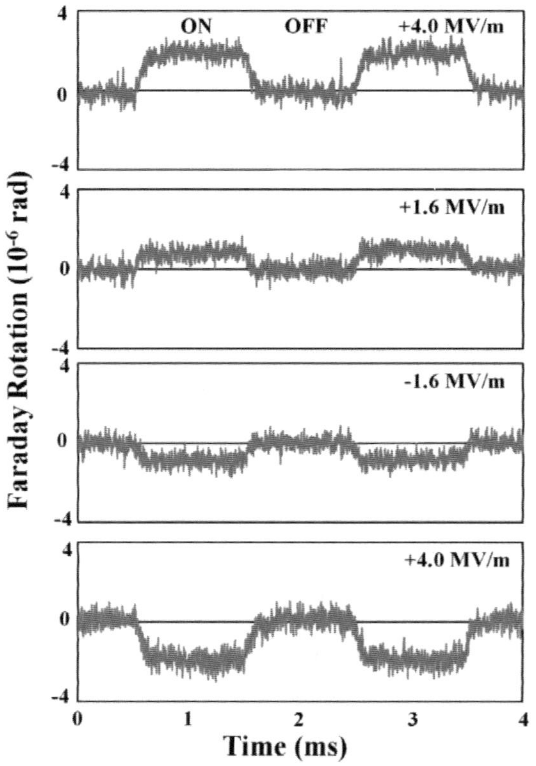

Figure 1. Electric-field dependence of the Faraday-rotation signals observed in the square-wave electric field of +4.0, +1.6, -1.6, and -4.0 MV/m at 290 K.

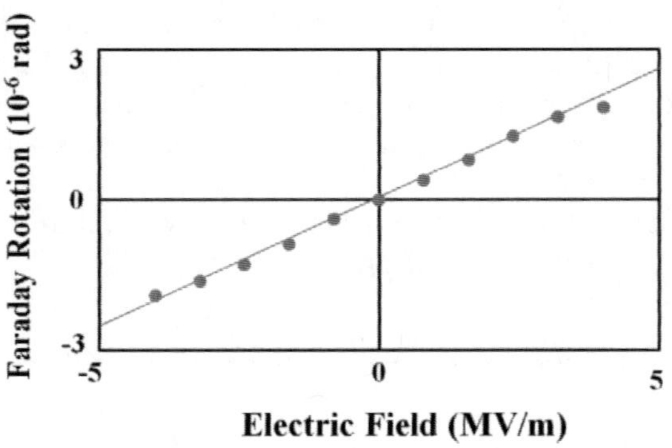

Figure 2. Electric-field dependence of the Faraday-rotation amplitude obtained from the millisecond signal at 290 K.

3.2. Measurement in the nanosecond region by pulse probe

The change of the electric-field induced Faraday-rotation signals observed in the nanosecond region at 280 K and 320 K is shown in figure 3, where the electric field is switched from -2.8 MV/m to +2.8 MV/m at t = 0. The rise time of the Faraday-rotation signal was observed to be ~5 ns. This time is considered to be determined by the rise time of the electric-field pulse. The Faraday-rotation signals are found below T_N, but are not found above T_N. Figure 4 shows the temperature dependence of the Faraday-rotation amplitude. It decreases with increasing temperature and disappears above T_N.

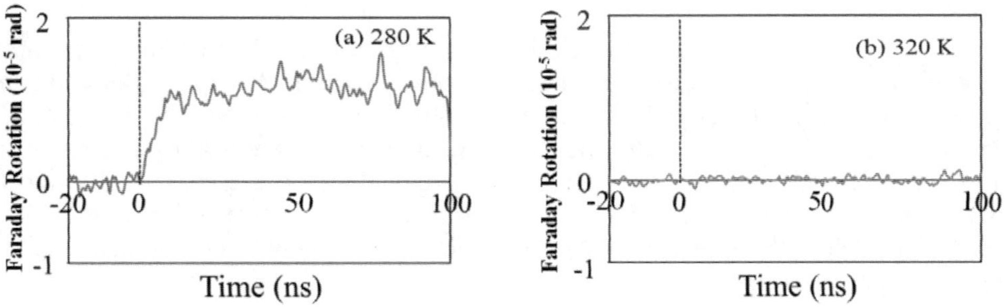

Figure 3. Faraday-rotation signals observed at (a) 280 K and (b) 320 K. The electric field is switched from -2.8 MV/m to +2.8 MV/m at t = 0.

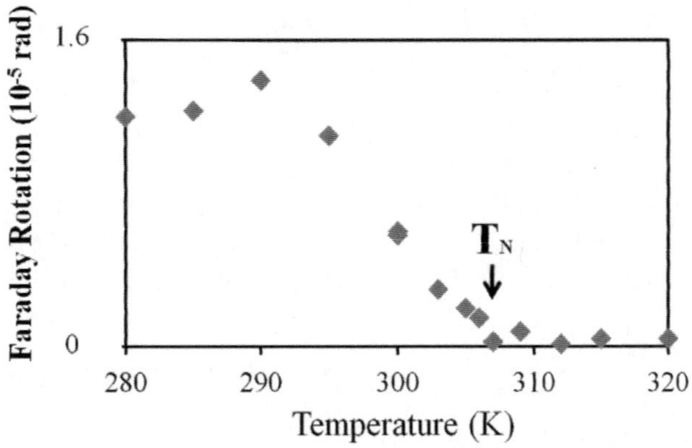

Figure 4. Temperature dependence of the Faraday-rotation amplitude.

The influence of the electric field on the magnetization in the sample can manifest itself differently in various regimes. The electric-field induced magnetization signal may be caused by the change in the magnetization within individual domain and/or the motion of domain walls in the electric field [11]. Figure 3(a) suggests that there is no domain-wall motion in the nanosecond region after the electric-field switching.

Figure 5 shows the electric-field dependence of the Faraday-rotation amplitude obtained from the nanosecond signal at 290 K. It linearly depends on the electric field.

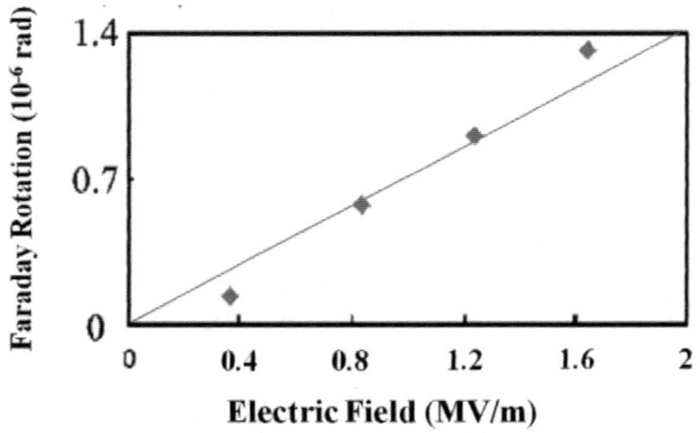

Figure 5. Electric-field dependence of the Faraday-rotation amplitude obtained from the nanosecond signal at 290 K.

4. Summary

We observed the dynamics of the electric-field induced magnetization in an antiferromagnet Cr_2O_3 by the Faraday-rotation measurement using a continuous-wave probe light in the millisecond and a pulse probe light in the nanosecond region. In the millisecond region, it was found that the Faraday-rotation amplitude linearly depends on the electric field. This result suggests the linear magnetoelectric effect. In the nanosecond region, the Faraday-rotation amplitude linearly depends on the electric field, decreases with increasing temperature, and disappears above T_N. In our experiment, the response time of the electric-field induced magnetization to the electric field could not be determined, but it was found that the response time is faster than 5 ns.

References

[1] Astrov D N 1960 *Sov. Phys. JETP* **11** 708
[2] Krichevstov B B, Pavlov V V, Pisarev R V and Gridnev V N 1996 *Phys. Rev. Lett.* **76** 4268
[3] Fiebig M, Fröhlich D, Krichevtsov B B and Pisarev R V 1994 *Phys. Rev. Lett.* **73** 2127
[4] Iyama A and Kimura T 2013 *Phys. Rev. B* **87** 180408(R)
[5] Folen V J, Rado G T and Stalder E W 1961 *Phys. Rev. Lett.* **6** 607
[6] Astrov D N 1961 *Sov. Phys. JETP* **13** 729
[7] Shtrikman S and Treves D 1963 *Phys. Rev.* **130** 986
[8] O'Dell T H and White E A D 1970 *Philos. Mag.* **22** 649
[9] Krichevtsov B B, Pavlov V V and Pisarev R V 1986 *JETP Lett.* **44** 607
[10] Krichevtsov B B, Pavlov V V and Pisarev R V 1988 *Sov. Phys. JETP* **67** 378
[11] Krichevtsov B B, Pisarev R V and Selitskii A G 1992 *Sov. Phys. JETP* **74** 565

Modulated photocurrent spectroscopies for characterization of the charge transport process in organic photovoltaics

Hiroki Nojima[1], Takashi Kobayashi[1,2], Takashi Nagase[1,2], and Hiroyoshi Naito[1,2]

[1] Department of Physics and Electronics, Osaka Prefecture University, 1-1 Gakuen-cho Naka-ku, Sakai, 599-8531, Japan

[2] The Research Institute for Molecular Electronic Devices (RIMED), Osaka Prefecture University, 1-1 Gakuen-cho Naka-ku, Sakai, 599-8531, Japan

E-mail: tkobaya@pe.osakafu-u.ac.jp (TK); naito@pe.osakafu-u.ac.jp (HN)

Abstract. Modified modulation photocurrent (MPC) measurements were carried out to investigate the carrier transport process in organic photovoltaics based on polymer and fullerene derivatives. In conventional MPC measurements, a photocurrent component modulating at the same frequency as the modulated incident light is recorded; however, in this study, a component modulating at twice the modulation frequency was detected. A linear process does not contribute to the detected signals; thus, this technique allows for focus on a process that converts incident photons into photocurrent in a nonlinear manner. An analysis using rate equations suggested that the nonlinearity resulted from the carrier density dependence of the internal electric field, which is the driving force of the carriers under the short-circuit condition. Furthermore, features due to both electron and hole transports were more clearly observed in the modified MPC measurements than in the conventional MPC measurements.

1. Introduction

The carrier transport process in organic photovoltaics (OPVs) has been extensively investigated in order to gain a deeper understanding of the operational mechanism and to obtain a guidance principle for further improvements of the power conversion efficiency (PCE) [1,2]. Various experimental techniques have been utilized for the investigations. In particular, modulation spectroscopy, e.g., impedance spectroscopy [3-5], photoinduced absorption measurements [6-9], and modulated photocurrent spectroscopy (MPC) [10-18], is a technique suitable for studying OPVs because of its good sensitivity for detecting signals in the same time range (>μs) as a carrier traveling to the electrodes or recombining. In MPC, OPVs are irradiated by light modulated at a certain frequency, f, and a photocurrent modulating at the same frequency is measured. By analyzing the modulation frequency dependence of the photocurrent, the carrier transport process can be investigated.

In modulation spectroscopy, it is technically possible to detect output signals modulating at frequencies other than f. For instance, a $2f$ signal is recorded in order to study a nonlinear process that generates a signal component in proportion to the square of the input or to eliminate contributions from a linear process [19,20]. In this study, we have applied $2f$ MPC spectroscopy to OPVs based on poly(3-hexylthiophene) (P3HT) and [6,6]-phenyl C_{61}-butyric acid methyl ester (PCBM). Although the photocurrent is largely proportional to the light intensity, $2f$ signals were clearly observed. A numerical analysis using rate equations suggested that the nonlinearity resulted from the carrier density dependence of the internal electric field in the semiconducting layer.

2. Experimental

In this work, OPV devices with the structure ITO/ZnO/P3HT:PCBM (100 nm)/MoO_3 (10 nm)/Al (50 nm) were prepared. The OPVs were fabricated in a glove box, which was filled with dry nitrogen gas, and subsequently sealed with a glass piece and fixed with epoxy. The details of the fabrication process can be found in Ref. 21. The current density–voltage (*J*–*V*) characteristics of the devices were measured using a source measuring unit under a simulated irradiation of 100 mW/cm^2 (AM 1.5 G). The PCE of the devices was about 2.4%. MPC measurements were carried out with a home-made setup consisting of a 488 nm semiconducting laser, transimpedance amplifier, and lock-in amplifier. The laser beam amplitude was modulated to be a sinusoidal wave, which consisted of a fixed ac component and a variable dc component. The modulation frequency was varied from 10 Hz to 2 MHz. A lock-in amplifier can measure the amplitudes of a signal component oscillating in-phase with respect to the modulated laser beam and a component oscillating with a 90° phase delay; in this paper, these are represented by the real and imaginary axes, respectively.

3. Results and discussion

Figure 1 shows an example of the modulation frequency dependence of the 1*f* MPC signals. The curves can be explained in terms of a simple Debye function [22] with a relaxation time of 0.63 μs, which may correspond to the time needed for the generated electrons or holes to reach one of the electrodes. Since the intensities of the 1*f* MPC signals are nearly proportional to the light intensity, the shape of the curves is virtually independent of the intensity of the laser dc component. Therefore, it seems to be impossible to obtain more information on the carrier transport process from the 1*f* MPC signals. On the other hand, the modulation frequency dependence of the 2*f* MPC signals is rich in structure, as shown in Figure 2. Although the origin of the two negative peaks observed around 10^3 and 10^5 Hz in Figure 2(a) cannot be understood intuitively, the data are expected to include more information on the carrier transport process. Note that the 2*f* MPC signals are two orders of magnitude smaller than the 1*f* MPC signals, but can still be recorded with a good signal-to-noise ratio using a conventional lock-in amplifier.

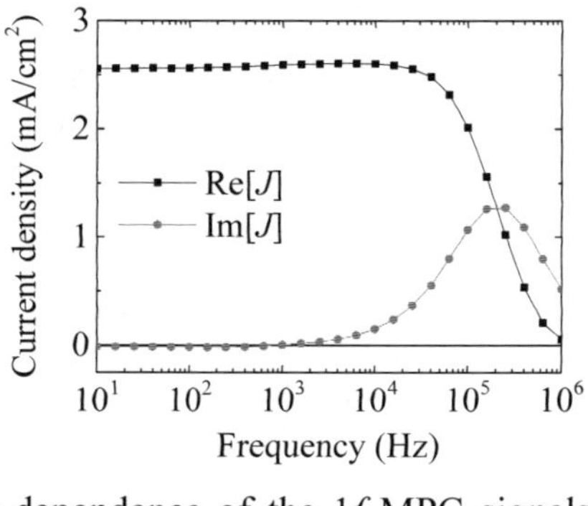

Figure 1. Modulation frequency dependence of the 1*f* MPC signals measured using a sinusoidally modulated laser with a peak-to-peak magnitude of 50 mW/cm^2 and an offset of 125 mW/cm^2.

To analyze the modulation frequency dependence, rate equations describing the temporal evolution of electron and hole densities (*n* and *p*) are considered:

$$\frac{dn(t)}{dt} = G(t) - \beta_e n(t), \tag{1}$$

$$\frac{dp(t)}{dt} = G(t) - \beta_h p(t). \tag{2}$$

In these equations, $G(t) \propto I_{dc} + I_{ac} \sin(2\pi f t)$ represents the carrier generation rate, where I_{dc} and I_{ac} are the intensities of the dc and ac light components, respectively. For OPVs under the short-circuit condition, most of the photogenerated carriers are extracted by the electrodes prior to recombination. In this case, the decay rates of electrons (β_e) and holes (β_h) are determined by the electron and hole mobilities (μ_e and μ_h), the internal electric field (F), and the film thickness (L). For simplicity, it is assumed that the electrons and holes travel the same distance, i.e., $L/2$. Then, $\beta_e = 2\mu_e F/L$ and $\beta_h = 2\mu_h F/L$ are obtained. Using $n(t)$ and $p(t)$, the short-current density J_{sc} can be written as

$$J_{SC}(t) = qF\{\mu_e n(t) + \mu_h p(t)\}, \tag{3}$$

where q is the elementary charge. If all of these constants, with the exception of $n(t)$, $p(t)$, and $G(t)$, were time-independent, the 2*f* MPC signals would be zero. Observing the 2*f* MPC signals, it is found that some of the constants are dependent on $n(t)$, $p(t)$, or both. In this study, only the carrier density dependence of the internal electric field was considered. According to the theory of inorganic semiconductors [23], the open-circuit voltage is equal to $(E_{Fe} - E_{Fh})/q$, where E_{Fe} and E_{Fh} are the quasi-Fermi levels for the electrons and holes and depend on $n(t)$ and $p(t)$, respectively. Under the short-circuit condition, the internal electric field is assumed to be the open-circuit voltage divided by the thickness of the semiconducting layer. Thus,

$$F = \frac{E_{Fe} - E_{Fh}}{qL} = \frac{k_B T}{qL} \ln\left\{\frac{n(t)p(t)}{n_i^2}\right\}, \tag{4}$$

where k_B is the Boltzmann constant, T is the temperature, and n_i is the intrinsic carrier density. Equation (4) introduces the nonlinear process into the model. $J_{sc}(t)$ was obtained by solving equations (1)–(4) numerically before being Fourier transformed into $J_{sc}(f)$.

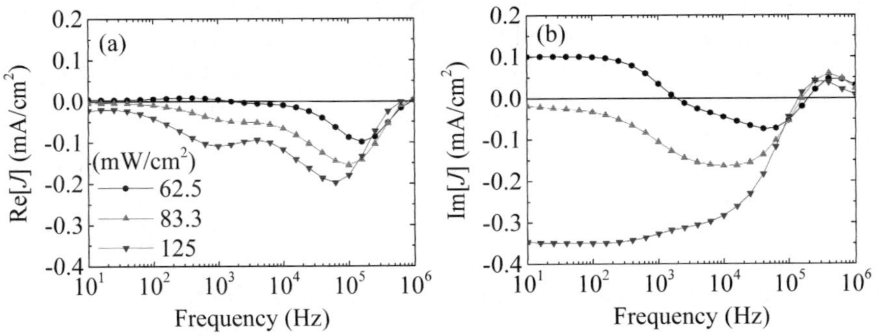

Figure 2. (a) Real and (b) imaginary parts of the modulation frequency dependence of the 2*f* MPC signals measured with a sinusoidally modulated laser with a peak-to-peak magnitude of 50 mW/cm^2 and an offset of 62.5, 83.3, or 125 mW/cm^2.

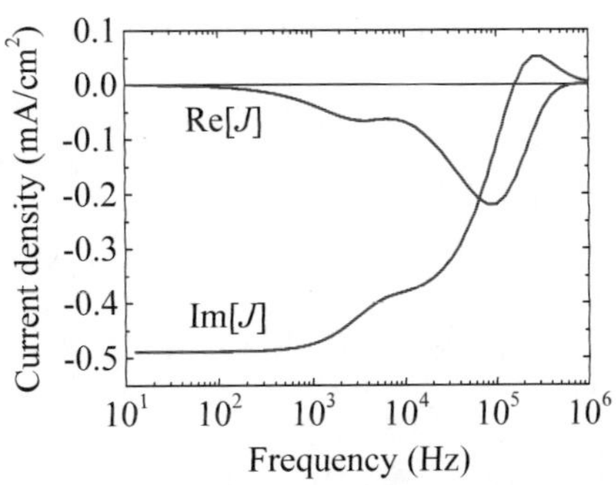

Figure 3. Modulation frequency dependence of the 2*f* MPC signals calculated with L = 100 nm, μ_e = 1 × 10^{-6} cm^{2}V^{-1}s^{-1}, and μ_h = 1 × 10^{-4} cm^{2}V^{-1}s^{-1}. Note that the same calculated result is obtained if the two mobility values are switched.

A calculated result is shown in Figure 3, which properly reproduces the experimental result obtained with an offset of 125 mW/cm^2. From the calculations, it is found that the two negative peaks can be attributed to the electrons and holes, although it is unknown which is due to the electrons. The separated negative peaks can be reproduced if there is a large difference between μ_e and μ_h. From the calculation, μ_e and μ_h are estimated to be around 10^{-6} $cm^2V^{-1}s^{-1}$ and 10^{-4} $cm^2V^{-1}s^{-1}$, respectively, or vice versa. With the current simple model, it is impossible to determine the mobilities more accurately and explain the modulation frequency dependence obtained with the weaker laser intensities, particularly in the lower frequency region ($<10^3$ Hz). This may be because the current model does not consider trapping and detrapping processes. However, this result demonstrated that $2f$ MPC measurements enable the separate study of electron and hole transport processes, e.g., mobility and driving force, in working OPVs.

4. Conclusions

$2f$ MPC measurements were applied to P3HT:PCBM-based OPVs. Unlike conventional $1f$ MPC measurements, a photocurrent component that was generated in proportion to the light intensity was suppressed in $2f$ MPC measurements. As a result, more detailed information on the carrier transport process was obtained. It was also demonstrated that the signatures due to the electron and hole transports appear in different frequency domains.

Acknowledgments

This work is supported in part by JSPS KAKENHI Grant Numbers JP17H01265 and JP18H03902.

References

[1] Brabec C, Scherf U, and Dyakonov V (ed.) 2014 *Organic Photovoltaics: Materials, Device Physics, and Manufacturing Technologies* 2nd Ed. (Weinheim: Wiley-VCH)
[2] Leo K (ed.) 2017 *Elementary Processes in Organic Photovoltaics* (Switzerland: Springer)
[3] Macdonald J R (ed.) 1978 *Impedance Spectroscopy: Emphasizing Solid Materials and Systems* (New York: Wiley-Interscience)
[4] Belmonte G G, Munar A, Barea E M, Bisquert J, Ugarte I and Pacios R 2008 *Org. Electron.* **9** 847
[5] Otsura T, Nakatsuka E, Nagase T, Kobayashi T and Naito H 2016 *J. Nanosci. Nanotechnol.* **16** 3349
[6] Wohlgenannt M, Ehrenfreund E and Vardeny Z V2006 Spectroscopy of Long-Lived Photoexcitations in π-Conjugated Systems *Photophysics of Molecular Materials* ed. G Lanzani (Weinheim: Wiley-VCH) ch 5
[7] Voigt S, Zhokhavets U, Al-Ibrahim M, Hoppe H, Ambacher O and Gobsch G 2008 *Phys. Stat. Sol. (b)* **245** 714
[8] Kobayashi T, Terada Y, Nagase T and Naito H 2011 *Appl. Phys. Express* **4** 126602
[9] Terada Y, Kobayashi T, Nagase T and Naito H 2014 *Thin Solid Films* **554** 209
[10] Particularly, in chemistry, MPC is also referred to intensity-modulated photocurrent spectroscopy.
[11] Oheda H 1981 *J. Appl. Phys.* **52** 6693
[12] Schumm G and Bauer G H 1989 *Phys. Rev. B* **39** 5311
[13] Byers J C, Ballantyne S, Rodionov K, Mann A and Semenikhin O A 2011 *ACS Appl. Mater. Interfaces* **3** 392
[14] Cho S, Kim K D, Heo J, Lee J Y, Cha G, Seo B Y, Kim Y D, Kim Y S, Choi S and Lim D C 2014 *Sci. Rep.* **4** 4306
[15] Set Y T, Heinemann M D, Birgersson E and Luther J 2013 *J. Phys. Chem. C* **117** 7993
[16] Kumar M, Dubey A, Reza K M, Adhikari N, Qiao Q and Bommisetty V 2015 *Phys. Chem. Chem. Phys.* **17** 27690
[17] Gao Y, Wise A J, Thomas A K and Grey J K 2015 *ACS Appl. Mater. Interfaces* **8** 285
[18] Neukom M, Züfle S, Jenatsch S and Ruhstallera B 2018 *Sci. Technol. Adv. Mater.* **19** 291
[19] Liptay W 1974 Dipole Moments and Polarizabilities of Molecules in Excited Electronic States *Excited States* ed. Lim E C (New York: Academic Press) p 129
[20] Sevastian L, Weiser G and Bässler H 1980 *Chem. Phys.* **61** 125
[21] Nakami S, Narioka T, Kobayashi T, Nagase T and Naito H 2017 *J. Phys.: Conf. Ser.* **924** 012009
[22] Kao K C 2004 *Dielectric Phenomena in Solids* (Amsterdam: Elsevier) p 92
[23] Sze S M and Ng K K 2007 *Physics of Semiconductor devices* 3rd Ed. (New Jersey: Wiley-Interscience) p 90

Time-resolved measurement of photocarrier generation in $CH_3NH_3PbI_3$ single crystals

Y Tanaka and I Akimoto

Department of Applied Physics, Wakayama University, Sakaedani 930, Wakayama 640-8510, Japan

akimoto@sys.wakayama-u.ac.jp

Abstract. A giant microwave response is transiently observed in $CH_3NH_3PbI_3$ single crystals under pulsed laser excitation at temperatures in both structural phases across the transition at 161 K. The response is caused by photocarriers generated in the specimen mounted in a microwave cavity. This detection technique is capable of measuring the carrier dynamics with a time-resolution of a few nanoseconds. Based on the time-resolved excitation spectrum near the absorption edge, we unveil that the photocarriers are generated by long-lived excitons as well as by fast band-to-band transition.

1. Introduction

Organo-metal halide perovskite systems have attracted increasing attention as light absorbers in solar cells [1]. Remarkable device performances [2] are originated form characteristic properties of these materials, such as a high optical absorption coefficient, free carrier generation, low trapping rate, a large carrier diffusion length, and photon recycling [3]. Regarding the structural phase transition, the rotational motion of organic cation comes to be restricted at the low temperature phase, affecting the dielectric constant significantly [4]. Dielectric fluctuation due to a collective orientational motion of cation is considered to be important to set up a low-trap platform for carrier transport in the high temperature phase [5]. The role of excitons for the device function has been a hot topic under investigation. The binding energy of free exciton in $CH_3NH_3PbI_3$ was figured out to be c.a. 13 meV in the low-temperature phase from the temperature dependence of steady-state photocurrent (PC) and photoluminescence (PL) with a single crystal [6] and the transient-absorption with a thin film [7].

In this study, we investigate photocarrier generation mechanisms in $CH_3NH_3PbI_3$ single crystals by a newly developed transient microwave response measurement. Microwave at the frequency of 9.6 GHz can probe overall photocarriers generated inside a crystal without electrodes and wire connection, in contrast to the ordinary surface-current measurements. Observed excitation spectrum of the signals is changed drastically across the phase transition temperature of T_c = 161 K in iodide, in accordance with the previous steady-state PC measurement [6]. From time-resolved excitation spectra, we unveil that the photocarriers are generated by long-lived excitons as well as by fast band-to-band transition in both structural phases.

2. Experiment

Single crystals of $CH_3NH_3PbI_3$ were synthesized in solution via inverse temperature recrystallization [8]. We confirmed that XRD peaks appeared at the expected angles for this material [8]. A crystal of typical size of $2.5\times2.5\times1.0$ mm^3 was mounted in a microwave cavity (Bruker, MD5W1) which is capable of adjusting the value of the quality factor (Q). The crystal was exposed to continuous

IOP Conf. Series: Journal of Physics: Conf. Series **1220** (2019) 012019 doi:10.1088/1742-6596/1220/1/012019

microwave and laser light pulses in the temperature range of 10 - 300 K. A transient microwave response was measured by an oscilloscope (Bruker, SpecJet in ELEXSYS E580) synchronized to the laser pulse. Wavelength of excitation light was selected near the absorption edge from a linearly-polarized tunable pulse laser (Spectra Physics, MOPO). Incident pulse energy was controlled in the range of 1 - 3 μJ. Temperature of sample was monitored by a K-type thermocouple on the cavity with assistance by a Cernox resistor mounted on the cryostat under cooling with liquid helium or nitrogen.

It is known that dielectric constant of the specimen is very large in the tetragonal phase above T_c and drastically decreases in the orthorhombic phase below T_c [4]. When a small sized dielectric material is installed in a microwave cavity, a loaded Q-value is influenced by the dielectric constant of the material. Actually the value monitored at a critical coupling was varied with temperatures reflecting the change of dielectric constant. Therefore, we measured a transient microwave response by adjusting the Q-value as low as of 800 (a time-resolution of 13 ns) at each temperature in order to perform measurement under the same condition.

3. Results and Discussion

A giant microwave response was observed under pulsed laser excitation depending on the excitation photon energy and temperature. Figure 1 shows the temporal profile of the microwave power response after the excitation laser pulse at various photon energies near the absorption edge, measured at temperatures, (a) 92 K in the orthorhombic phase and (b) 284 K in the tetragonal phase. Each time profile consists of the fast and slow decay components. The fast decay component within 0.1 μs changes sharply at the low energy side. Overall characteristics of the response are shown in coloured contour maps (Fig. 2) at (a) 92 K and (b) 284 K, as functions of time after a pulsed excitation and excitation wavelength. Besides the fast component at the high intensity (red-yellow) parts, a long-lived component continued over 1 μs is remarkably recognized as a broad peak around 1.64 eV at 92 K or around 1.52 eV at 284 K,.

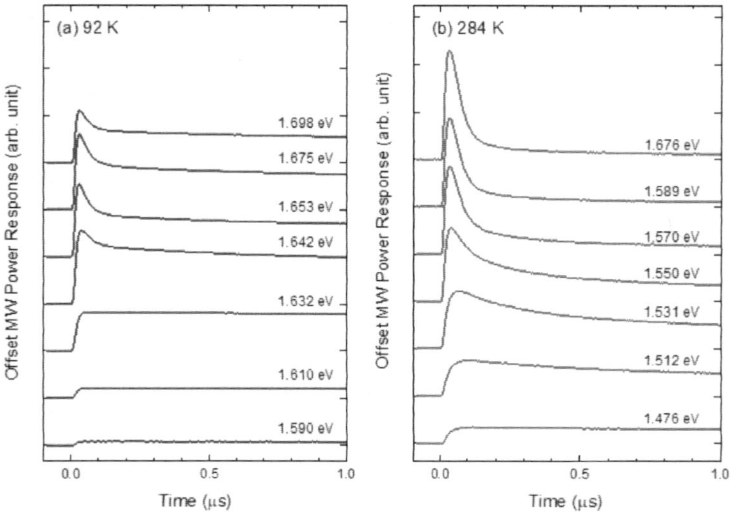

Figure 1 Microwave (MW) power response excited at various excitation photon energies, measured at temperatures of (a) 92 K in the orthorhombic phase and (b) 284 K in the tetragonal phase.

These signals appeared irrelevantly to an applied magnetic field, in contrast to the signals in the cyclotron resonance [9]. Present signals can be attributed to the transient carriers injected by optical excitation as follows. In general, a loaded Q-value of a microwave cavity depends on a dielectric property of an inserted specimen. If the real or imaginary part of the dielectric constant of the specimen is changed, the loaded Q-value is modified [10]. When the optical conductivity, which is directly relating to the imaginary part of the dielectric constant, instantaneously increases by photocarrier generation in the specimen, the Q-value suddenly decreases and then a part of irradiated microwave power cannot enter the cavity at the moment. As a result, a transient response of microwave power in a balanced detector is observed. Therefore, a dynamical photocarrier generation in the specimen can be monitored as a transient microwave power response in present method.

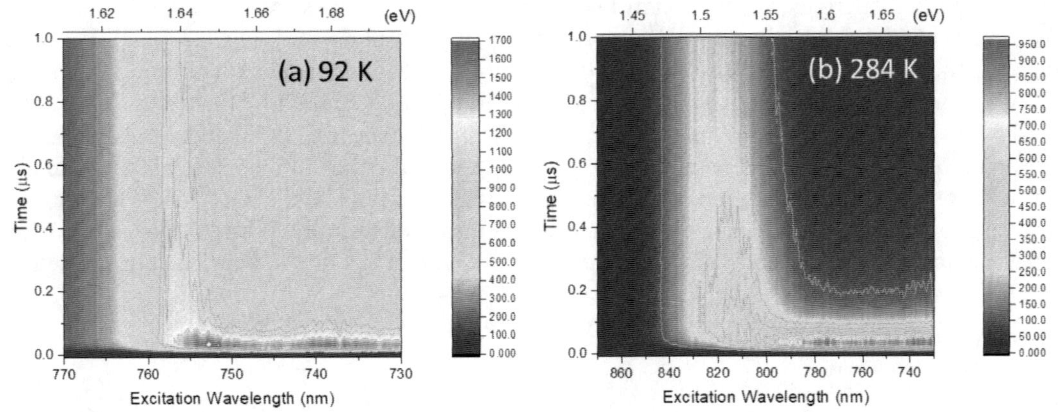

Figure 2. A contour map of the transient microwave power response at (a) 92 K and (b) 284 K, as functions of time after a pulsed excitation and excitation wavelength. The excitation photon energy was indicated in the upper horizontal axis.

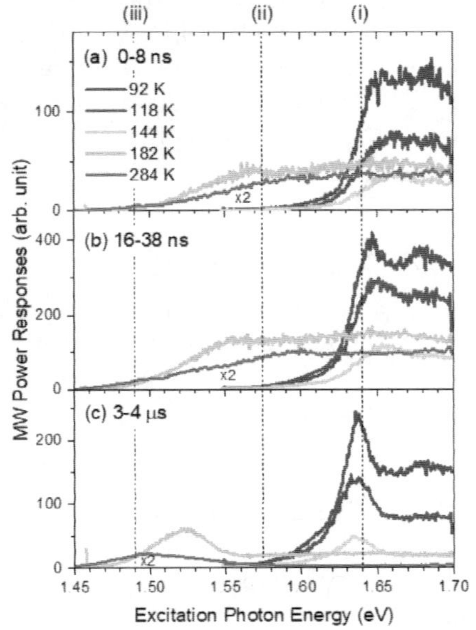

Figure 3. Microwave power responses as a function of the excitation photon energy at various temperatures of 90, 120, 150, 180, and 290 K, averaged in time window (a) 0 - 8 ns, (b) 16 - 38 ns, and (c) 3 - 4 μs. The curves at 284 K were magnified two times. The vertical broken lines indicate the photon energies of (i) absorption edge in the orthorhombic phase, (ii) absorption edge in the tetragonal phase, and (iii) edge of photocurrent at room temperature, where the values were taken from ref. [6].

Time-resolved excitation spectrum provides dynamical and spectroscopic aspects of the carrier generation. Figure 3 shows the excitation spectra as a function of the excitation photon energy at various temperatures averaged in three time-windows, (a) 0 - 8 ns, (b) 16-38 ns, and (c) 3-4 μs. The vertical broken lines indicate the energies of (i) absorption edge in the orthorhombic phase (1.640 eV), (ii) absorption edge in the tetragonal phase (1.575 eV), and (iii) PC edge at room temperature (1.459 eV), where the values were taken from ref. [6]. At the time immediately after the excitation pulse (Fig. 3a), the excitation spectra at 92, 118, and 144 K correspond to the intrinsic absorption of the specimen in the low temperature phase. Those at 182 and 284 K indicate responses to the excitation photon energy covering the absorption in the high temperature phase.

At the temporal peak (Fig. 3b) and the later delay (Fig. 3c) times, the excitation spectra in the low temperature phase exhibited peaks at 1.646 eV and at 1.635 eV, respectively, indicating peak shift depending on the delay time. Each spectrum appears as a superposition of an isolated band and a continuum, referring to the dull-shaped edge of absorption spectrum which was resolved into an exciton band and band-to-band transition in literatures [7, 11]. The energies at half of maximum of spectrum in Fig. 3a and at peak in Fig. 3c were plotted in Fig. 4 as a function of temperature, including peaks at 1.535 eV at 182 K and 1.492 eV at 284 K (Fig. 3c) in the high temperature phase. These characteristic photon energies exhibit a significant jump of values across the T_c, which is qualitatively in agreement with previous reports on the energies of exciton and band gap [11, 12]. The energy difference of 13.5 meV at 10 K was in consistent with previously reported exciton binding energy in the orthorhombic phase [6, 7]. Therefore, the long-lived isolated band is attributed to carriers generated by dissociation of long-lived excitons. The lifetime longer than 0.5 μs may indicate a carrier dynamics dominated by exciton recombination process [13] or a contribution of delayed excitons due to shallow traps as discussed in colloidal quantum dot system of $CsPbBr_3$ [14]. The long-lived responses extended to the lower energy side below absorption edge in the tetragonal phase (Fig.3a-c) are probably related to recent report of a circular photogalvanic effect due to Rashba spin splitting [15].

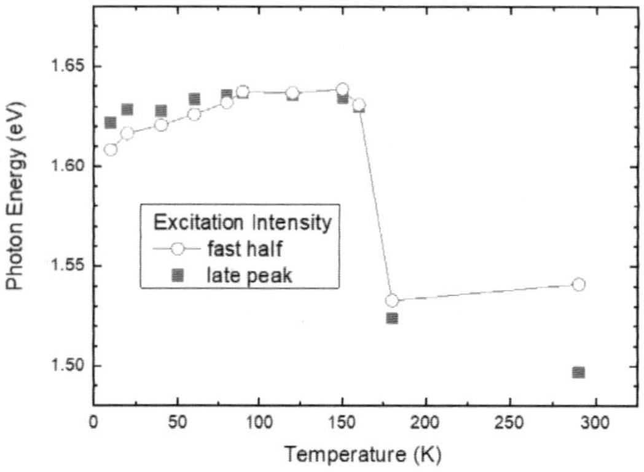

Figure 4. Temperature dependence of characteristic photon energies, the photon energy at half maximum of excitation spectra (open circles) extracted from Fig. 3a and the photon energy at the peak of the excitation spectra (open squares) extracted from Fig. 3c.

4. Conclusion

We investigated the photocarrier generation in a $CH_3NH_3PbI_3$ single crystal by measuring a transient microwave response. Time-resolved excitation spectra at different time windows unveiled that the photocarriers are generated through long-lived excitons as well as fast band-to-band transition. Such microwave detection is highly sensitive to study a photocarrier generation without wire connection, involving carrier dynamics measurement with a time-resolution of a few nanoseconds.

References

[1] Kojima A, Teshima K, Shirai Y, and Miyasaka T 2009 *J. Am. Chem. Soc.* **131** 6050
[2] Best Research-Cell Efficiencies at a homepage of NREL, "https://www.nrel.gov/pv/assets/images/efficiency-chart-20180716.jpg"
[3] Yamada Y, Yamada T; Phuong L; Maruyama N, Nishimura H, Wakamiya A, Murata Y, Kanemitsu Y 2015 *J. Am. Chem. Soc.* **137** 10456
[4] Onoda-Yamamura N, Yamamuro O, Matsuo T, and Suga H 1992 *J. Phys. Chem. Solids* **53** 935
[5] Kanno S, Imamura Y, Saeki A, and Hada M 2017 *J. Phys. Chem.* C **121** 14051
[6] Phuong L Q, Nakaike Y, Wakamiya A, and Kanemitsu Y 2016 *J. Phys. Chem. Lett.* **7** 4905
[7] Yang Y, Yang M, Li Z, Crisp R, Zhu K, and Beard M C 2015 *J. Phys. Chem. Lett.* **6** 4688
[8] Saidaminov M I, Abdelhady A L, Murali B, Alarousu E, Burlakov V M, Peng W, Dursun I, Wang L, He Y, Maculan G, Goriely A, Wu T, Mohammed O F, Bakr O M 2015 *Nature Commun.* **6** 7586
[9] Akimoto I and Naka N 2017 *Appl. Phys. Exp.* **10** 061301
[10] Porţeanu H E, 2005 *Rev. Sci. Instrum.* **76** 035106
[11] Ruf F, Magin A, Schultes M, Ahlswede E, Kalt H, and Hetterich M 2018 *Appl. Phys. Lett.* **112** 083902
[12] Tahara H, Endo M, Wakamiya A, and Kanemitsu Y 2016 *J. Phys. Chem.* C **120** 5347
[13] Chen X, Lu H, Yang Y, and Beard M C 2018 *J. Phys. Chem. Lett.* **9** 2595
[14] Wang Y, Zhi M, and Chan Y 2017 *J. Phys. Chem.* C **121** 28498
[15] Niesnera D, Hauckb M, Shresthac S, Levchukc I, Mattc G J, Osvetc A, Batentschukc M, Brabec C, Weberb H B, and Faustera T 2018 *PNAS* **115** 9509

IOP Conf. Series: Journal of Physics: Conf. Series **1220** (2019) 012020 doi:10.1088/1742-6596/1220/1/012020

Generation processes of superfluorescence of biexcitons in CuCl quantum dots by one- and two-photon resonant excitation

K Kawamura[1], T Yoshida[1], M Nasu[1], J Ishihara[1], A Ishikawa[2] and K Miyajima[1*]

[1]Department of Applied Physics, Graduate School of Science, Tokyo University of Science, 6-3-1 Niijuku, Katsushika-ku, Tokyo 125-8585, Japan
[2]Department of Science for Advanced Materials, Graduate School of Engineering, University of Yamanashi, 4-3-11 Takeda, Kofu-city, Yamanashi 400-8511, Japan

*Corresponding Author

E-mail address: miyajima@rs.tus.ac.jp

Abstract. We investigated the influence of excitation processes of a biexciton on the generation of superfluorescence, which is pulsed radiation from coherently coupled two-level systems, from biexcitons in CuCl quantum dots. Two excitation processes of biexcitons were examined: resonant two-photon excitation of biexcitons, and resonant one-photon excitation of excitons. The excitation density dependence of the time-resolved photoluminescence was measured. The results showed that for one-photon excitation, a stronger peak intensity, longer delay time and wider pulse width were observed compared to resonant two-photon excitation. The high density of the excited dots in the one-photon excitation process results in the strong pulse, while an initial population of excitons results in a suppression of coherent coupling.

1. Introduction

Superfluorescence (SF) is defined as cooperative spontaneous emission from coherently-coupled many excited two-level systems [1]. In the generation process of SF, complete population inversion of many two-level systems is formed by a strong excitation pulse. Their dipole moments are then synchronized by interaction with a radiation field, which is triggered by a spontaneous emission. They then cooperatively emit photons. Therefore, to generate SF, the distance between any given pair of two-level systems should be shorter than the emission wavelength, and a long dephasing time is required to keep synchronization during emission. Until now, most studies on SF reported mainly on atoms and molecules [2]-[4]. However, in recent years, observation of SF in semiconductors has been reported [5],[6]. An exciton or biexciton in a semiconductor has a shorter radiative lifetime compared to in atom or molecule, which induces ultrashort pulses. In addition, semiconductors have an advantage with regard to applications to new light sources owing to the possibility of current injections and miniaturization. Semiconductor quantum dots (QDs) assembly is an appropriate system for generating SF because the fabrication of high-density QD assembly is possible and a QD has quantized energy levels and long dephasing time.

We reported on SF from CuCl QDs assembly [7][8]. CuCl is a suitable material for studying the fundamental properties of excitons and biexcitons because of their large exciton and biexciton binding energies [9][10]. Owing to latter, CuCl QDs assembly has a two-photon absorption band for biexcitons which is below the exciton absorption band [11]. Therefore, a complete population inversion between the biexciton and exciton states can be achieved initially by resonant two-photon excitation of the biexcitons, as shown in Figure 1(a). Under this excitation process, the generation of ultrashort pulsed SF and the transition from a spontaneous emission to the SF are observed for photoluminescence (PL) of biexcitons in CuCl QDs [7][8]. Furthermore, in the case of resonant excitation of the exciton in which a biexciton originates via a relaxation from two excitons in a QD as shown in Figure 1(b), pulsed emissions from the biexcitons were observed [7]. However, the influence of the excitation processes of the biexciton for generating SF has not been investigated. In this study, we measured time-resolved biexciton luminescence under one- and two-photon excitation and their excitation density dependences are discussed. Differences in the actual excited dot density and the initial population density of the excitons and biexcitons affected the time-profiles of the SF.

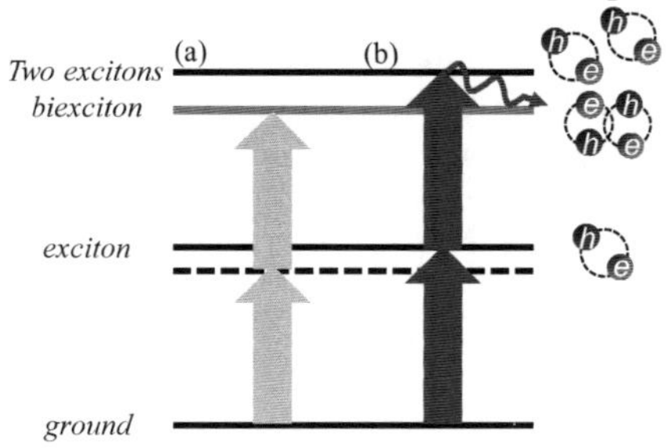

Figure 1. Energy diagram of resonant two-photon excitation process of biexcitons via virtual level (yellow arrows) (a) and one-photon excitation processes of excitons to generate biexcitons through a relaxation from 2 excitons (blue arrows) (b).

2. Experimental setup

We fabricated CuCl QDs embedded in NaCl single crystal using the transverse Bridgman method and the annealing process [12]. The annealing process was performed at 580 °C for 24 hours. The average radius of the CuCl QDs was estimated as ~4.9 nm from the exciton energy obtained by PL excitation measurement. The sample was then cleaved as a plate with a thickness of 1.21 mm and was placed in a holder in a cryostat. During subsequent optical measurement, the sample temperature was kept at 3 K.

The optical Kerr gate method was used for time-resolved PL measurements. A picosecond pulsed laser beam was produced by a regenerative amplified mode-lock Ti:sapphire laser (wavelength, 800 nm; repetition rate, 1 kHz). The laser beam was then divided into two components using a half mirror: one for the gate light and the other for the pump light of an optical parametric amplifier (OPA). The excitation light was obtained from the fourth harmonic generation of a signal beam of the OPA using two β-BaB_2O_4 crystals. The excitation light was focused on the sample with a deformed shape (length, 1870 μm; width, 41 μm) using a cylindrical lens, then, the PL from an edge of the sample was collected. The excitation energy was 3.210 eV for one-photon excitation and 3.185 eV for resonant in two-photon excitation. These values were determined from the PL excitation spectrum of the biexcitons.

3. Results and discussion

The excitation density dependences on the time profiles of the biexcitonic PL under one-photon and two-photon excitation are shown in Figure 2(a) and (b), respectively, with scattered excitation light. In both cases, the time profiles changed from exponential decays to the pulsed emissions.

Figure 3 shows the excitation density dependences of the delay time (a), the pulse width (b), and the peak intensity (c). Here, the delay time was defined as the time at the peak of the time profile, and the pulse width was defined as the full-width at half-maximum. When the excitation density increases, a transition from spontaneous emission to SF occurs, which results in the increase in the delay time and a large superlinearity of the peak intensity. A further increase in the excitation density results in a dominance of the time profile by the SF, which results in a decrease in the delay time [8]. The ideal dependence of the peak intensity is the square of the density of the excited dots. In addition, the pulse width is expected to decrease. From Figure 3, the time profiles for one-photon excitation clearly exhibit the aforementioned characteristics. In contrast, the time profiles for two-photon excitation exhibited unexpected behaviors of the delay time and pulse width; they increased gradually for high excitation density regions. However, because the peak intensity exhibited a similar behavior to that reported in a previous report [8], the generation of the SF is concluded. The origin of the unexpected behavior has not been resolved. It is suggested that the observed time profile consists of a few different SF components. Therefore, for a high excitation density, a new SF component with a long delay time appeared in addition to the main SF component, which results in a long delay time and wide pulse width of the whole profile.

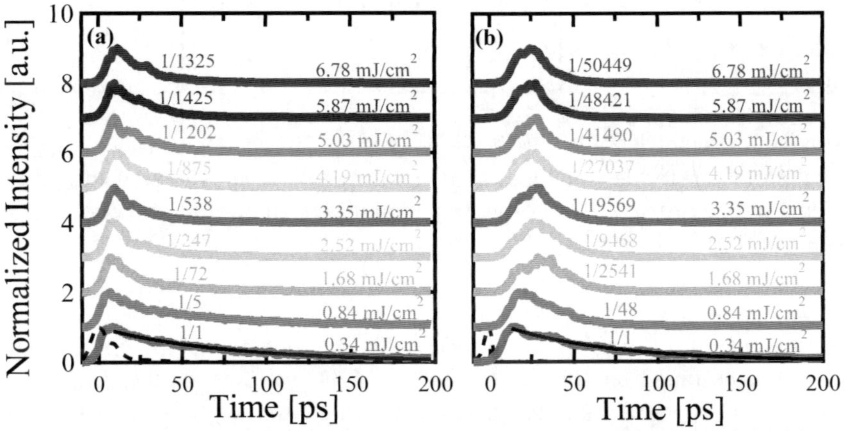

Figure 2. Excitation density dependences of time profiles at the peak of the biexcitonic PL obtained under (a) one- and (b) two-photon excitation processes, with photon energies of 3.210 eV and 3.185 eV, respectively. The black broken curves represent the scattered excitation light. The solid black curve shows a fitting by the exponential function. All spectra were normalized by their maximum intensities.

In this case, we focus on the differences between the excitation processes. In Figure 3 (a), the delay time for the one-photon process was approximately 20 ps longer compared to the two-photon process. The generation time for the biexciton from two electron-hole pairs under band-to-band excitation was reported as ~4 ps at 2 K [13]. Therefore, in the case of resonant excitation of the excitons, the biexcitons can be generated more quickly. Therefore, it is concluded that the long delay time resulted not from a relaxation process for two excitons, but from the suppression of coherent coupling of biexcitons. In addition, the pulse width for a one-photon process is wider than that for a two-photon excitation process, except for an excitation density over 4 mJ/cm^2. This result also supports the idea of a suppression of coherent coupling of biexcitons. We think that an initial population inversion ratio is important for the coherent coupling.

The two-photon process establishes a complete population inversion, while the one-photon process generates biexcitons and excitons simultaneously. For the correlated QDs with three energy levels (ground, exciton and biexciton levels), super-radiant coupling is dominant under a population inversion. However, when the exciton density is higher than the biexciton density, sub-radiant coupling becomes dominant. In the case of one-photon excitation, the initial population of the exciton, which is not negligible for the high excitation density, suppresses the super-radiant coupling. Our theoretical calculation based on the semiconductor luminescence equations [14] showed that the exciton population induces a sub-radiant process of the biexcitons. This agrees with the differences in the delay time and pulse width observed for one- and two-photon excitation. On the other hand, the peak intensity for one-photon excitation was stronger than that for two-photon excitation. In the case of the former, the absorption probability of the excitation light is larger than for the case of resonant two-photon excitation. Consequently, the actual density of the excited dots became larger, which results in a strong peak intensity. In addition, one-photon excitation exhibited a stronger power of 5.3 compared to a value of 3.5 for two-photon excitation. When the biexciton density exceeds the exciton density, SF components can be generated. This strong nonlinearity originates from the increase in the number of the cooperative two-level systems and the increase in the population inversion.

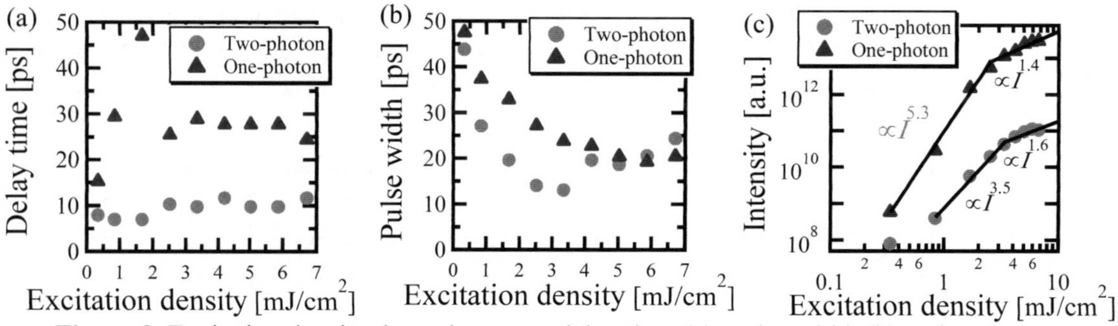

Figure 3. Excitation density dependences on delay time (a), pulse width (b) and peak intensity (c) in the time profiles for two-photon excitation (red circles) and one-photon excitation (blue triangles).

4. Conclusions

We investigated the influence of the excitation processes of biexcitons, which are resonant two-photon excitation of biexcitons and resonant one-photon excitation of the excitons, on the generation of SF in CuCl QDs. SF under one-photon excitation processes exhibited a stronger peak intensity, longer delay time and wider pulse width compared to the case of two-photon excitation. The strong peak intensity resulted from the large absorption probability of the excitation light. The actual density of the excited QDs is high for the one-photon excitation process. Furthermore, the longer delay time and wider pulse width indicates the inhibition of synchronization between dipole moments, resulting from the initial population density of the biexcitons and excitons. Our results indicate that the excitation process strongly influences on generating SF.

References

[1] Bonifacio R and Lugiato L A 1975 *Phys. Rev.* A **11** 1507
[2] Skribanowitz N, Herman I P, MacGillivray J C and Feld M S 1973 *Phys. Rev. Lett.* **30** 309
[3] Gibbs H M, Vrehen Q H F and Hikspoors H M J 1977 *Phys. Rev. Lett.* **39** 547
[4] Florian R, Schwan L O and Schmid D 1984 *Phys. Rev.* A **29** 2709
[5] Dai D C and Monkman A P 2011 *Phys. Rev.* B **84** 115206
[6] Cong K, Wang Y, Kim J, Noe G T, McGill S A, Belyanin A and Kono J 2015 *Phys. Rev.* B **91** 235448
[7] Miyajima K, Kagotani Y, Saito S, Ashida M and Itoh T 2009 J. Phys.:*Condens. Matter* **21** 195805
[8] Miyajima K, Kumagai Y and Ishikawa A 2017 *J. Phys. Chem.* C **121** 27751
[9] Saito K, Hasuo M, Hatano T and Nagasawa N 1995 *Solid State Commun.* **98** 33
[10] Tang Z K, Yanase A, Segawa Y, Matsuura N, and Cho K 1995 Phys. Rev. B **52** 2640
[11] Sato G, Akatsu T and Miyajima K 2016 *Mater. Res. Express* **3** 025002
[12] Miyajima K, Akatsu T and Itoh K 2018 *J. Crystal Growth* **490** 84
[13] Yano S, Yamamoto A and Goto T 1998 *Phys. Rev.* B **57** 7203
[14] Kira M and Koch S W 2006 *Phys. Rev.* A **73** 013813

IOP Conf. Series: Journal of Physics: Conf. Series **1220** (2019) 012021 doi:10.1088/1742-6596/1220/1/012021

Stimulated Raman scattering in anatase TiO_2

M Watanabe

Department of Interdisciplinary Environment, Graduate School of Human and Environmental Studies, Kyoto University, Kyoto 606-8501, Japan

watanabe.masayuki.4v@kyoto-u.ac.jp

Abstract. We report an experimental study of stimulated Raman scattering in anatase TiO_2 at 4K under both resonance and off-resonance conditions. Efficient first- and higher-order stimulated Raman emission of the lowest optical mode was observed, especially for the resonance case. From the dependence of the first-order Stokes emission on laser intensity, we estimate values of the Raman gain coefficient which are large compared with those reported for other crystalline materials. The large gain values are attributed to the narrow linewidth of the Raman line at low temperatures.

1. Introduction

Stimulated Raman scattering (SRS) is one of the nonlinear optical processes driven by coherent interactions between light and elementary excitations of materials [1]. Over the years, many applications such as Raman lasers [2] and attosecond pulse generation [3] have been developed out of this process. Spectroscopic applications of SRS have also been developed to obtain valuable information on the materials under investigation. It was pointed out that the major criteria for a material with efficient SRS are a narrow Raman linewidth and a high Raman scattering cross section [4]. Beattie and Gilson [5] observed a very intense Raman line in anatase TiO_2 which is well known as a photocatalyst material. Thus anatase is expected to be a good candidate for SRS. We shall report an investigation of SRS in anatase under off-resonant and resonant excitation.

2. Experimental

The samples used were single crystals of anatase TiO_2 grown by chemical transport reactions. They were kept in a continuous flow helium cryostat. For our measurements of stimulated Raman emission, an optical parametric amplifier (1.94 eV or 3.14 eV, 10 Hz repetition rate, 2 ns duration) pumped by the third harmonic of a Nd:YAG laser was used as the excitation source. The laser beam was focused into the sample by a lens of 25-cm focal length. Light scattered at nearly the right angle to the incident beam was analyzed with a double-grating monochromator equipped with a CCD detector. Spectral resolution of our setup was 0.2 (0.6) cm^{-1} for 1.94 (3.14) eV excitation.

3. Results and discussion

Figure 1 shows the spontaneous Raman spectrum of anatase TiO_2 measured at 5K. A cw He-Ne laser, 1.96 eV was used as the excitation source. Since the sample is not oriented, all the Raman-active vibrational modes with different polarizability tensors ($1A_{1g}$+$2B_{1g}$+$3E_g$) are detected [6]. It is found that the spectrum of the lowest E_g phonon at 137 cm^{-1} is quite sharp and intense, indicating a large scattering cross section for this mode [5]. These properties are the major criteria for a high gain Raman

IOP Conf. Series: Journal of Physics: Conf. Series **1220** (2019) 012021 doi:10.1088/1742-6596/1220/1/012021

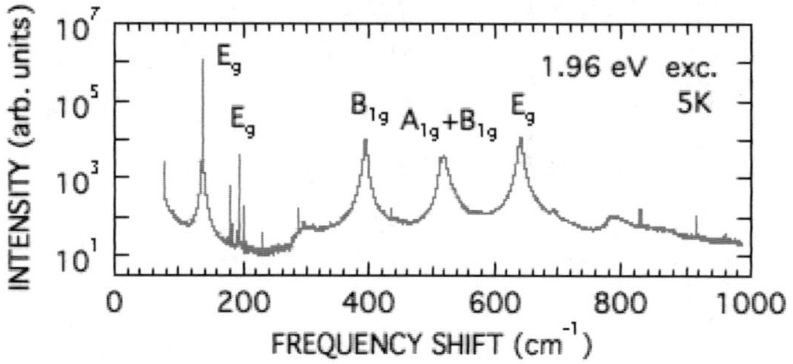

Figure 1. Spontaneous Raman spectrum of anatase at 5K excited with the 632.8 nm line of a cw He-Ne laser. The spectrum is plotted on a semi-log scale. Vibrational mode symmetries of the Raman lines are indicated. The other weak lines are plasma lines of the laser.

material [4].

Figure 2 shows the stimulated Raman spectra at 4 K and its dependence on incident laser intensity. Since the laser photon energy, 1.94 eV was well below the band-gap energy of anatase, 3.2 eV [7], the spectra were measured under the off-resonant excitation. At low excitation intensity, as shown in figure 2 (a), the spectrum is similar to the spontaneous Raman spectrum (figure 1). In addition, a weak anti-Stokes component (AS) due to the lowest E_g mode is seen. At higher intensities, the first-order Stokes emission of the E_g mode is strongly amplified through nonlinear coupling with the laser wave (figure 2(b)). As the excitation intensity is increased further, new Stokes lines appear, as shown in figure 2(c). There are intense lines at shifts of two and three times the frequency of the E_g phonon. Such higher-order Stokes components have been often observed in SRS experiments [1]. When the first-order Stokes emission becomes sufficiently intense, it acts as a source for the amplification of a light wave at the second-order Stokes energy. This process is iterated generating higher-order Stokes waves. It is found that the anti-Stokes component is also amplified.

Next we investigate the SRS process in resonant-scattering regime. Figure 3 shows the observed Raman spectra for incident laser energy of 3.14 eV which is slightly below the band-gap energy. As in the case of the off-resonant excitation, Stokes emission of the E_g mode is strongly amplified and the

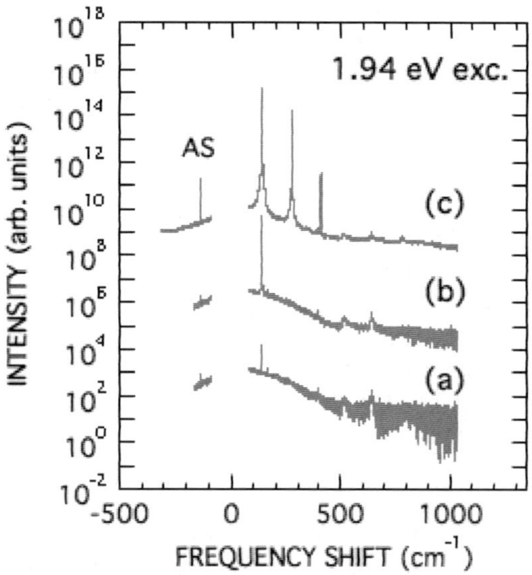

Figure 2. Stimulated Raman spectra at 4 K under off-resonant excitation for laser intensities equal to (a) 300, (b) 740 and (c) 2000 MW/cm^2, respectively. Each spectrum is offset by three decades for clarity.

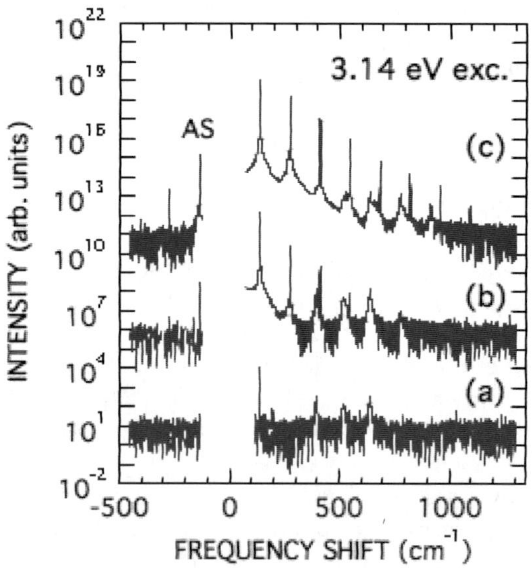

Figure 3. Stimulated Raman spectra at 4 K under resonant excitation for laser intensities equal to (a) 80, (b) 370 and (c) 1200 MW/cm^2, respectively. Each spectrum is offset by five decades for clarity.

number of the Stokes components increases with increasing excitation intensity. We observe the Stokes emission up to the eighth-order at laser intensity of 1.2 GW/cm^2 as shown in figure 3(c). Thus the number of the Stokes components in figure 3(c) is larger than that in figure 2(c) in spite of the low laser intensity compared with that in figure 2(c), indicating the resonant enhancement of Raman scattering efficiency.

Besides the sharp Stokes components, there are several bands in the range above 700 cm^{-1} in figures 3(b) and 3(c). These bands arise from spontaneous Raman scattering caused by the Stokes components. For example, the band at 777 cm^{-1} corresponds to the spontaneous Raman emission of the 640 cm^{-1}-mode (E_g symmetry) induced by the first-order Stokes emission. This fact indicates that the conversion efficiency of laser into the Stokes emission is high. Several anti-Stokes components are also observed in figures 3 (b) and 3(c). These components are produced by Raman-induced four-wave mixing processes [1].

The observed intensities of the first- and second-order Stokes emission are plotted as functions of the laser intensity in figure 4 for 1.94 eV excitation and in figure 5 for 3.14 eV excitation. For comparison, the intensity of the 395 cm^{-1}-mode (B_{1g}) is also shown. For both excitation energies, while the B_{1g} emission shows a linear dependence on incident laser intensity over the whole intensity range investigated (as shown by thin solid lines), the Stokes components show different behaviors depending on the laser intensity. With reference to [8], four different regions can be distinguished for the first-order Stokes emission : (A) at low intensities, the Stokes emission increases linearly with laser intensity (thin solid lines), indicating that spontaneous Raman scattering dominates ; (B) amplification of the Stokes emission occurs ; (C) substantial growth of the Stokes emission is observed, as shown by dotted lines (the lines are guides to the eye.) ; (D) the Stokes emission saturates under the resonant excitation (figure 5).

SRS is observed in the regions (B),(C) and (D). The threshold for SRS occurs at laser intensity of 340 (80) MW/cm^2 for 1.94 (3.14) eV excitation. We can estimate values of the Raman gain coefficient g by fitting an equation $I_{\mathrm{Stokes}}=(\mathrm{const})Ie^{gIl}$ to the first-order Stokes data in the region (B) [9]. Here I is the incident laser intensity and l is an effective interaction length. The bold solid curves in figures 4 and 5 were obtained by the fitting. Taking l = 0.5 mm (the sample size), the Raman gain coefficient is estimated to be 0.12 (0.14) cm/MW for 1.94 (3.14) eV excitation at 4 K (the accuracy of the gain values is approximately ±60% due to the possible errors in estimating I and l). These values are large

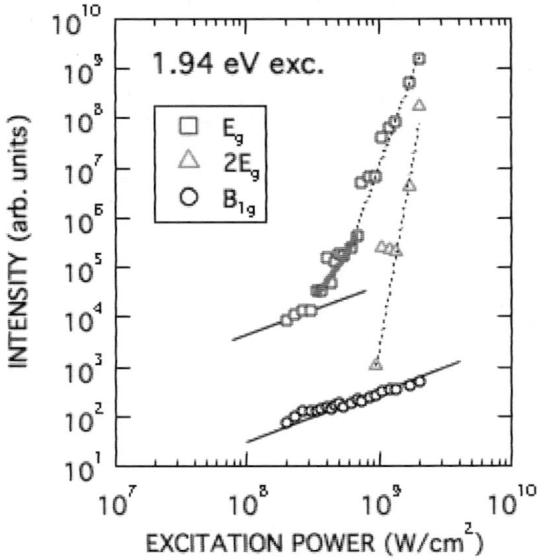

Figure 4. Experimental intensities of several Stokes lines for incident laser energy of 1.94 eV as functions of excitation power.

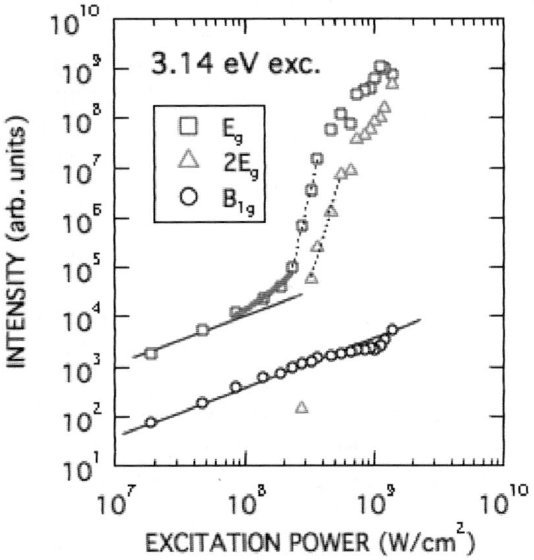

Figure 5. Same as figure 4, with incident laser energy of 3.14 eV.

IOP Conf. Series: Journal of Physics: Conf. Series **1220** (2019) 012021 doi:10.1088/1742-6596/1220/1/012021

compared with reported values for other crystalline materials. For example, gain values of 0.04 cm/MW for the 521 cm^{-1}-mode in Si at 77K [9] and 0.044 cm/MW for the 1066 cm^{-1}-mode in $NaNO_3$ at room temperature [10] have been reported. The gain coefficient is inversely proportional to the Raman linewidth [1]. The linewidth of the lowest E_g mode in anatase at 4 K is below our instrumental resolution (0.2 cm^{-1} full width at half-intensity-maximum) and is much smaller than the linewidths of 0.8 cm^{-1} for Si and of 2.0 cm^{-1} for $NaNO_3$. Thus it is concluded that the observed large gain is (at least partly) due to the narrow Raman linewidth. In fact, we did not observe SRS in anatase at room temperature where the Raman line of the E_g mode broadened considerably (about 7 cm^{-1}). It is seen that the estimated gain coefficient in resonance condition is almost equal to that in off-resonance condition. We have shown that intense photoexcitation with energies slightly below the band-gap energy induced resonant two-photon transitions at low temperatures [7]. Two-photon absorption at the laser frequency may suppress the Raman gain in resonance condition. Since the gain is inversely proportional to the cube of the Stokes frequency [1], the factor also contributes to the suppression.

In the region (C), the first-order Stokes emission increases sharply with increasing excitation intensity. The results are similar to those found in SRS experiments for liquid O_2 and N_2 [8]. In [8], the authors ascribed the results to the onset of Raman oscillation caused by feedback of Stokes emission via Rayleigh scattering. Since our sample is not a parallel-sided slab, it is unlikely that the onset of oscillation arises from reflection at the sample surfaces. Then it is inferred that the feedback via Rayleigh scattering is the cause of oscillation, as in the case of [8]. It is to be noted that substantial growth of the second-order Stokes emission is also observed in the region (C), as shown by dotted lines (the lines are guides to the eye.).

Under the resonant excitation (figure 5), at higher laser intensities above 1 GW/cm^2, the Stokes emission saturated and further increase in laser intensity resulted in reduction of the emission intensity due to damage to the crystal. As mentioned above, resonant two-photon transition occurs under intense photoexcitation at 3.14 eV, thus producing many electron-hole pairs in the crystal [7]. Then the two-photon excitation is expected to act as a competing nonlinear process in the region (D).

4. Conclusion

We have studied stimulated Raman effect in anatase TiO_2 at 4 K under the off-resonance and resonance conditions. It is found that anatase exhibits highly efficient first-order stimulated Stokes emission associated with the E_g mode at 137 cm^{-1}. We also observe the high-order Stokes and anti-Stokes emission, when we increase the incident laser intensity, especially for the resonant case. The large gain values estimated from the stimulated Raman emission are assigned to the narrow linewidth of the Raman line of the E_g mode at low temperatures. Our results show that anatase has large Raman nonlinearity and suggest the possibility of stimulated Raman effect to control the material property through excitation of intense coherent lattice vibrations.

References

[1] Kaiser W and Maier M 1972 *Laser Handbook* vol 2, ed F T Arecchi and E O Schulz-Dubios (Amsterdam: North-Holland) p 1077
[2] Spence D J 2017 *Prog. Quantum Electron.* **51** 1
[3] Agostini P and DiMauro L F 2004 *Rep. Prog. Phys.* **67** 813
[4] Murray J T, Powell R C and Peyghambarian N 1996 *J. Lumin.* **66&67** 89
[5] Beattie I R and Gilson T R 1968 *Proc. R. Soc. Lond. A* **307** 407
[6] Ohsaka T, Izumi F and Fujiki Y 1978 *J. Raman Spectrosc.* **7** 321
[7] Watanabe M and Hayashi T 2005 *J. Lumin.* **112** 88
[8] Grun J B, McQuillan A K and Stoicheff B P 1969 *Phys. Rev.* **180** 61
[9] Ralston J M and Chang R K 1970 *Phys. Rev. B* **2** 1858
[10] Karpukhin S N and Stepanov A I 1986 *Sov. J. Quantum. Electron.* **16** 1027

Observation of optical Stark effect between 1*s* - 2*p* exciton levels in CuCl single crystal

S Efumi, Y Uchibori, J Ishihara and K Miyajima*

Department of Applied Physics, Graduate School of Science, Tokyo University of Science, 6-3-1 Niijuku, Katsushika-ku, Tokyo 125-8585, Japan

*Corresponding Author

E-mail address: miyajima@rs.tus.ac.jp

Abstract. We measured the transient reflection spectra induced by infrared light for a CuCl single crystal using pump-probe spectroscopy. A Stark shift of the 1*s* exciton was observed using light with a near resonant energy between the 1*s* and 2*p* exciton levels. The amount of energy shift due to the Stark effect was obtained from the absorption spectra calculated using the Kramers–Kronig transform. We investigated the energy shift of 1*s* excitons with respect to the excitation density. This is the first published report on the Stark effect of 1*s* exciton associated with the internal energy levels of the excitons in CuCl crystal.

1. Introduction

When high intensity light such as a laser beam is incident on a material sample, a dressed state is generated in which an electron level of an atom is strongly coupled with a photon. The dressed state can be observed as an energy shift by the optical Stark effect during light irradiation, and the magnitude of the shift is determined by the intensity of the incident light and the detuning of the energy. In semiconductors, an exciton dressed state can be generated as in the case of an atomic system, because excitons have hydrogen atom-like energy. To observe the optical Stark effect in semiconductors, non-resonant light is generally used to avoid the complication of the optical response associated with the incoherent effect of excitons or carriers excited in the case of resonant excitation [1,2]. In addition, the optical Stark effect related to the internal levels of the exciton is quite useful because of the coherent control of atom-like levels in solid-states that it affords. For Cu_2O and GaAs quantum well, the optical Stark effect of 2*p* exciton has been reported for infrared light resonant for 1*s*–2*p* transition energy [3,4].

In this study, we focused on an exciton in CuCl single crystal. CuCl is a direct transition semiconductor, and the exciton with the lowest energy (named as Z_3 exciton) has a small Bohr radius and a large exciton binding energy (197 meV [4]). The internal energy levels related to the Rydberg states were observed by the absorption spectrum and the photoluminescence (PL) excitation spectrum [5]. In addition, transient absorptions from 1*s* to 2*p* and 3*p* exciton states, respectively, were observed via pump-probe spectroscopy [6]. The transition energy from 1s to 2p states was 169.5 meV. For CuCl, the Stark shift of the exciton has been reported when the photon energy of the pump light was near resonant to the transition energy from an exciton to biexciton states [7]. However, the Stark shift induced by the pump light resonant to the energy between the 1*s* and 2*p* levels of the exciton has not been

IOP Conf. Series: Journal of Physics: Conf. Series **1220** (2019) 012022 doi:10.1088/1742-6596/1220/1/012022

reported. We observed the optical Stark effect by time-resolved reflection spectroscopy. An energy diagram of this effect which was measured in this study is shown in Figure 1.

A reduction of reflectance originated from the irradiation of the pump light. From the absorption spectrum obtained by Kramers-Kronig transform (K - K transform), a red shift of the peak energy was confirmed, in addition to a reduction of the absorption coefficient and broadening of the spectral width at the pump light irradiation.

2. Experiment setup and method

The CuCl single crystal was fabricated by a vapor phase growth method. From theoretical calculations of the reflection spectrum to fit the experimental results, the damping constant of the exciton 1*s* level and the exciton dead layer were determined to be Γ_{1s} = 0.75 meV and l = 11 nm, respectively. Transient changes in the reflection spectra were measured by pump-probe spectroscopy using dual optical parametric amplifiers (OPAs) pumped by a regeneratively amplified pulse from a mode-locked Ti:sapphire laser. Time width of correlation between the pulses from two OPAs was 1.5 ps. The probe pulse was obtained by self-phase modulation, using a YAG crystal, of sun-frequency light (wavelength of 500 nm) of a signal and pump light from an OPA. A bandpass filter was used to select the wavelength near the 1s exciton energy in CuCl. The pump pulse in the mid-infrared region was obtained by difference frequency generation, using an $AgGaS_2$ crystal, of the signal and idler light from the other OPA.

The pump energy was tuned to 150 meV, 165 meV and 175 meV. The sample was positioned in a holder in a cryostat and temperature was kept at 3 K. The spot size of the pump light was 221 μm and that of the probe pulse was 215 μm in diameter on the sample surface. The relative delay times for the two pulses was controlled by a mechanical delay line. The reflection spectra were measured using a spectrometer (focal length: 500 mm; grating: 2400 g/mm; spectral resolution: 0.3 meV) equipped with a charge-coupled device (CCD).

3. Results and discussions

Figure 2 shows the contour image of the time-resolved reflection spectra at the excitation energy of 165 meV. A change in the reflectance was observed at the delay time of 0 ps. It is concluded that this is due to the Stark effect because the time width of the change in the reflectance is approximately equal to the pump pulse width. Figure 3 (a) shows the reflection spectrum (blue curve) at the time origin and that at −65 ps (black curve) without the influence of the pump light. It was confirmed that the reflectance decreased due to irradiation of the pump light.

In order to investigate the energy shift caused by the Stark effect in detail, the reflection

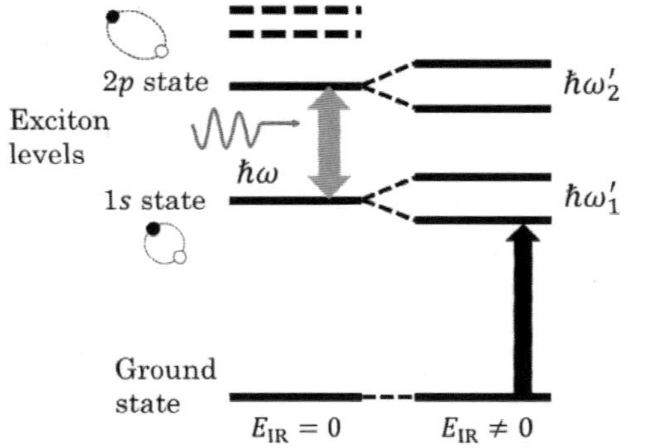

Figure 1 Energy diagram of Stark effect. The eigenvalues split by the Stark effect are represented by equation (1) and (2) in text, respectively.

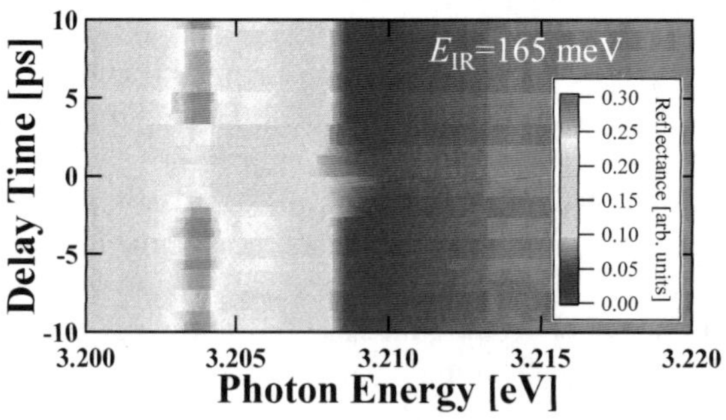

Figure 2 Contour image of reflection spectra near 1*s* exciton energy of 3.202 eV. The pump photon energy was 165 meV and a density of 366 μJ/cm^2.

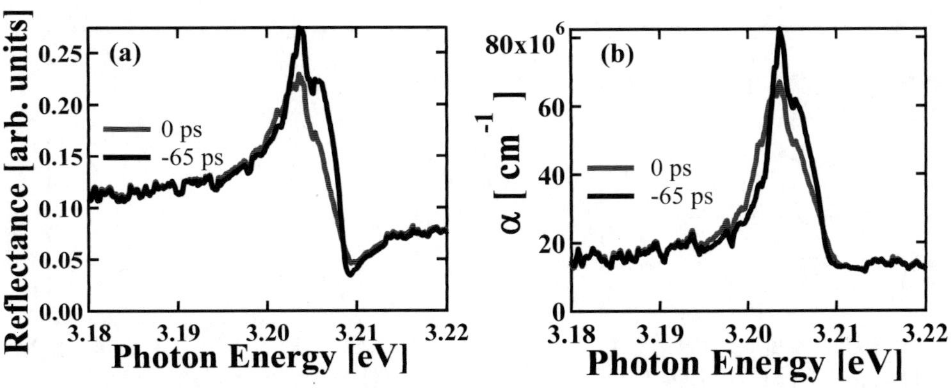

Figure 3 Reflection spectrum at delay time of 0 ps (blue) and -65 ps (black) (a), absorption spectrum at delay time of 0 ps (blue) and delay time -65 ps (black), which were obtained from the K-K transform of the reflection spectrum (b).

spectra were converted into absorption spectra using the Kramers-Kronig (K-K) transform as shown in Figure 3 (b). The blue and black solid curves are the absorption spectra at the time origin and the −65 ps, respectively. We observed a reduction of the absorption coefficient and a broadening of the spectral width. Figure 4 shows the changes in the absorption spectra at pump energies of 150, 165, and 175 meV (detuning of 19.5, 4.5 and −5.5 meV) with a pump density of 254, 366 and 315 μJ/cm^2, respectively. The positive peaks at 3.201 eV implies a red-shift of the 1*s* exciton energy level. In contrast, the negative change in the absorption from 3.202 to 3.210 eV was difficult to analyze. It is suggested that the complexity of this change is due to the crystallinity of the sample, e.g. the damping constant and the thickness of the dead layer. As a result of the Stark effect, the exciton energy splits into two levels (upper and lower levels). The red-shift shown in Figure 4 indicates the appearance of the lower levels. Simultaneously, the blue shift which originates from the upper level is expected. The oscillator strengths from the ground state to the lower and upper levels depend on the detuning energy; the oscillator strength of the lower level becomes strong for positive detuning, and vice versa [8]. When the photon energy of the pump light is increased, the negative peak in the vicinity of 3.207 eV decreases. It then disappeared at $\hbar\delta = -5.5$ meV, which suggest the appearance of the upper level. However, due to the complexity of the change in the absorption spectra, the peak energy and its strength could not be identified. Therefore, from the red-shift due to the lower level at the photon energy of 150 meV, we discuss the optical Stark effect.

Figure 5 shows the excitation density dependence of the Stark shift when light with an energy of 150 meV is used. The eigenvalues in the exciton dressed state are given by equations [8] (1) and (2),

$$\hbar\omega_1' = \hbar\left(\omega_1 + \frac{\delta}{2}\right) \pm \frac{\hbar}{2}\sqrt{\delta^2 + \Omega_0^2} \quad (1)$$

$$\hbar\omega_2' = \hbar\left(\omega_2 - \frac{\delta}{2}\right) \pm \frac{\hbar}{2}\sqrt{\delta^2 + \Omega_0^2} \quad (2)$$

where δ is the detuning from the resonance energy, Ω_0 is the resonant Rabi frequency which is given by $\Omega_0 = |\mu_{12}E_0|/\hbar$ using the electric field intensity E_0, the transition matrix element μ_{12}. $\hbar\omega_1$ and $\hbar\omega_2$ are the energies of the two states, respectively, and $\hbar$ is the Dirac constant. By using the Stark shift due to a lower level of obtained from the experiment, the transition dipole moment $|\mu_{12}|$ between 1*s* and 2*p* exciton levels is obtained from the equations (1) including the resonant Rabi frequency with $\hbar\omega_1 =$ 3.202 eV, $\hbar\delta =$19.5 meV. Using the result of the excitation density dependence of the Stark shift, the transition dipole moment was obtained for the respective excitation density, and by obtaining the average value, the transition dipole moment $|\mu_{12}|$ was determined as 2.97 ± 0.09 eÅ. Using the energy shift due to the Stark effect, the transition dipole moment between the 1*s* and 2*p* exciton levels in the CuCl single crystal was obtained.

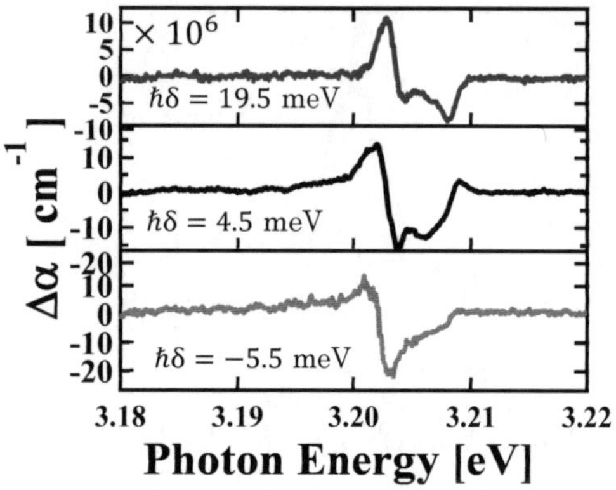

Figure 4 Absorption difference spectra for each detuning. The excitation density is 254, 366 and 315 μJ/cm^2 for detuning $\hbar\delta$ =19.5, 4.5 and −5.5 meV, respectively.

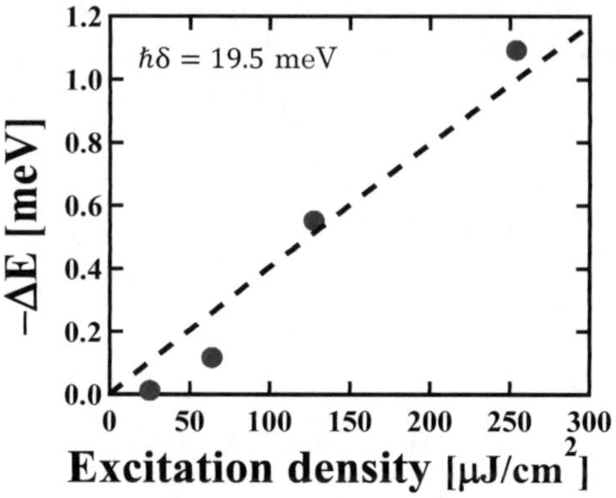

Figure 5 Excitation density dependence of optical Stark shift at the detuning $\hbar\delta$ =19.5 meV. The black dashed line shows the behavior of energy shift in the dressed model in the range of excitation density used.

4. Conclusions

The optical Stark effect between the 1*s* and 2*p* exciton levels in CuCl was observed by time-resolved reflection spectrum measurement. A reduction of the reflectance with a time width approximately equal to the pump pulse width was observed from the time-resolved reflection spectrum. Furthermore, from the absorption spectrum obtained from the Kramers-Kronig transform (K-K transform), it was confirmed that the peak energy was red shifted, the absorption amount decreased, and the spectrum width broadened during the time period of pump light irradiation. The transition dipole moment between the 1*s* and 2*p* exciton levels in CuCl were obtained from the excitation density dependence of the Stark shift at the time of pump light irradiation. The future task is to consider the influence of the coupled 2*p* level on the 1*s* level and to study the behavior of the Stark effect in the near resonance region by improving the quality of the sample.

References

[1] Mysyrowicz A, Hulin D, Antonetti A, Migus A, Masselink T W, and Morkoc H 1986 *Phys. Rev. Lett.* **56**, 2748.
[2] Hulin D and Joffre M, 1990 *Phys. Rev. Lett.* **65**, 3425.
[3] Frohlich D, Nothe A and Reimann K, 1985 *Phys. Rev. Lett.* **55**, 1335.
[4] Wagner M, Schneider H, Stehr D, Winnerl S, Andrews M. A, Schartner S, Strasser G, and Helm M 2010 *Phys. Rev. Lett.* **105**, 167401.
[5] Saito K, Hasuo M, Hatano and T, Nagasawa 1995 *Solid State Commun.* **94**, 33.
[6] Miyajima K, Sakaniwa K, and Sugawara M, 2016 *Phys. Rev. B* **94**, 195209.
[7] Shimano R and Gonokami-Kuwata M 1994 *Phys. Rev. Lett.* **72**, 530.
[8] Wagner M, Schneider H, Stehr D, Winnerl S, Andrews M. A, Schartner S, Strasser G, and Helm M 2011 *Phys. Status. Solidi. B* **248**, 4, 859.

IOP Conf. Series: Journal of Physics: Conf. Series **1220** (2019) 012023 doi:10.1088/1742-6596/1220/1/012023

Temperature-dependent carrier injection routes under optical excitation in high-purity diamond crystals

S Hamabata[1], I Akimoto[1] and N Naka[2]

[1] Department of Material Science and Chemistry, Wakayama University, Sakaedani 930, Wakayama 640-8510, Japan
[2] Department of Physics, Kyoto University, Kitashirakawa Oiwakecho, Sakyo-ku, Kyoto 606-8502, Japan

akimoto@sys.wakayama-u.ac.jp

Abstract. We have investigated carrier injection routes in high-purity diamond crystals in a wide temperature range using the time-resolved cyclotron resonance method. By analysing the excitation spectra, we find that carriers are injected under optical excitation above the energy threshold for exciton generation with emission of a transverse acoustic phonon at 10 K and 80 K, whereas with absorption of a transverse optical phonon at 300 K. The excitation power dependence clearly shows that the carriers are generated by thermal dissociation of excitons at 300 K.

1. Introduction

Diamond is the last frontier among the group IV semiconductors, with the large indirect energy gap of 5.47 eV at 300 K. Because of the outstanding properties, such as the high dielectric breakdown voltage, long spin relaxation time, and multivalley structure of the conduction band, diamond is an attractive material for next generation devices in the fields of power electronics, spintronics, and valleytronics [1, 2]. As in the case of pure silicon, which had been applied to a light-triggered thyristor as a successful power device, clarifying the optical routes for carrier injection in pure diamond is an urgent issue. In particular, detailed understanding of the coexistent state of carriers and excitons is necessary for the development of diamond devices working at room temperature, as the excitons in diamond are highly stabilized by the large exciton binding energy [3, 4]. This situation is characteristic of diamond, in contrast to the currently relevant materials, e.g., silicon or gallium arsenide, in which excitons are unstable at room temperature due to their small exciton binding energies [5].

For diamond, it is known that neutral excitons are primarily generated under optical excitation [3, 4] and they subsequently equilibrate with charge carriers [6, 7]. In this study, we investigate the ultraviolet spectral aspect of carrier injection in a high-purity diamond crystal using the time-resolved cyclotron resonance (TRCR) method. This measurement principle was already demonstrated for pure silicon [8] and pure diamond [9] at 10 K, in which two carrier injection routes (the band-to-band transition and dissociation of excitons) were unveiled and the exciton binding energy of diamond was renewed [9]. Here, we extend the measurements up to device-operation temperatures.

2. Experiment

We used a high-purity diamond crystal grown by the chemical-vaper-deposition (CVD) method in dimensions of $1.2 \times 4.5 \times 0.07$ mm^3, where the largest plane was oriented to the (001) crystal plane.

IOP Conf. Series: Journal of Physics: Conf. Series **1220** (2019) 012023 doi:10.1088/1742-6596/1220/1/012023

The sample was carefully mounted on a 2×8 mm^2 face of a right-angle prism using a small amount of vacuum grease. The prism was attached to a truncated quartz rod of ϕ 3.0 in diameter, and the rod was inserted into a quartz tube of ϕ 3.2/ ϕ 5.0 in inner/outer diameter (see the inset in Fig. 1). This tube was inserted in a dielectric cavity containing a sapphire tube (Bruker, MD5W1) in a cryostat, with an adjustable quality factor for microwave at 9.6 GHz. The [001] crystal axis was tilted by 40 degrees in the (110) plane with respect to the direction of the external magnetic field.

We measured temporal power responses of microwave resonating with the cyclotron motion of carriers using time-resolved paramagnetic resonance equipment (Bruker, ELEXSYS E580). The cavity quality factor was set at 400, corresponding to a time resolution of 7 ns. The sample was irradiated by laser pulses through the optical window of the cryostat, the sapphire tube of the dielectric cavity, and the prism. The wavelength ranged from 219.4 nm to 226.4 nm, where the light source was an optical parametric oscillator (Spectra-Physics, MOPO, 10 Hz, 5 ns) pumped by the third harmonics of infrared light from a Nd:YAG laser. We obtained excitation spectra by measuring the signal intensity divided by the simultaneously measured laser power as a function of the excitation wavelength. The incident laser power was reduced to the range of a few μJ by an optical attenuator, and further reduced by neutral density filters for measurements of the excitation power dependence.

3. Results and discussion

Figure 1 shows CR spectra at temperatures of 10 K and 80 K. Four peaks are separated at 10 K as previously reported [10], that are resonant to the light hole (lh), heavy hole (hh), and two electrons (e1, e2) belonging to distinct valleys under the magnetic field. The spectrum obtained at 80 K was broader than that at 10 K, probably owing to the enhanced rate of acoustic phonon scattering obeying the $T^{3/2}$ rule, where T is the temperature [11]. The spectrum at 300 K was much broader (not shown in Fig. 1), and the microwave response was still measureable. Hereafter, we focus on the CR intensity at 0.16 T which is resonant to the electron (e2).

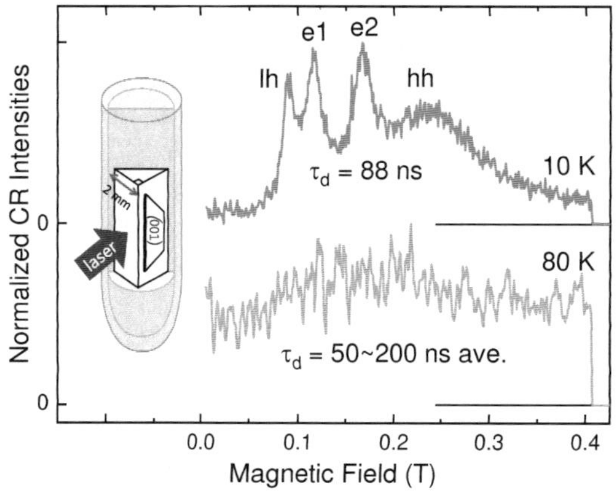

Figure 1. CR spectrum at 10 K at a delay time of 88 ns and that at 80 K averaged for 50–200 ns. The excitation photon energy was 5.550 eV. The inset shows schematic of the mounted sample.

Figure 2 shows the normalized CR time profiles at 10, 80, and 300 K. Under excitation at the photon energy of 5.520 eV (a) above the exciton absorption edge, signals peaked at delay times of 24, 48, and 16 ns at 10, 80, and 300 K, respectively. The signals decayed with lifetimes of 60, 425, and 125 ns at 10, 80, and 300 K, respectively. The lifetime was defined as the time for the 1/e of the peak intensity. We consider that the delayed peaks appeared owing to incubation of exciton dissociation, depending on the exciton lifetime and density [12]. The longest delay time at 80 K is considered to be caused by the exciton lifetime longest at approximately100 K [13], whereas the shorter delay time at 300 K is caused by the increase in the exciton density due to the higher absorption coefficient at higher temperatures [3]. On the other hand, under excitation at the photon energy of 5.620 eV (b), where carriers are directly generated by band-to-band transitions in addition to exciton dissociation [9], faster responses appeared within 40 ns. The signals peaked at 12, 16, and 8 ns at 10, 80, and 300 K, respectively, and decayed with lifetimes of 60, 300, and 104 ns at 10, 80, and 300 K, respectively.

IOP Conf. Series: Journal of Physics: Conf. Series **1220** (2019) 012023 doi:10.1088/1742-6596/1220/1/012023

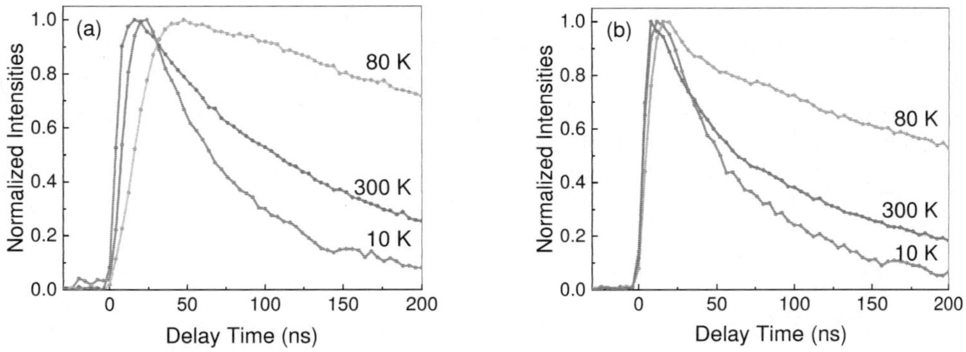

Figure 2. Normalized time profiles of the CR signal resonant to e2 at 0.161 T at 10, 80, and 300 K. The excitation photon energy was (a) 5.520 eV and (b) 5.620 eV.

Figure 3 shows the TRCR excitation spectra obtained at 10, 80, and 300 K. The time windows were chosen as (a) 80–280 ns at 10 K, (b) 352–552 ns at 80 K, and (c) 156–356 ns at 300 K, after the signal decayed to the 1/e of the peak intensity. For such late times, we observed that carriers were dominantly generated by dissociation of excitons [9]. The excitation spectrum at 80 K was similar to that at 10 K in terms of the onset energies. The onset energy at 5.493 eV coincides with the exciton generation edge assisted by emission of a transverse acoustic (TA) phonon ($E_{ex} + \hbar\omega_{TA}$), where E_{ex}= 5.406 eV is the exciton energy and $\hbar\omega_{TA}$ = 87 meV is the TA phonon energy [14] (See the inset). The second onset at 5.547 eV is assigned to exciton generation assisted by emission of a transverse optical (TO) phonon ($E_{ex} + \hbar\omega_{TO}$), where $\hbar\omega_{TO}$ = 141 meV is the TO phonon energy [14]. Because of the large phonon energies in diamond, the excitation spectra exhibited onsets at the same positions and no trace of phonon absorption up to 80 K.

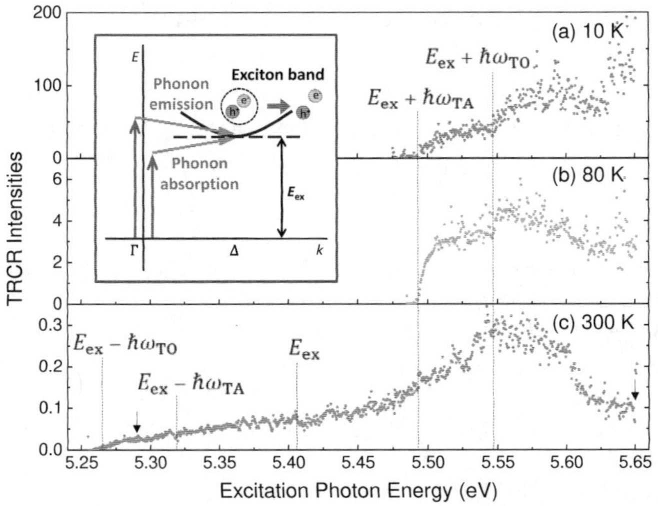

Figure 3. CR excitation spectra at (a) 10 K, (b) 80 K, and (c) 300 K. Vertical broken lines indicate the energy positions of $E_{ex} \pm \hbar\omega_{TA}, E_{ex} \pm \hbar\omega_{TO}$ and E_{ex} . The downward arrows indicate the excitation energies for the power dependence measurements shown in Fig. 4. The inset shows schematic of phonon-assisted transitions to the exciton band.

On the other hand, the signal at 300 K arose at the lower energy side with the onset at 5.265 eV. The onset energy coincides with the threshold for exciton generation assisted by absorption of a TO phonon ($E_{ex} - \hbar\omega_{TO}$) (See the inset). The onset energy for TA phonon absorption ($E_{ex} - \hbar\omega_{TA}$) is higher as indicated by the vertical broken line (the second one from the left) in Fig. 3(c). The above results make apparent that carriers are generated via excitons at 300 K with phonon absorption in addition to phonon emission. The measured excitation power dependence in Fig. 4 indicates that the CR signal intensity at 300 K proportionally varies with the excitation intensity, both under excitation

IOP Conf. Series: Journal of Physics: Conf. Series **1220** (2019) 012023 doi:10.1088/1742-6596/1220/1/012023

at 5.29 eV for phonon absorption and at 5.65 eV for phonon emission. Therefore, the carriers are generated dominantly via a one-body process, such as thermal ionization of excitons, instead of a two-body process due to collision of excitons at 10 K [7]. Finally, we would like to note that the excitation spectra in Fig. 3(a-c) deviated from a function $\sqrt{E - E_{ex} - \hbar\omega_i}$, which is characteristic of phonon-assisted exciton transition, above 5.58 eV. This is caused by the reduction of the CR signal under a high carrier density condition owing to a high absorption coefficient [4].

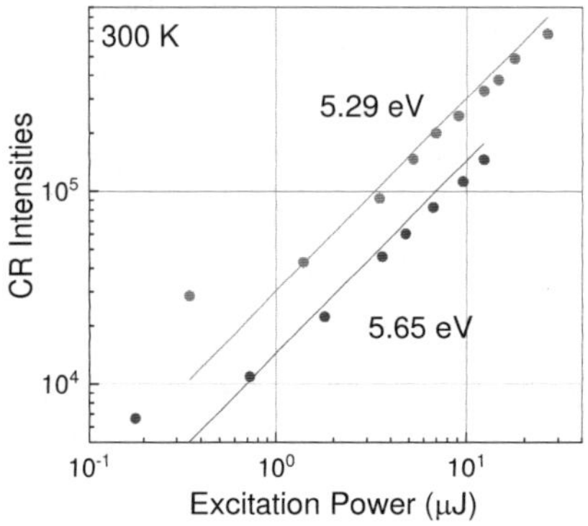

Figure 4. Excitation power dependence of the CR signal at 300 K, measured under excitation photon energies at 5.29 and 5.65 eV. The signals were integrated over time after excitation.

4. Conclusion

We investigated the excitation spectra of cyclotron resonance signals due to electrons in a high-purity diamond crystal up to 300 K. Based on the spectroscopic analysis, we found that carriers are generated by dissociation of excitons generated only by phonon emission at 10 and 80 K, and by phonon emission and absorption at 300 K. This indicates that it is possible to inject carriers via thermal ionization of excitons, even under excitation at low photon-energy region (5.265–5.493 eV, i.e., 235.5–225.7 nm in wavelength) with an absorption coefficient smaller than 1 cm^{-1} [3]. We believe that this knowledge is useful for device application, as bipolar carriers are uniformly injected over the entire crystal in such a condition.

Acknowledgements

We thank J. H. Kaneko (Hokkaido University) for providing the diamond sample grown by the CVD method. This work was supported by JSPS KAKENHI (Grant Nos. 15K05129 and 17H02910) and the Murata Science Foundation.

References

[1] Bottegoni F, Zucchetti C, Conte S D, Frigerio J, Carpene E, Vergnaud C, Jamet M, Isella G, Ciccacci F, Cerullo G, and Finazzi M 2017 *Phys. Rev. Lett.* **118** 167402.
[2] Cheng J L, Rioux J, Fabian J, and Sipe J E 2011 *Phys. Rev.* B **83** 165211.
[3] Clark C D, Dean P J, and Harris P V 1964 *Proc. Roy. Soc.* A **277** 312.
[4] Dean P J and Male J C 1964 *Proc. Roy. Soc.* A **277** 330.
[5] Sze S M 1969 *Physics of Semiconductor Devices* (New York: Wiley-Interscience).
[6] Gourley P L and Wolfe J P 1982 *Phys. Rev.* B **25** 6338.
[7] Naka N, Morimoto H, and Akimoto I 2016 *physica status solidi* A **10** 2551.
[8] Akimoto I and Naka N 2017 *Appl. Phys. Exp.* **10** 061301.
[9] Akimoto I, Hamabata S, Kaneko J H, and Naka N (in preparation).
[10] Naka N, Fukai K, Handa Y, and Akimoto I 2013 *Phys. Rev.* B **88** 035205.
[11] Akimoto I, Handa Y, Fukai K, and Naka N 2014 *Appl. Phys. Lett.* **105** 032102.
[12] Naka N, Fukai K, Handa Y, and Akimoto I 2014 *J. Lumin.* **152** 93.
[13] Naka N, Omachi J, Sumiya H, Tamasaku K, Ishikawa T, and Kuwata-Gonokami M 2009 *Phys. Rev.* B **80** 035201.
[14] Jacoboni C and Reggiani L 1983 *Rev. Mod. Phys.* **55** 645.

Temperature dependence of photoluminescence properties of water-soluble ZnSe quantum dots

Yong-Shin Lee, Taichi Taniguchi, and DaeGwi Kim *

Department of Applied Physics, Graduate School of Engineering, Osaka City University, 3-3-138, Sugimoto, Sumiyoshi-ku, Osaka 558-8585, Japan

*E-mail: tegi@a-phys.eng.osaka-cu.ac.jp

Abstract. We have investigated the optical properties of ZnSe quantum dots (QDs) prepared by a hydrothermal method. The photoluminescence (PL)-decay profiles become slower with an increase in temperature up to 160 K, contrary to an ordinary behaviour due to thermal quenching. The temperature dependence of the PL-decay profile is explained by a three-state model consisting of a ground state and two excited states of the lower-lying bound-exciton and higher-lying dark-exciton states. The analysis of the temperature dependence of the decay time indicates that the dark-exciton state contributes to the PL-decay process in the ZnSe QDs.

1. Introduction

Many studies have been conducted from a scientific viewpoint to understand the intrinsic nature of physical/chemical properties of semiconductor quantum dots (QDs) as well as from an interest in the application to new functional materials. The band-gap energy of ZnSe is 2.7 eV, so that high-quality ZnSe QDs have potentials for applications in photonic devices in the blue and UV regions. Furthermore, ZnSe QDs have an environmental advantage of being Cd-free.

So far, high-quality ZnSe QDs were synthesized in organic solvents [1] and also in aqueous solutions [2, 3], and the PL-wavelength was successfully changed by preparing alloy QDs of ZnCdSe [1, 2] and ZnSeS [4]. For further improvement of PL properties, core/shell QD structures of ZnSe/ZnS [5] and ZnSe/CdSe [6] were also prepared. However, the main concerns of previous studies on ZnSe QDs were the synthesis and PL properties at room temperature. In order to understand the PL mechanism, it is necessary to conduct systematic studies of the temperature dependence of the PL spectra and PL-decay profiles. In the present work, we have investigated the temperature dependence of absorption, PL, and PL-decay profiles in colloidal ZnSe QDs.

2. Experiments

In the synthesis of ZnSe QDs by a hydrothermal method, ZnSe precursor solution was prepared by adding freshly prepared NaHSe aqueous solution to a N_2-saturated $Zn(ClO_4)_2 \cdot 6H_2O$ aqueous solution in the presence of *N*-acetyl-L-cysteine (NAC) as a ligand [3]. Then, 10 mL of the precursor solution was loaded into an autoclave. The autoclave was maintained at 200 °C for a desired time and then cooled down to room temperature. In order to investigate the temperature dependence of the optical properties, we dispersed colloidal QDs into polymer films as follows. The sample solution was mixed with a polyvinyl alcohol aqueous solution. Then, the mixed solution was spread on a glass, and the excess water was evaporated by heating at 60 °C for 8 hours.

IOP Conf. Series: Journal of Physics: Conf. Series **1220** (2019) 012024 doi:10.1088/1742-6596/1220/1/012024

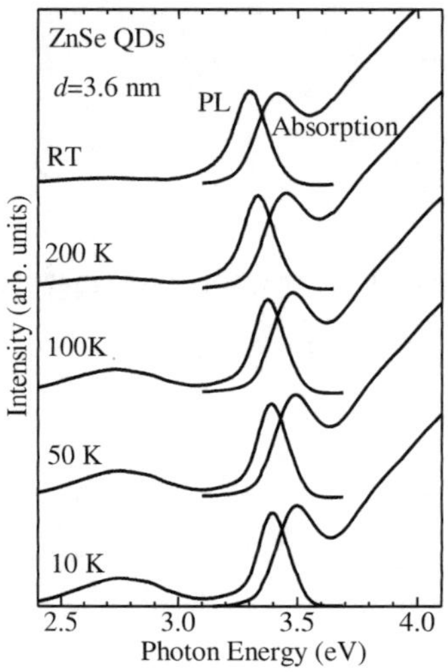

Figure 1. Temperature dependence of absorption and PL spectra for ZnSe QDs.

Absorption spectra were measured using a double-beam spectrometer with a resolution of 0.2 nm. For PL measurements, the 325 nm line of a He-Cd laser was used as the excitation light source, and the emitted PL was analyzed with a single monochromator with a spectral resolution of 0.5 nm. For measurements of PL-decay profiles, third-harmonic-generation (THG) light (355 nm) of a laser-diode pumped yttrium aluminium garnet (YAG) laser with a pulse duration of 20 ns and a repetition of 10 kHz was used as the excitation light. The PL-decay profiles were obtained by a time-correlated single-photon counting method. The sample temperature was controlled using a closed-cycle helium-gas cryostat.

3. Results and discussion

Figure 1 shows the temperature dependence of the absorption and PL spectra of ZnSe QDs with 3.6 nm in diameter. The absorption- and PL-peak energies shift to the higher energy side with a decrease in temperature. Figure 2(a) shows the temperature dependence of the lowest absorption peak (exciton) energy. It is well known that the temperature dependence of the exciton energy in a direct-gap semiconductor can be described by Varshni's law [7]. The solid curve indicates the calculated result for the temperature dependence of the absorption energy of a ZnSe bulk crystal. This result indicates that the observed temperature dependence results from the intrinsic property of the ZnSe crystal.

Figure 2(b) shows the temperature dependence of the integrated intensity of the band-edge PL band. The intensity of the band-edge PL band at room temperature is almost 50 % of that at 10 K, which

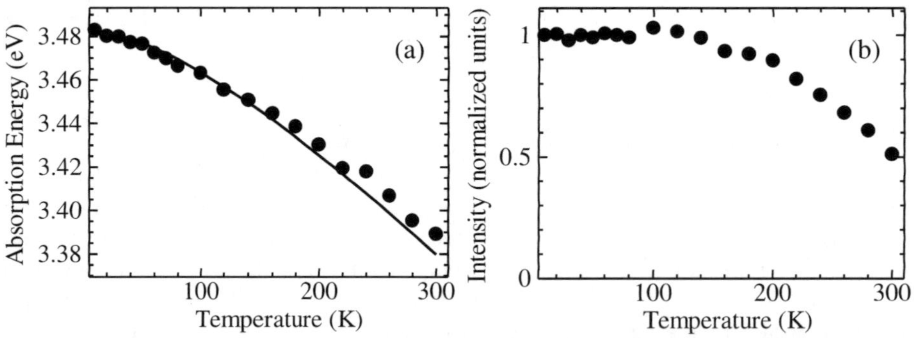

Figure 2. (a) Temperature dependence of the lowest absorption peak energy. (b) Temperature dependence of the integrated intensity of the band-edge PL.

IOP Conf. Series: Journal of Physics: Conf. Series **1220** (2019) 012024 doi:10.1088/1742-6596/1220/1/012024

means that the thermal quenching effect is not large. This result demonstrates that the influence of the nonradiative process is remarkably suppressed in the ZnSe QDs.

Figure 3(a) shows the temporal profiles of the band-edge PL at 10, 70, 120, and 200 K. In this temperature range, temporal profiles become slightly slower with an increase in temperature. In the usual case, as the temperature is increased, the nonradiative-decay rate related to the thermal quenching becomes larger, so that the PL intensity and the PL decay time are decreased. In fact, as shown in figure 2(b), the band-edge PL intensity gradually decreases with an increase in temperature. Nevertheless, the observed decay profiles become longer, which is contrary to the above scenario. The temperature dependence of the PL-decay profiles in ZnSe QDs can be quantitatively explained by a three-state model consisting of a ground state and two excited states: a lower-lying bound-exciton state and a higher-lying free-exciton state ("dark-exciton state") having an optically inactive triplet nature [8]. With an increase in temperature, the thermal population of the higher-lying dark-exciton state is significant, and the decay time becomes longer.

In order to discuss the PL-decay profiles quantitatively, we analyzed them by a Kohlrausch-Williams-Watts-type function [8]. The best fits were obtained with a combination of one monoexponential and one stretched exponential functions:

$$A_1 \exp\left(-t/\tau_1\right) + A_2 \exp\left[-\left(t/\tau_2\right)^{\beta}\right] \quad (1)$$

Here, the parameters A_1 and A_2 represent the relative weights of the exponential component and stretched exponential decay one. From the stretched-exponential PL-decay component, we can obtain an average decay time $\langle\tau_2\rangle=(\tau_2/\beta)\cdot\Gamma(1/\beta)$ as introduced in reference 9. Figure 3(b) shows the temperature dependence of $\langle\tau_2\rangle$ and the fast decay time of τ_1. The slow decay time of $\langle\tau_2\rangle$ becomes slower with an increase in temperature up to 160 K, while the value of τ_1 gradually increases. The intensity ratio of the slow component ($\langle\tau_2\rangle$) to the fast one (τ_1) is ~0.7 and over at every temperature. Thus, in order to discuss the temperature dependence of the PL decay profiles, we adopt the decay time of $\langle\tau_2\rangle$ as an intrinsic decay time because the fast component (τ_1) would reflect relaxation processes of excitons and carriers.

Finally, we quantitatively discuss the temperature dependence of $\langle\tau_2\rangle$ based on the three-state model. The decay time based on the above-mentioned three-state model is given as [8]:

$$\tau = \frac{1 + \frac{g_T}{g_B}\exp\left(-\frac{\Delta E}{k_B T}\right)}{\frac{1}{\tau_B} + \frac{g_T}{g_B}\frac{1}{\tau_T}\exp\left(-\frac{\Delta E}{k_B T}\right)} \quad (2)$$

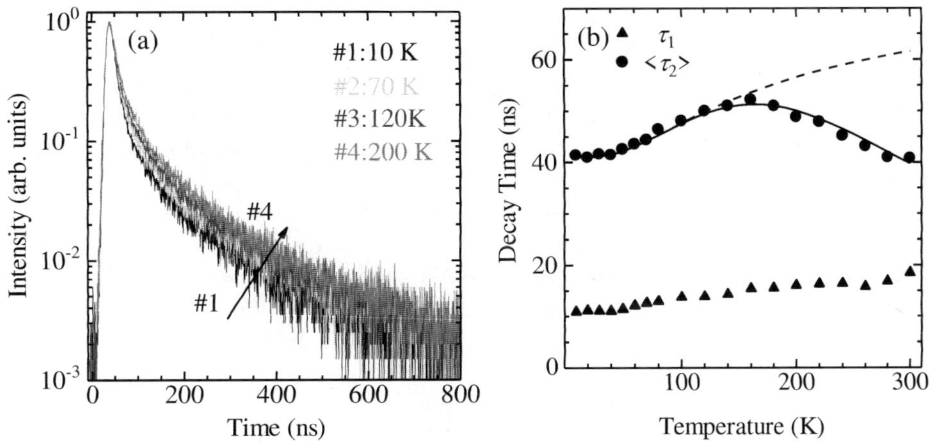

Figure 3. (a) Temperature dependence of temporal profiles of the band-edge PL. (b) Temperature dependence of the decay time of τ_1 (closed triangles) and $\langle\tau_2\rangle$ (closed circles).

Here, $1/\tau_T$ ($1/\tau_B$) and g_T (g_B) denote the radiative-decay rate and the density of states of the dark-exciton state (the bound exciton state), respectively. The dashed curve in figure 3(b) indicates the calculated result for the temperature dependence of τ using equation (2) with τ_T=600 ns, τ_B =41 ns, ΔE=15 meV, and g_T/g_B=1. The calculated result quantitatively explains the experimental one below 160 K. At temperatures higher than 160 K, the calculated result still increases gradually, while $<\tau_2>$ begins to decrease. The discrepancy between the calculated and the experimental results can be attributed to the influence of the nonradiative recombination process because equation (2) represents only the "*radiative*" decay rate in the three-state model. Thus, we take into account the temperature dependence of the nonradiative decay rate, $1/\tau_{nr}(T)$, in the form of a well-known thermal-activation type, which is given as:

$$\frac{1}{\tau_{\text{nr}}(T)} = \frac{1}{\tau_{\text{nr}}(T=0)} \exp\left(-\frac{E_{\text{a}}}{k_{\text{B}}T}\right) \tag{3}$$

where E_a represents the thermal-activation energy for the nonradiative process. We calculated the temperature dependence of $\tau(T)$, which is defined as:

$$\frac{1}{\tau(T)} = \frac{1}{\tau_{\text{r}}(T)} + \frac{1}{\tau_{\text{nr}}(T)} \tag{4}$$

where $1/\tau_r$(T) is given by equation (2). The solid curve in figure 3(b) represents the calculated result with use of the fitting parameters of $\tau_{nr}(T=0)$=7.5 ns and E_a=70 meV. The calculated result quantitatively explains the temperature dependence of $<\tau_2>$. Thus, a slight decrease in $<\tau_2>$ at temperatures higher than 180 K is consistently explained by taking into account the nonradiative process. Consequently, the semi-quantitative explanation of the temperature dependence of the PL decay time demonstrates the contribution of the dark-exciton state to the PL process in the ZnSe QDs.

4. Conclusion

The optical properties of size-controlled ZnSe QDs prepared by the hydrothermal method were investigated. The temperature dependence of the decay time of the slow component exhibits the following anomalous behaviour: The decay time increases with an increase in temperature up to 160 K, then further increase of temperature leads to a decrease of the decay time. Using the three-state model consisting of the ground state, lower-lying bound-exciton state, and higher-lying dark-exciton state, the increase of the decay time is reasonably explained. This demonstrates that the dark-exciton state is the origin of the anomalous temperature dependence of the decay time.

References

[1] Zhong X, Han M, Dong Z, White T J and Knoll W 2003 *J. Am. Chem. Soc.* **125** 8589
[2] Deng Z, Lie F L, Shen S, Ghosh I, Mansuripur M and Muscat A J 2009 *Langmuir* **25** 434
[3] Lee Y S, Nakano K, Bu H B and Kim D 2017 *Appl. Phys. Express* **10** 065001
[4] Shavel A, Gaponik N and Eychmüller A 2004 *J. Phys. Chem.* B **108** 5905
[5] Dong B, Cao L, Su G and Liu W 2010 *Chem. Commun.* **46** 7331
[6] Zhong X, Xie R, Zhang Y, Basché T and Knoll W 2005 *Chem. Mater.* **17** 4038
[7] Varshni Y P 1967 *Physica* **34** 149
[8] Kim D, Mishima T, Tomihira K and Nakayama M 2008 *J. Phys. Chem.* C **112** 10668
[9] Schöps O, Thomas N L, Woggon U and Artemyev M V 2006 *J. Phys. Chem.* B **110** 2074

IOP Conf. Series: Journal of Physics: Conf. Series **1220** (2019) 012025 doi:10.1088/1742-6596/1220/1/012025

Emission characteristics for a single CdSe quantum dot on an optical nanofiber at cryogenic temperatures

K Muhammed Shafi, Wei Luo, Ramachandrarao Yalla and Kohzo Hakuta

Center for Photonic Innovations, University of Electro-Communications, Chofu, Tokyo 182-8585, Japan

E-mail: shafi@cpi.uec.ac.jp

Abstract. We investigate the emission characteristics of a single CdSe quantum dot on an optical nanofiber at cryogenic temperatures from the viewpoint of quantum photonics. We show that the charged exciton (trion) of the quantum dot is a promising quantum emitter for both aspects of spectral and temporal characteristics. The nanofiber/quantum-dot system may give a promising work bench for the future quantum photonic network.

1. Introduction

In recent years, various quantum nanophotonic systems with quantum emitters have been developed in the context of single photon manipulation. A key point of such systems is to confine the light field modes to a tiny area around λ^2, where λ is the wavelength of the emitterd photons. In this view, one promising approach would be to use optical nanofibers (ONFs), tapered optical fibers with sub-wavelength waist diameters. In ONFs, guided modes at the nanofiber waist can be confined to an area with a size of wavelength. Various novel quantum optical processes have been demonstrated so far using hybrid systems of ONFs and quantum emitters [1].

Regarding quantum emitters, one promising candidate would be core/shell quantum dots (QDs) in colloidal solutions, which can realize wide tunability with high quantum efficiency (QE). Although it has been recognized that one weak point of such QDs is a rather long exciton decay time [2], the weak point can be lifted by using thick shell or bulky gradient-shell structures for the CdSe core/shell QDs. Using such QD structures at cryogenic temperatures, it is shown that the QDs are observed not only in neutral conditions, but also in charged conditions [3], where a charged exciton (trion) transition reveals a one order shorter radiative decay time than that for the neutral exciton transition.

So far, however, all above works have been carried out under microscopes for QDs cast on a glass plate. Such situation is far from real photonic applications. In this paper, we extend the work to a hybrid system of an ONF and a single thick shell CdSe QD, which is operated at cryogenic temperatures. This situation can be readily extended to photonic applications in fiber networks. By observing emitted photons through the fiber guided modes, we systematically investigate emission characteristics for one specific QD deposited on a nanofiber for exciton and trion transitions. We show that the trion at cryogenic temperatures would be a promising candidate as a quantum emitter for the ONF and QD hybrid system.

IOP Conf. Series: Journal of Physics: Conf. Series **1220** (2019) 012025 doi:10.1088/1742-6596/1220/1/012025

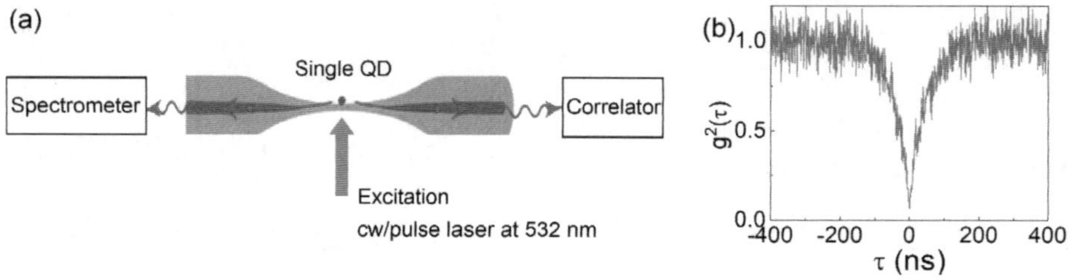

Figure 1. (a) Experimental setup. (b) Photon correlation characteristics of a single QD

2. Experimental setup

ONFs were fabricated by tapering single mode optical fibers using a heat and pull technique. The nanofiber waist diameter was set to 320 nm so that fluorescence photons from an emitter on the ONF surface can be efficiently channeled into the fiber guided modes with an efficiency around 20% [4]. We used thick shell CdSe/ZnS core/shell QDs in toluene solution synthesized by NS Materials Inc. Quantum efficiency (QE) of the QD in solution was measured to be 85%. A single QD was deposited on an ONF using a needle-dispenser with the QD solution [5].

Regarding the cryogenic system, we used a custom designed optical cryostat with a pulse-tube refrigerator. The ONF with a single QD was installed in a sample chamber, which was fixed to a cold plate at the bottom of the cryostat. The QD on a nanofiber was cooled down to 3.7 K through buffer-gas cooling by filling the sample chamber with He-gas. Optical transmission of ONF did not show any drop by lowering the temperature to cryogenic temperatures.

Figure 1(a) illustrates a measurement scheme for the emission characteristics of a single QD on a nanofiber. The QD was excited perpendicularly to the nanofiber with a cw or pulsed laser at a wavelength of 532 nm. Emissions were measured through optical fibers. One side of the ONF was introduced to an optical multichannel analyzer (OMA) spectrometer with two gratings of 300 and 1,800 lines/mm. The other side of the ONF was used for photon counting/correlation measurements with two single-photon counting modules (SPCMs). The photon arrival times at the two SPCMs were recorded using a time tagging module (Pico-Harp 300) to obtain photon counting-rates and photon correlations. We measured the photon correlation function $g^{(2)}(\tau)$ of the QD on the nanofiber to confirm the single QD emission. The measured $g^{(2)}(\tau)$ is displayed in Fig. 1(b), revealing a $g^{(2)}(0)$ value close to zero.

3. Results and discussion

Figure 2(a) displays the QD emission spectrum at 3.7 K measured at an excitation intensity of 30 W/cm^2 with an integration time of 60 s. Well resolved four peaks are observed. The observed energy separation between peaks 1 and 2 was 15.2 meV (5.0 nm in wavelength), leading to the assignment of the two peaks as zero phonon lines (ZPLs) for neutral exciton and charged exciton (trion), respectively [6]. Spectral width of each peak was 1.3 nm FWHM. Peaks $3'$ and 3 were the longitudinal optical phonon (LO-phonon) replica of the exciton and trion ZPL, respectively. The spectrum in the top and bottom of Fig. 2(a) are observed at a different time. It is observed that occurrence probability of the exciton and trion changes by the observation timing. For the top case, neutral condition is much more probable than the charged condition.

Since the spectral profile for the QD emission is governed by the spectral diffusion process, the emission spectrum was measured for shorter integration times to examine the intrinsic spectral width. Fig. 2(b) shows the spectrum observed with an integration time of 1 s at 3.7 K. Although S/N-ratio is not good, one can readily recognize a sharp peak at the trion ZPL wavelength. The observed width is 0.12 nm FWHM ($\sim 350\ \mu eV$), which is the spectral resolution limit for the

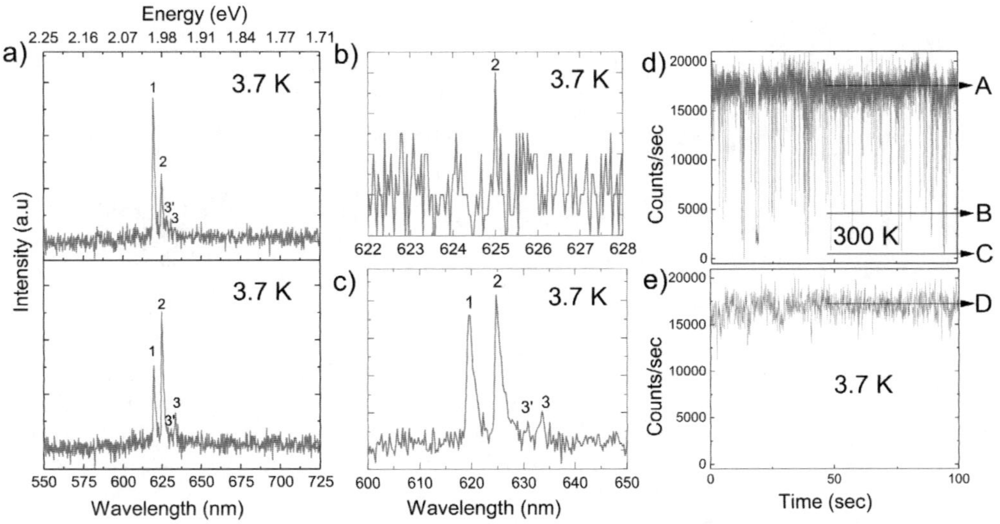

Figure 2. (a) Emission spectrum of single QD at 3.7 K with an integration time of 60 s. Peak 1 and 2 correspond to ZPLs for exciton and trion, respectively. Peak 3′ and 3 are the LO-phonon replica of exciton and trion, respectively. Two spectra displayed in top and bottom panels were measured at different timings. (b) Emission spectrum at 3.7 K with an integration time of 1 s. (c) Emission spectrum at 3.7 K with an integration time of 270 s. (d) Total photon count rate measured at room temperature. (e) Total photon count rate measured at 3.7 K

present OMA spectrometer. It means that the intrinsic spectral width for the QD is much narrower than the observed width.

We found that the occurrence probability reaches a stationary condition for integration time longer than 100 s. Fig. 2(c) shows a spectrum measured with an integration time of 270 s at 3.7 K. One can see exciton and trion ZPLs with corresponding LO-phonon replica. Spectral widths are broader than those observed with an integration time of 60 s. From the peak intensity ratio, we estimated the stationary occurrence probability for neutral and charged condition at 3. 7 K to be 46% and 54%, respectively. Figs. 2(d, e) display photon count rate measured at the end of ONF for a period of 100 s at room temperature and at cryogenic temperature of 3.7 K, respectively. At room temperature, one can clearly see three steps in the count rate, and the three steps are marked by A, B, and C. Peaks A and B can be attributed to neutral exciton emission and trion emission respectively [7]. Based on the observed step behavior, we estimated the exciton QE to be close to 100% from a fact that the QE for the mixture of exciton and trion is 85%. By lowering the temperature below 50 K, the step behavior in the count rate disappeared and the count rate showed a single step behavior. Measured example at 3.7 K is displayed in Fig. 2(e). Note that the count rate at 3.7 K marked by D is equal to that of exciton at room temperature. It means that the QE of trion increased by lowering the temperature and reached a value close to 100%.

Figures 3(a-c) show temporal behaviours at 3.7 K obtained by measuring temporal correlations between the exciting laser pulse and emitted photons. Repetition rate of the laser pulse was 0.5 MHz and the measurement time was 200 s for each decay measurement. Measurements were carried out using two types of interference filters (IF). One filter was a narrow band filter with an FWHM of 2 nm which could resolve exciton and trion peaks separated by 5.0 nm by adjusting the filter tilt-angle. The other filter which had an FWHM of 10 nm was

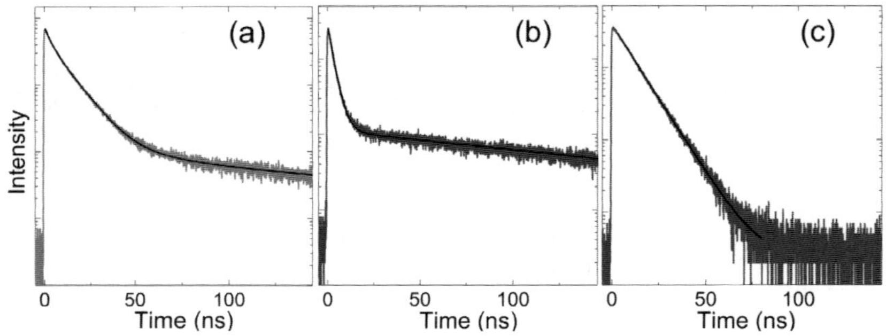

Figure 3. Semi-log plot of temporal behaviors for a single QD on an ONF measured at 3.7 K. (a) Decay profile measured using a 10 nm IF filter. (b) Exciton decay profle measured using a 2 nm IF filter. (c) Trion decay profile measured using a 2 nm IF filter

used for measuring emissions from both exciton and trion simultaneously.

Figure 3(a) displays a decay profile measured with the 10 nm FWHM IF filter. One can recognize multiple exponential decay curves. Figs. 3(b, c) display decay profiles measured for exciton and trion separately using the 2 nm FWHM IF filter, respectively. In Fig. 3(b), one can clearly see a double exponential behavior with a fast decay time of 3.2 ns and a slow decay time of 140 ns. The observed double exponential behavior can be well explained by a fine structure model of band edge exciton consisting of two states, termed as dark and bright states [2]. On the other hand, the trion decay profile in Fig. 3(c) shows clearly a single exponential decay process with a decay time of 11 ns. This means that the trion transition can effectively be treated as a simple two level transition.

4. Conclusion

We have investigated photoemission characteristics for a single CdSe QD on an ONF at cryogenic temperatures. From the observed results, we conclude that charged exciton (trion) is superior to the neutral exciton as a quantum emitter. The trion shows a simple two level transition and reveals a more than one order faster response time than the effective exciton response time. Regarding the QE value, although the trion QE is low at room temperature, it can be improved by lowering the temperature to a value close to 100%. Although the occurrence probability of trion is 54% for the present QD, the probability can be improved close to 100% by introducing a photo-assisted technique to the synthesis of QDs as reported in Ref 8. The ONF/QD system with trion may give a promising work bench for the future quantum photonic network.

Acknowledgements

The work was supported by the Japan Science and Technology Agency (JST) through the Strategic Innovation Program. Authors thank NS Materials Inc. for providing the QD samples.

References

[1] Nayak K P, Sadgrove M, Yalla R, Kien F L and Hakuta K 2018 *J. Opt.* **20** 073001
[2] Biadala L, Louyer Y, Tamarat P and Lounis B 2009 *Phys. Rev. Lett.* **103** 037404
[3] Fernee M J, Littleton B N and Dunlop H R 2009 *ACS Nano* **3** 3762
[4] Yalla R, Kien F L, Morinaga M and Hakuta K 2012 *Phys. Rev. Lett.* **109** 063602
[5] Yalla R, Nayak K P and Hakuta K 2012 *Opt. Express* **20** 2932
[6] Louyer Y, Biadala L, Tamarat P and Lounis B 2010 *Appl. Phys. Lett.* **96** 203111
[7] Gomez D E, Embden J, Mulvaney P, Fernee M J and Dunlop H R 2009 *ACS Nano* **3** 2281
[8] Rinehart J D, Schimpf A M, Weaver A L, Cohn A W and Gamelin D R 2013 *Am. Chem. Soc.* **135** 18782

IOP Conf. Series: Journal of Physics: Conf. Series **1220** (2019) 012026 doi:10.1088/1742-6596/1220/1/012026

Temperature dependence of photoluminescence dynamics of exciton-exciton inelastic scattering in a GaAs/AlAs multiple-quantum-well structure

Yuichiro Miyazaki and Masaaki Nakayama

Department of Applied Physics and Electronics, Graduate School of Engineering, Osaka City University, Sugimoto, Sumiyoshi-ku, Osaka 558-8585, Japan

nakayama@a-phys.eng.osaka-cu.ac.jp

Abstract. We have investigated the photoluminescence (PL) dynamics of exciton-exciton inelastic scattering at temperatures from 10 to 60 K. It was found that the energy dependence of the PL decay rate is scaled by that of the group velocity of the photon-like lower polariton (LP) at each temperature, taking account of the broadening factor of the polariton state. This fact demonstrates that the PL decay rate is dominated by the photon-like LP which is the final state of the exciton-exciton inelastic scattering process. The broadening factor is proportional to temperature, which indicates the influence of acoustic phonon scattering on the LP state.

1. Introduction

The photoluminescence (PL) properties of exciton-exciton inelastic scattering have attracted attention in physics and applications from the 1970s [1]. Wide-gap semiconductors such as ZnO [2-4], CdS [5], GaN [6,7], and CuI [8,9] have been samples for the investigation because of high stability of the excitonic system. In addition, the PL from exciton-exciton inelastic scattering was also observed in GaAs/AlAs multiple-quantum-well (MQW) structures because of the enhancement of the exciton binding energy by the quantum confinement effect [10,11]. In the exciton-exciton inelastic scattering process, one exciton is scattered into a higher exciton state with a hydrogenic quantum number $n \geq 2$ including the continuum state ($n=\infty$), while the other is scattered onto a photon-like lower polariton (LP) branch under the energy and momentum conservation. The photon-like LP is converted to a photon, the so-called P emission. Thus, the polariton nature of the P emission is one of key characteristics of the exciton-exciton inelastic scattering process. In previous works, it was reported that the energy dependence of the decay rate of the P emission is scaled by that of the LP group velocity in ZnO thin films [4] and GaAs/AlAs MQWs [11]. This suggests the polariton nature of the P emission. However, the measurement temperature was limited to a low temperature around 10 K; namely, there has been no systematic investigation of the PL dynamics of the P emission.

In this work, we have investigated the temperature dependence of the PL dynamics of the P emission in a GaAs (15 nm)/AlAs (15 nm) MQW from the viewpoint of the polariton characteristics. We discuss the energy dependence of the P-emission decay rate at temperatures from 10 to 60 K in comparison with that of the group velocity of the photon-like LP, taking account of the broadening factor of the polariton state.

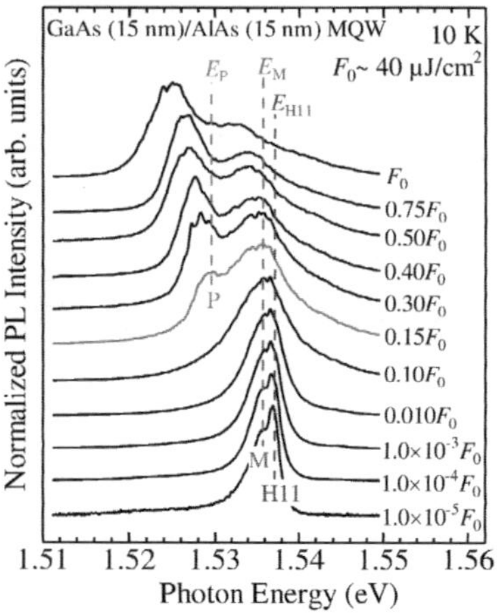

Figure 1. Steady-state PL spectra at various excitation fluences at 10 K in the GaAs (15 nm)/AlAs (15 nm) MQW, where the maximum fluence is F_0 ~ 40 μJ/cm^2 and the intensity of each PL spectrum is normalized by the maximum one.

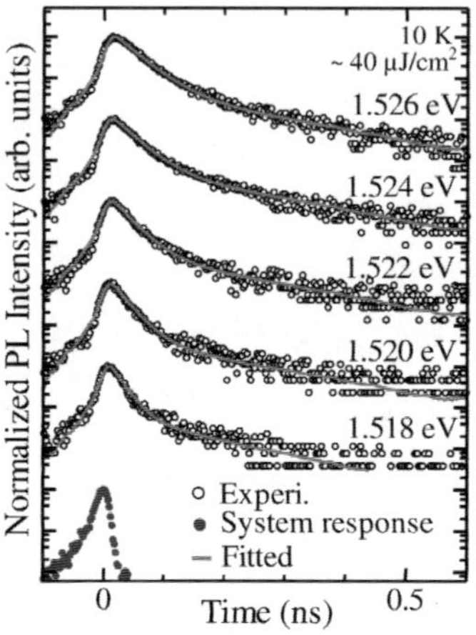

Figure 2. Detection-energy dependence of the PL-decay profile at ~40 μJ/cm^2 at 10 K, where the open and solid circles indicate the experimental PL results and system response, respectively, and the intensity of each PL-decay profile is normalized by the maximum one. The solid curve indicates the fitted result using a convolution method for the system response and $I_{PL}(t)$ given by Eq. (2).

2. Experimental details

The sample used was a GaAs (15 nm)/AlAs (15 nm) MQW with 20 periods grown on a (001) GaAs substrate by molecular beam epitaxy. The excitation light source was a mode-locked Ti:sapphire laser with a pulse duration of 110 fs and a repetition rate of 76 MHz. The peak energy of the excitation light was fixed at 1.698 eV. Time-resolved PL spectra were measured with a streak camera system with time and spectral resolutions of 16 ps and 0.20 nm, respectively. We also observed steady-state PL spectra with a cooled charge coupled device attached to a 32-cm single monochromator with a resolution of 0.15 nm. In the PL measurements, the sample was set into a closed-cycle helium cryostat.

3. Results and discussion

Figure 1 shows the excitation fluence dependence of the steady-state PL spectrum at 10 K in the GaAs (15 nm)/AlAs (15 nm) MQW, where the vertical dashed lines labelled E_{H11} and E_M indicate the first quantized heavy-hole (H11) exciton and biexciton energies, respectively, which are taken from Ref. [10]. At the lowest excitation fluence, the H11-PL band is dominant. With an increase in excitation fluence, the relative intensity of the M-PL band continuously increases. The low energy tail of the M-PL band is peculiar to the biexciton PL: the so-called inverse-Boltzmann-type line shape. At an excitation fluence of $0.15F_0$, where F_0 was ~40 μJ/cm^2, a new PL band labelled P appears with a threshold-like nature. The energy spacing between the P-PL peak and H11-exciton energy is 7.6 meV corresponding to the energy difference between the n=2 (2s) and n=1 (1s) H11 excitons [12]. This result indicates that the P-PL band originates from the P emission due to exciton-exciton inelastic scattering because of the following reason. The P-emission energy is given by [2]

$$E_P = E_{H11,n=1} - (E_{H11,n\geq2} - E_{H11,n=1}) - 3\sigma k_B T_{eff}, \quad (1)$$

where the subscript n indicates the hydrogenic quantum number, T_{eff} is an effective temperature, and σ is a positive constant smaller than unity. At the threshold, the term of $3\sigma k_B T_{eff}$ is negligible at 10 K [11]. The vertical dashed line labelled E_P indicates $E_p=E_{H11,n=1}-(E_{H11,n=2}-E_{H11,n=1})$, which agrees with the P-PL peak energy at the threshold. The P-PL peak energy shifts toward the low energy side with an increase in excitation fluence from the threshold, which reflects an increase in T_{eff} in Eq. (1). The

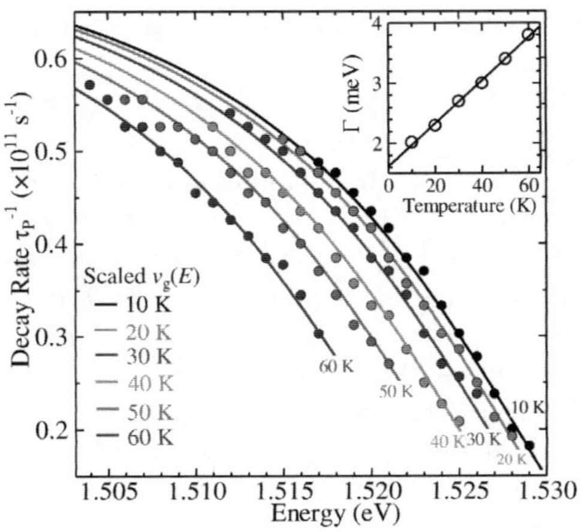

Figure 3. Detection-energy dependence of the decay rate τ_P^{-1} (solid circles) of the P emission at temperatures from 10 to 60 K in the GaAs (15 nm)/AlAs (15 nm) MQW. The solid curve indicates the scaled energy dependence of the group velocity $v_g(E)$ of the photon-like LP at each temperature. The inset shows the temperature dependence of the broadening factor Γ (open circles) used as the fitting parameter. The solid line indicates the least-square fitting result.

energy shift of the P-PL peak is continuous; therefore, contribution of the exciton state with $n \geq 3$, which will result in a discontinuous energy shift because of the binding energy difference, is not considerable.

Figure 2 shows the detection-energy dependence of the PL-decay profile (open circles) at ~40 μJ/cm^2 at 10 K, where the solid circles indicate the system response. It is obvious that the decay profile consists of the fast and slow components. The fast component was observed in the excitation fluence region higher than the threshold for the P emission. This fact indicates that the fast component results from the P emission. The slow component originates from the biexciton PL because the P- and M-PL bands overlap with each other as shown in Fig. 1. The PL-decay profile, $I_{PL}(t)$, including a rise component is written as

$$I_{\mathrm{PL}}(t) = -I_{\mathrm{rise}}(-t/\tau_{\mathrm{rise}}) + I_{\mathrm{P}}(-t/\tau_{\mathrm{P}}) + I_{\mathrm{M}}(-t/\tau_{\mathrm{M}}), \quad (2)$$

where the subscript P (M) indicates the P (M) PL and $I_{rise}=I_P+I_M$. The solid curve depicts the fitted result using a convolution method for the system response and $I_{PL}(t)$. The fitted result well explains the PL-decay profile. The estimated P-emission decay times are τ_P=21, 24, 26, 30, and 36 ps at 1.518, 1.520, 1.522, 1.524, and 1.526 eV, respectively. Note that τ_P systematically increases with an increase in energy. In contrast, the M-PL decay time, τ_M, is independent of energy: 0.15±0.02 ns.

In Fig. 3, the black solid circles indicate the detection-energy dependence of the P-emission decay rate, τ_P^{-1}, at 10 K. The P emission is caused by conversion from the photon-like LP, which is the final state in the exciton-exciton inelastic scattering process, to a photon. In the framework of the polariton picture, the P-emission decay rate can be written as $\tau_P^{-1}(E)=Av_g(E)$ [4,11], where the inverse of the scaling factor, $1/A$, corresponds in principle to the propagation length of the photon-like LP for the photon conversion. The LP group velocity is defined as $v_g(E) = (1/\hbar) dE_{\mathrm{LP}}(k_{//})/dk_{//}$, where $E_{LP}(k_{//})$ is the LP dispersion relation and $k_{//}$ is the in-plane wave vector along the interface in the MQW. The LP dispersion relation was calculated using the following equation [11]:

$$\hbar^2 c^2 k_{//}^2 / E^2 = \varepsilon_{\mathrm{b}} + f / [E_{\mathrm{H11}}^2(k_{//}) - E^2 - iE\Gamma], \quad (3)$$

where ε_b is the background dielectric constant, c is the light speed in vacuum, f is the oscillator strength, $E_{\mathrm{H11}}(k_{//}) = E_{\mathrm{H11}} + \hbar^2 k^2 / (2M_{//})$, $M_{//}$ is the in-plane exciton mass, and Γ is the broadening factor. The oscillator strength is given by $f=2E_T\varepsilon_b\Delta_{LT}$ [13], where E_T is the transverse exciton energy corresponding to E_{H11} and Δ_{LT} is the energy difference between the longitudinal and transverse excitons. The values of ε_b and $M_{//}$ are 10.86 [14] and 0.177m_0 [11] for GaAs, where m_0 is the free electron mass. The value of Δ_{LT} in the MQW was estimated from the experimental GaAs-layer thickness dependence [15]: 0.31 meV. The value of Γ at 10 K was taken from the full width at half maximum of the H11-exciton band in the PL-excitation (PLE) spectrum [10], 2.0 meV, on the assumption that the PLE line shape of the lowest-lying H11 exciton is consistent with the absorption spectrum. The black solid curve indicates the scaled energy dependence of the calculated group velocity, $v_g(E)$, at 10 K: A=8.4 cm^{-1}.

The detection-energy dependences of the P-emission decay rate (solid circles), τ_P^{-1}, at temperatures of 20, 30, 40, 50, and 60 K are also depicted in Fig. 3. In the calculation of $v_g(E)$, the broadening factor was treated as a fitting parameter because we have no available data. The temperature dependence of the H11 exciton energy was calculated using the Varshni equation: $E_{H11}(T)=E_{H11,10K}-\alpha T^2/(T+\beta)$, where α=5.405×10^{-4} eV/K and β=204 K for GaAs [16]. The scaling factor A should be independent of temperature because $1/A$ corresponds to the propagation length of the photon-like LP as described above. Thus, the value of A was fixed at 8.4 cm^{-1} obtained at 10 K. It is evident that $\tau_P^{-1}(E)$ is systematically scaled by $v_g(E)$ in the temperature range from 10 to 60 K. This fact clearly demonstrates the polariton nature of the P emission. Note that the scaling factor A is a phenomenological value at present. The inset in Fig. 3 shows the temperature dependence of the broadening factor. The solid line indicates the least-square fitting result, which indicates that the broadening factor is proportional to temperature: 36 μeV/K. The proportional relation indicates that acoustic phonon scattering is the dominant effect on the temperature dependence of the broadening factor of the polariton state [17]. The quantitative discussion of the proportional constant is beyond the scope of this paper.

4. Conclusions

We have revealed that the detection-energy dependence of the P-emission decay rate is systematically scaled by the energy dependence of the group velocity of the photon-like LP, which is the final state of the exciton-exciton inelastic scattering process, in the temperature range from 10 to 60 K. This fact demonstrates the polariton nature of the P emission. In addition, it was confirmed that the broadening factor is proportional to temperature, which indicates the influence of acoustic phonon scattering on the broadening factor of the polariton state.

Acknowledgment

This work was supported by JSPS KAKENHI Grant Number 18K03494.

References

[1] For a review, Klingshirn C and Haug H 1981 *Phys. Rep.* **70** 315
[2] Klingshirn C 1975 *Phys. Status Solidi* B **71** 547
[3] Tang Z K, Wong G K L, Yu P, Kawasaki M, Ohtomo A, Koinuma H and Segawa Y 1998 *Appl. Phys. Lett.* **72** 3270
[4] Wakaiki S, Ichida H, Mizoguchi K, Kim D, Kanematsu Y and Nakayama M 2011 *Phys. Status Solidi* C **8** 116
[5] Bohnert K, Schmieder G and Klingshirn C 1980 *Phys. Status Solidi* B **98**, 175
[6] Kurai S, Kawabe A, Sugita T, Kubo S, Yamada Y, Taguchi T and Sakai S 1999 *Jpn. J. Appl. Phys.* **38** L102
[7] Tanaka H, Ando M, Uemura T and Nakayama M 2006 *Phys. Status Solidi* C **3** 3512
[8] Tanaka I and Nakayama M 2002 *J. Appl. Phys.* **92** 3511
[9] Ichida H, Kanematsu Y, Mizoguchi K, Kim D, and Nakayama M 2007 *Phys. Rev.* B **76** 085417
[10] Nakayama M, Hirao T and Hasegawa T 2009 *J. Appl. Phys.* **105** 123525
[11] Furukawa Y, Takeuchi H and Nakayama M 2014 *J. Phys. Soc. Jpn.* **83** 054709
[12] Andreani L C and Pasquarello A 1990 *Phys. Rev.* B **42** 8928
[13] Klingshirn C F 2007 *Semiconductor Optics* 3rd ed. (Berlin: Springer) p.73
[14] Madelung O 2004 *Semiconductors: Data Handbook* 3rd ed. (Berlin: Springer) p.117
[15] Ivchenko E L, Kochereshko V P, Kopev P S, Kosobukin V A, Uraltsev I N and Yakovlev D R 1989 *Solid State Commun.* **70** 529
[16] Blakemore J S 1982 *J. Appl. Phys.* **53** R123
[17] Rudin S, Reinecke T L and Segall B 1990 *Phys. Rev.* B **42** 11218

Preparation of ZnSe-ZnS alloy quantum dots by a hydrothermal method and their optical properties

Hisaaki Nishimura, Yuxin Lin, Yuki Kunimasa, and DaeGwi Kim *

Department of Applied Physics, Graduate School of Engineering, Osaka City University, 3-3-138, Sugimoto, Sumiyoshi-ku, Osaka 558-8585, Japan

*tegi@a-phys.eng.osaka-cu.ac.jp

Abstract. We prepared colloidal $ZnSe_xS_{1-x}$ alloy quantum dots (QDs) and investigated their optical properties. $ZnSe_xS_{1-x}$ QDs were successfully prepared by a hydrothermal method, which was confirmed by the results of X-ray structural analysis. When the alloy composition was decreased from $x = 1$ to 0, the absorption energy continuously shifted to the high energy side. Band-edge photoluminescence (PL) was the main PL band in ZnSe QDs, whereas only defect-related PL band was observed in ZnS QDs. The band-edge PL was clearly observed as the main PL band for $x = 0.52$ to 1, which shifted to the high energy side by decreasing x.

1. Introduction

In semiconductor quantum dots (QDs), absorption and photoluminescence (PL) energies can be controlled by QD size owing to quantum confinement effect [1]. However, PL quantum yield (QY) in QDs having small sizes and high surface to volume ratio is usually low because of the influence of nonradiative recombination process and surface defects. Therefore, it is still a challenge to prepare semiconductor QDs having high PL QY in a wide energy range merely by controlling QD size. In alloy semiconductors, it is possible to control the band-gap energy over a wide range by changing alloy composition. Therefore, it is expected that optical properties of the alloy semiconductor QDs can be controlled not only by QD size but by alloy composition as well.

To date, hot-injection method is widely used not only for preparation of alloy QDs but for synthesis of various types of QDs as well, including II-VI and IV-VI semiconductors [2, 3]. DeGroot *et al.* succeeded in synthesizing ZnCdSe alloy QDs using the hot-injection method [4]. On the other hand, preparation of water-soluble QDs is necessary for the application of QDs in bioscience fields. The QDs synthesized by the hot injection method can disperse in water by ligand exchange [5]. Therefore, direct synthesis of water-soluble QDs having good optical properties is still an important issue even now. Cheng *et al.* synthesized water-soluble ZnCdTe QDs by a reflux method [6]. The alloy composition was controlled by changing the molar ratio of Zn^{2+} and Cd^{2+} in precursor solutions. In this study, ZnSe-ZnS ($ZnSe_xS_{1-x}$) alloy QDs those are cadmium-free, were prepared by a hydrothermal method. Because $ZnSe_xS_{1-x}$ QDs are an alloy of two wide band-gap semiconductors of ZnSe and ZnS, they can be expected to be applied as an optical functional material in the near ultraviolet region.

2. Experiments

$ZnSe_xS_{1-x}$ QDs were prepared by a hydrothermal method. First, solutions of ZnSe and ZnS as two precursors were prepared separately. The precursor solution of ZnSe or ZnS was prepared by mixing *N*-

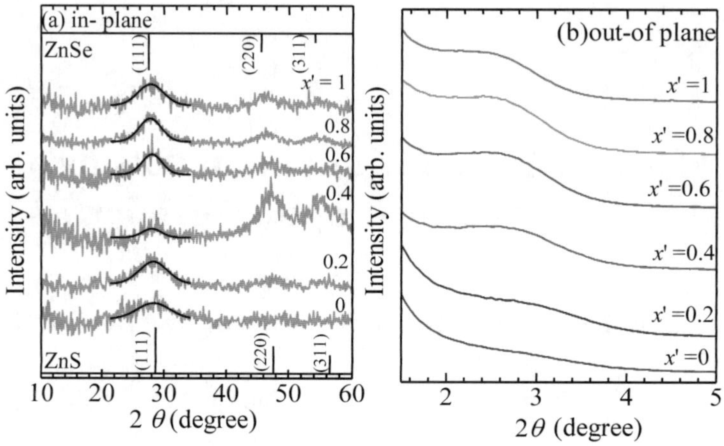

Figure 1. (a) Wide-angle in-plane XRD pattern. (b) Small-angle out-of-plane XRD pattern.

acetyl-*L*-cysteine as a ligand, and $Zn(ClO_4)_2 \cdot 6H_2O$ as a Zn^{2+} ion source, whereas NaHSe or Na_2S as a Se^{2-} or S^{2-} ion source [7]. Then, a precursor solution of $ZnSe_xS_{1-x}$ was prepared by mixing ZnS and ZnSe precursor solutions having a molar ratio between 0 and 1. Finally, 10 ml of the precursor solution was transferred into an autoclave and heated at 200 °C using an oil bath to prepare QDs. We define x' as the mixing ratio of ZnS and ZnSe precursor solutions.

To perform X-ray diffraction (XRD) measurements, multilayer structures of QDs were fabricated by layer-by-layer (LBL) assembly in which QDs and positively charged polyelectrolytes of poly (diallyldimethylammonium chloride) were alternately deposited [8]. The periodicity in the laminating direction was obtained from small-angle out-of-plane XRD measurements, and average diameter of QDs was estimated [8]. In addition, the actual alloy composition, x, was estimated from the analysis of wide-angle in-plane XRD measurements. Absorption and PL spectra were recorded using a JASCO V-650 UV/Vis spectrophotometer and a JASCO FP-8300 spectrofluorometer, respectively.

3. Results and discussion

Figure 1(a) shows the result of wide-angle X-ray structural analysis. For ZnSe QDs ($x' = 1$), diffraction peaks were observed at 27.2°, 46.1° and 54.5°, which corresponds to the (111), (220), and (311) planes in ZnSe crystal of zincblende structure. However, for ZnS QDs ($x' = 0$), the signals corresponding to (220) and (311) planes are very weak, and only the diffraction peak corresponding to the (111) plane was clearly observed at 28.5°. Thus, the lattice constant was calculated using the Bragg's law based on the diffraction angle of the (111) plane in each QD sample. The actual alloy composition, x, was estimated on the basis of Vegard's law and is listed in Table 1. Figure 1(b) shows the result of small-angle out-of-plane XRD. The diffraction peak of ZnSe QDs ($x = 1$) was observed at 2.56°, indicating

Table 1. Lattice constant, alloy composition (x), and QD diameter of ZnSe-ZnS alloy QDs prepared for a reaction time of 60 min by heating precursor solutions having the mixing ratio (x') of ZnS and ZnSe precursors.

x'	0	0.2	0.4	0.6	0.8	1
Lattice constant (nm)	0.539	0.550	0.554	0.555	0.560	0.570
x	0	0.36	0.52	0.58	0.76	1
QD diameter (nm)	3.0	3.1	3.3	3.3	3.3	3.5

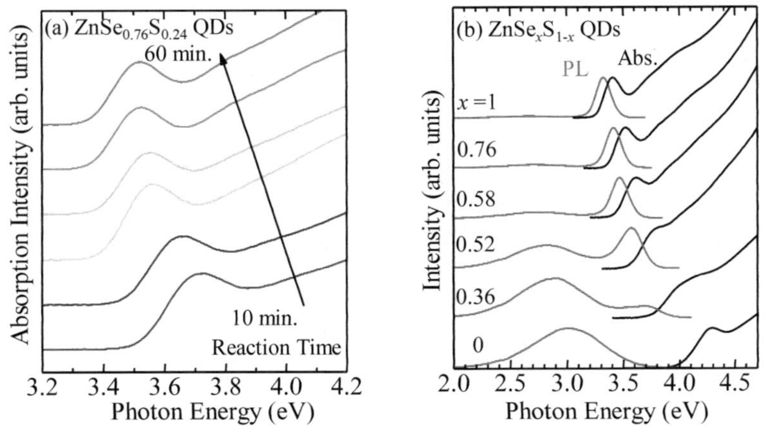

Figure 2. (a) Absorption spectra in $ZnSe_{0.76}S_{0.24}$ QDs prepared with various reaction times. (b) Absorption and PL spectra of $ZnSe_xS_{1-x}$ alloy QDs with x=0, 0.36, 0.52, 0.58, 0.76, and 1.

that the ZnSe QD multilayers had a structural regularity with a periodicity of 3.5 nm in the lamination direction. The results of the prepared alloys are listed in Table 1.

Figure 2(a) shows absorption spectra of $ZnSe_{0.76}S_{0.24}$ QDs prepared for different reaction time. The absorption peak was red-shifted, by increasing the reaction time, indicating that the average QD size was increased. Similar results were obtained for QD samples having other alloy compositions. For each kind of alloy QDs, the QD size became the maximum at a reaction time of 60 min.

Next, we discuss the dependence of optical properties on the alloy composition in $ZnSe_xS_{1-x}$ QDs prepared for 60 min. Figure 2(b) shows absorption and PL spectra of $ZnSe_xS_{1-x}$ alloy QDs having x = 0, 0.36, 0.5 2, 0.58, 0.76, and 1. In the absorption spectra for x = 0 (x = 1), the absorption peak was observed on the higher energy side than the band-gap energy of 3.7 eV (2.7 eV) in the ZnS (ZnSe) bulk crystal, demonstrating the formation of ZnS (ZnSe) QDs. The absorption peak shifted to the high energy side by increasing the molar ration of ZnS, suggesting the increase in the band-gap energy. In PL spectra of alloy QDs having x = 0.58 to 1, the band-edge PL was observed as the main PL band, while defect-related PL was dominant in QDs having x = 0 to 0.52.

The alloy-composition dependence of the absorption energy was discussed subsequently. The band-gap energy of alloy semiconductors generally has a bowing characteristic with respect to the alloy composition. Ebina *et al.* reported that the x-dependence of the band-gap energy in a bulk crystal of $ZnSe_xS_{1-x}$, $E_g(x)$, can be given by

$$E_g(x)=E_g(\mathrm{ZnS}) + [E_g(\mathrm{ZnSe}) - E_g(\mathrm{ZnS}) - b]\, x + b\, x^2 \quad (1)$$

where a bowing parameter of b is 0.63 [9]. The solid curve in figure. 3 represents the calculated result of $E_g(x)$. The open circles denote the experimental results of the x-dependence of the absorption energy in $ZnSe_xS_{1-x}$ QDs, which was estimated from the energy minimum of the second-derivative spectrum. The x-dependence of the absorption is similar to that of the band-gap energy as shown by the solid curve. The large energy difference between them is caused by the quantum confinement effect.

The lowest exciton energy in II-VI semiconductor QDs is well explained by the electron-hole individual confinement model [10]. In alloy QDs, the effective mass of electrons and holes as well as the dielectric constant depend on the alloy composition. Thus, the exciton energy in a spherical alloy QD having a radius R and an alloy composition x is given approximately by [11]

$$E(R,x)=E_g(x)+\frac{\hbar^2}{2\mu(x)}\left(\frac{\pi}{R}\right)^2-1.8\frac{e^2}{\varepsilon(x)R}, \quad (2)$$

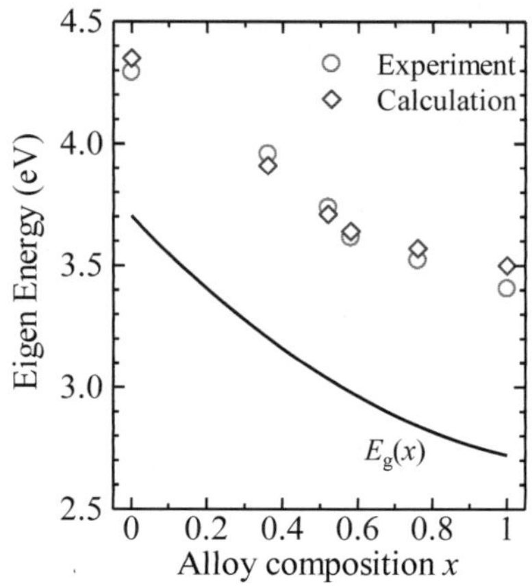

Figure 3. Alloy-composition dependence of the absorption energy Solid curve: Calculated result of $E_g(x)$. Open circles: Experimental results of the x-dependence of the absorption energy in $ZnSe_xS_{1-x}$ QDs. Open diamonds: Calculated results using equation (2).

where $\mu(x)$ and $\varepsilon(x)$ denote the x- dependence of the reduced mass and dielectric constant of the exciton, respectively. The open diamonds denote the calculated results of how the optical transition energy depend on x of $ZnSe_xS_{1-x}$ QDs using equation (2). Because the detailed x-dependence of $\mu(x)$ and $\varepsilon(x)$ in $ZnSe_xS_{1-x}$ has not yet been clarified, we assume that $\mu(x)$ and $\varepsilon(x)$ vary linearly with x between the values in ZnSe and ZnS bulk crystals. The value of R obtained from the analysis of out-of-plane XRD as shown in Tabel.1 was substituted. The calculated results agree well with the experimental values. These results demonstrate that the quantum confinement effects in $ZnSe_xS_{1-x}$ QDs can quantitatively explain the x-dependence of the absorption energy. In addition, alloy QDs of $ZnSe_xS_{1-x}$ can be successfully prepared by the hydrothermal method.

4. Conclusion

$ZnSe_xS_{1-x}$ alloy QDs were successfully prepared by the hydrothermal method. In the absorption spectrum of the sample prepared with $x = 0$ ($x = 1$), the absorption peak was observed on the higher energy side than the band gap energy of 3.7 eV (2.7 eV) in the ZnS (ZnSe) bulk crystal, demonstrating the formation of ZnS (ZnSe) QDs. By decreasing the alloy composition from $x = 1$ to 0, the absorption energy continuously shifted to the high energy side. Band-edge PL is the main PL band in ZnSe QDs, whereas only defect-related PL band was observed in ZnS QDs. The band-edge PL was dominant for $x = 0.52$ to 1, which shifted to the high energy side by decreasing x.

References

[1] Brus L 1991 *Appl. Phys.* A **53** 465
[2] Murray C B, Norris D J and Bawendi M G 1993 *J. Am. Chem. Soc.* **115** 8706
[3] Li H, Chen D, Li L, Tang F, Zhang L and Ren J 2010 *Cryst. Eng. Comm.* **12** 1127
[4] DeGroot M, Rösner H and Corrigan J 2006 *Chem. Eur. J.* **12** 1547
[5] Reiss P, Bleuse J and Pron A 2002 *Nano Lett.* **2** 782
[6] Cheng J, Li D, Cheng T, Ren B, Wang G and Li J 2014 *J. Alloy Com.* **589** 539
[7] Lee Y S, Bu H B, Taniguchi T, Takagi T, Sobue S, Yamada H, Iwaki T and Kim D 2016 *Chem. Lett.* **45** 878
[8] Kim D, Tomita S, Ohshiro K, Watanabe T, Sakai T, Chang I and Kim H 2015 *Nano Lett.* **15** 4343
[9] Ebina A, Fukunaga E and Takahashi T 1974 *Phys. Rev.* B **10**, 2495
[10] Brus L 1986 *J. Phys. Chem.* **90**, 2555
[11] Kim D, Nabeshima A and Nakayama M 2005 *Jpn. J. Appl. Phys.* **44**, 1514

IOP Conf. Series: Journal of Physics: Conf. Series **1220** (2019) 012028 doi:10.1088/1742-6596/1220/1/012028

Absorption and photoluminescence properties of CdSe quantum dots prepared by hydrothermal method

TaeGi Lee, Kaito Noguchi, Hisaaki Nishimura, and DaeGwi Kim*

Department of Applied Physics, Graduate School of Engineering, Osaka City University, 3-3-138, Sugimoto, Sumiyoshi-ku, Osaka 558-8585, Japan

*tegi@a-phys.eng.osaka-cu.ac.jp

Abstract. Herein, we investigated the preparation and optical properties of water-soluble CdSe quantum dots (QDs). CdSe QDs with a narrow size distribution were hydrothermally prepared by reacting Cd^{2+} with NaHSe in the presence of *N*-acetyl-L-cysteine as ligand. Furthermore, photoluminescence quantum yield increased to ~47% when a ZnS shell was applied to prepare the CdSe/ZnS core/shell QDs.

1. Introduction

In semiconductor quantum dots (QDs), both the absorption and photoluminescence (PL) energies can be controlled by the size of the QD owing to the quantum size effect. Semiconductor QDs have been expected to be new fluorescent materials because they can absorb light with a wide range of wavelengths and excellent in anti-photobleaching properties as compared with organic dyes. To date, several studies have been conducted on the preparation and optical properties of semiconductor QDs [1]. In 1993, Murray *et al.* reported the successful preparation of CdSe QDs with a narrow size distribution and high PL efficiency using chemical reactions in an organic solvent [2]. CdSe QDs have become model materials for studying the optical properties of QDs [3, 4].

To apply QDs to bio-imaging application, water-soluble QD is necessary. QD synthesized by the hot injection method can be dispersed in water via the ligand exchange process [5, 6]. A more straightforward method is to synthesize the QDs in an aqueous solution [7]. However, such water-soluble CdSe QDs have a problem of low PL quantum yield (QY) and broad absorption and PL spectral widths. Thus, directly synthesizing water-soluble CdSe QDs with good optical properties is still a challenge.

Qian *et al.* successfully prepared mercaptopropionic acid (MPA)-capped CdSe QDs with a PL QY of 25% using a microwave-assisted hydrothermal method [8]. Schumacher *et al.* focused on CdSe/ZnS core/shell QDs that are more biocompatible than CdSe-CdS QDs, and synthesized CdSe/ZnS QDs with a PL QY of 13% [9]. In our previous work, we successfully prepared CdTe and ZnSe QDs with a narrow size distribution and high PL QY through the hydrothermal method using *N*-acetyl-L-cysteine (NAC) as ligand [10-13]. In the present work, we investigated the preparation and optical properties of water-soluble CdSe QDs, prepared via the hydrothermal method. Furthermore, we prepared CdSe/ZnS core/shell QDs to improve the PL properties of QDs by covering the CdSe QDs with a ZnS-shell layer.

2. Experimental

IOP Conf. Series: Journal of Physics: Conf. Series **1220** (2019) 012028 doi:10.1088/1742-6596/1220/1/012028

CdSe QDs were prepared using the hydrothermal method. The CdSe precursor solution was prepared by mixing NAC as ligand, $Cd(ClO_4)_2 \cdot 6H_2O$ as a Cd-ion source, and NaHSe as an Se-ion source. The following molar ratios and concentrations of monomers were used Cd:Se:NAC = 1.0:0.3:6.0 and $[Cd^{2+}]$ = 20 mM. To prepare the QDs, 10 ml of the precursor solution was transferred into an autoclave and heated at 180°C using an oil bath for a specified period, and then cooled to room temperature. We prepared CdSe/ZnS core/shell QDs to improve the PL properties. First, ZnS precursor solution was prepared as follows. $Zn(ClO_4)_2 \cdot 6H_2O$ (0.53 mmol) and NAC (2.54 mmol) were dissolved in 20 mL DI water, and the pH was adjusted to 5.0. After adding $Na_2S \cdot 9H_2O$ (0.16 mmol), the precursor solution was adjusted to pH 6.0. The ZnS precursor solution was added to the already prepared CdSe QD solution, and the CdSe/ZnS QDs were prepared by continuously adding ZnS solution while heating at 120 °C.

Absorption and PL spectra were measured using JASCO V-650 UV-Vis spectrophotometer and FP-8300 spectrofluorometer, respectively. A laser diode (375 nm, Hamamatsu PLP M10306-27) with a pulse duration of 50 ps and a repetition rate of 100 kHz was used as excitation light to measure the PL-decay profiles. The decay profiles were obtained using a time-correlated single-photon counting method. X-ray diffraction (XRD) measurements were performed using Rigaku SmartLab with Cu Kα radiation while transmission electron microscopy (TEM) images were obtained using JEOL JEM-2100.

3. Results and Discussion

Figure 1(a) shows the absorption and PL spectra of the CdSe QDs synthesized with reaction times of 3, 10, 20, 40, and 80 min. The absorption-peak energy in each sample is shifted to the higher-energy side than the band gap energy of 1.74 eV in the CdSe bulk crystal. This high-energy shift suggests the formation of CdSe QDs. Furthermore, it is noted that the synthesized CdSe QDs have a narrow size distribution since the absorption peak is clearly observed in all the samples. As the reaction time increased, the onset and peak energies of the absorption spectra were shifted to the lower-energy side. This is because the quantum confinement effect decreases with increasing QD size. Donegá *et al.* reported that the lowest exciton energy in CdSe QDs with a diameter d is given by [14] .

$$E(d) = 1.858 + \frac{1}{0.220d^2 + 0.008d + 0.373} \quad (\mathrm{eV}). \tag{1}$$

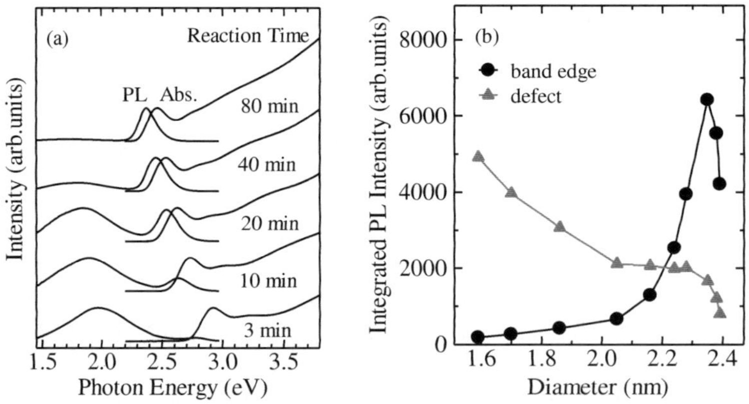

Figure 1. (a) Absorption and PL spectra of CdSe QDs synthesized with the reaction times of 3, 10, 20, 40, and 80 min. (b) Dependence of PL intensity of the band-edge PL (solid circles) and defect-related PL band (solid triangles) on the QD diameter.

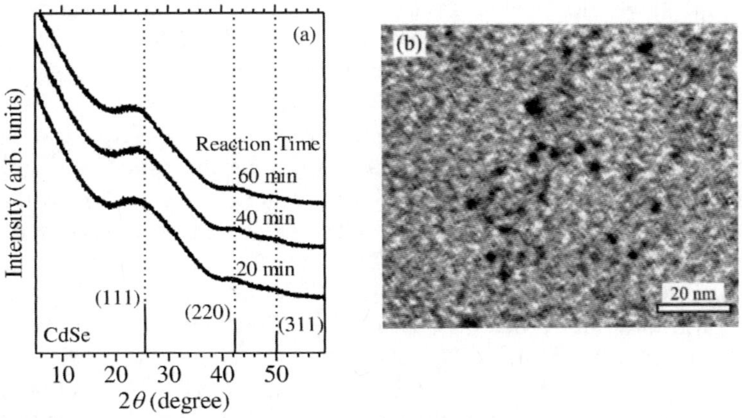

Figure 2. (a) XRD patterns of CdSe QDs synthesized with reaction times of 20, 40, and 60 min. (b) TEM image of the CdSe QDs synthesized with a reaction time of 60 min.

From the absorption-peak energy, the mean diameter can be estimated to be 1.6, 1.9, 2.1, 2.2, and 2.4 nm for CdSe QDs synthesized with reaction times of 3, 10, 20, 40, and 80 min, respectively, on the basis of equation (1).

Band-edge PL and defect-related PL were observed in the PL spectrum, wherein the intensity ratio of both PLs strongly depends on the QD size. Figure 1(b) shows the dependence of the PL intensity for the band-edge PL (solid circles) and defect-related PL (solid triangles) on the QD diameter. The band-edge PL was observed as the main PL band in samples with 2.3 nm diameter or larger, whereas the defect-related PL band was observed as the main PL band in samples with 2.2 nm diameter or smaller. The ratio of the surface area to volume decreases as the QD diameter increases. This is considered to cause the suppression of the defect-related PL. Moreover, the decrease in the intensity of the PL in CdSe QDs of 2.3-nm or larger diameters is considered to be due to the thermal decomposition of the ligand [11, 15].

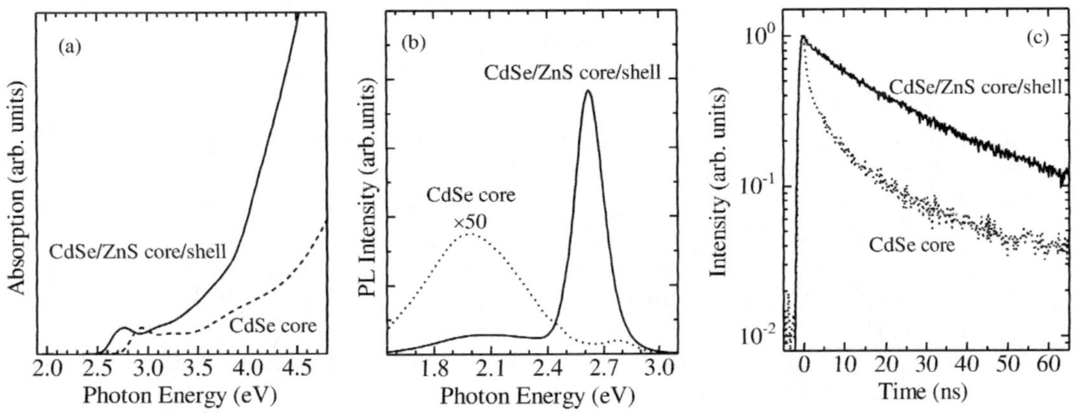

Figure 3. (a) Absorption and (b) PL spectra, and (c) PL-decay profiles of the band-edge PL in CdSe core and CdSe/ZnS core/shell QDs.

To confirm the formation of CdSe QDs, X-ray structural analysis and TEM measurements were performed. Figure 2(a) shows the XRD patterns of CdSe QDs synthesized with reaction times of 20, 40, and 60 min. The three diffraction peaks correspond to the (111), (220), and (311) planes of the cubic zinc-blende CdSe lattice structure. Therefore, we conclude that the synthesis of CdSe QDs with a zinc blende structure was successful. Figure 2(b) shows a TEM image of the CdSe QDs synthesized with a reaction time of 60 min. Based on the TEM image, the average diameter of the QDs was obtained as ~2.4 nm. This agrees well with the diameter estimated from the absorption-peak energy.

The CdSe/ZnS core/shell QDs were prepared by covering the surface of the CdSe QD with a ZnS layer to improve the PL characteristics. Figure 3(a) shows the absorption spectra of the CdSe-core and the CdSe/ZnS core/shell QDs. In the core/shell QDs, a new absorption structure was confirmed around 4.1 eV, which is higher than the band gap energy of 3.7 eV in the bulk ZnS crystal. This is considered to be a signal demonstrating the formation of ZnS shell. Furthermore, it can be seen that the lowest absorption-peak energy in the core/shell QDs was shifted to the low-energy side as compared with the absorption peak in the core QDs. This is due to the decrease in the effective quantum confinement energy after the ZnS shell is formed. Thus, based on the absorption characteristics, it can be concluded that the CdSe/ZnS core/shell QDs were successfully prepared.

Figure 3(b) shows the PL spectra of the CdSe core and the CdSe/ZnS core/shell QDs. It can be seen clearly that the PL characteristics were remarkably improved by the formation of the ZnS layer. The CdSe/ZnS core/shell QDs exhibit a high PL QY of ~47%. To the best of our knowledge, thus far, this PL QY is the highest reported for CdSe-based QDs synthesized in aqueous solution. Figure 3(c) shows the PL-decay profiles of the band-edge PL in the CdSe core and the CdSe/ZnS core/shell QDs. It can be seen that the decay profile in the core/shell QDs is prolonged as compared with the decay profile in core QDs. When the ZnS shell is formed, the non-radiative recombination process on the surface of the core QDs is suppressed, resulting in higher PL QY and longer decay time.

4. Conclusion

We successfully synthesized water-soluble NAC-capped CdSe QDs using the hydrothermal synthesis method. The average size of the CdSe QDs was controllable by the reaction time. In addition, by covering the surface of the core QD with a ZnS layer, the PL characteristics improved drastically. As a result, we successfully prepared CdSe/ZnS core/shell QDs with a PL QY as high as ~47%.

References

[1] Klimov V I 2004 *Semiconductor and Metal Nanocrystals* (Marcel Dekker Inc.)
[2] Murray C B, Norris D J and Bawendi M G 1993 *J. Am. Chem. Soc.* **115** 8706
[3] Donegá C M, Bode M and Meijerink A 2006 *Phys. Rev.* B **74** 085320
[4] Klimov V I, Ivanov S A, Nanda J, Achermann M, Bezel I, Mcguire J A and Piryatinski A 2007 *Nature* **447** 441
[5] Bruchez Jr M, Moronne M, Gin P, Weiss S and Alivisatos A P 1998 *Science* **281** 2013
[6] Reiss P, Bleuse J and Pron A 2002 *Nano Lett.* **2** 781
[7] Rogach A L, Kornowski A, Gao M, Eychmüller A and Weller H 1999 *J. Phys. Chem.* B **103** 3065
[8] Qian H, Li L and Ren J 2005 *Mater. Res. Bull.* **40** 1726
[9] Schumacher W, Nagy A,Waldman W J and Dutta P K 2009 *J. Phys. Chem.* C **113** 12132
[10] Bu H, Kikunaga H, Shimura K, Takahasi K, Taniguchi T and Kim D 2013 *Phys. Chem. Chem. Phys.* **15** 2903
[11] Watanabe T, Takahashi K, Shimura K, Bu H, Kim H D and Kim D 2017 *Bull. Chem. Soc. Jpn.* **90** 52
[12] Lee Y S, Bu H, Taniguchi T, Takagi T, Sobue S, Yamada H, Iwaki T and Kim D 2016 *Chem. Lett.* **45** 878
[13] Lee Y S, Nakano K, Bu H and Kim D 2017 *Appl. Phys. Express* **10** 065001
[14] Donegá C M and Koole R 2009 *J. Phys. Chem.* C **113** 6511
[15] Zhao D, He Z, Chan W H and Cho M M F 2009 *J. Phys. Chem.* C **113** 1293

Temperature dependence of photoluminescence properties of water-soluble CdS quantum dots

Kunio Shimura, TaeGi Lee, and DaeGwi Kim*

Department of Applied Physics, Graduate School of Engineering, Osaka City University, 3-3-138, Sugimoto, Sumiyoshi-ku, Osaka 558-8585, Japan

*tegi@a-phys.eng.osaka-cu.ac.jp

Abstract. We investigated the size dependence of band-edge photoluminescence (PL) dynamics for CdS quantum dots (QDs). The temperature dependence of the PL-decay profiles of CdS QDs with an average diameter of 3.7–6.0 nm was measured. The PL-decay profiles became longer as the temperature increased. Further, it was found that the temperature dependence of the PL-decay profiles depends greatly on the QD size. These experimental results can be understood by considering that the magnitude of the splitting energy between the bright- and dark-exciton states depends on the QD size and becomes larger as the QD size becomes smaller.

1. Introduction

The observation of an optically passive state, the so-called dark-exciton state, is a characteristic photoluminescence (PL) property of semiconductor quantum dots (QDs) [1,2]. Theoretical and experimental studies indicate that the splitting energy (Δ_{ST}) between optically active bright- and dark-exciton states, which is less than 1 meV in the bulk crystal, increases from several meV to several tens of meV owing to the quantum confinement effect in QDs [3, 4]. CdSe QDs have been extensively studied as a model material for QD studies [5, 6]. Crooker *et al.* quantitatively explained the temperature dependence of the PL dynamics using a three-state model comprising a ground state and two excited bright- and dark-exciton states [7]. Compared to CdSe QDs, it is theoretically predicted that CdS QDs have a larger Δ_{ST} value [8, 9], and the contribution of dark excitons to the PL processes is considered to be large.

Previously, we prepared water-soluble CdS QDs in which the band-edge PL is observed as the main PL band [9]. In addition, film samples containing CdS QDs with an average grain size of 4.2 nm were prepared, and the temperature dependence of the PL dynamics was measured. We found that the dark-exciton state has a large influence on the PL processes, even at room temperature (RT), because Δ_{ST} has a value of 40 meV for CdS QDs with a diameter of 4.2 nm [9]. Since the value of Δ_{ST} largely depends on the QD size, it is important to investigate the size dependence of the PL dynamics. In this study, we investigate the size dependence of the band-edge PL dynamics in CdS QDs.

2. Experiments

CdS QDs were prepared by injecting H_2S gas into aqueous solutions containing $Cd(ClO_4)_2$ and sodium hexametaphosphate (HMP). The sizes of the CdS QDs were controlled using a size-selective photo-etching technique [9]. The QD surface was modified by the addition of $Cd(ClO_4)_2$ after adjusting the pH of the solutions to the alkaline region. The sample solution was mixed with polyvinyl alcohol (PVA) aqueous solutions, and the final solution was spread on glass. The excess water was evaporated by heating the glass to 80 °C for 2 h to obtain film samples, i.e., CdS QDs dispersed in PVA films.

For PL measurements, the 325-nm line of a He-Cd laser was used as the excitation-light source. Third-harmonic-generation light (355 nm) from a laser-diode-pumped yttrium aluminum garnet laser with a pulse duration of 20 ns and a repetition of 10 kHz was used as the excitation light for measuring the PL-decay profiles.

3. Results and discussion

Figure 1 shows the absorption and PL spectra of CdS QD solution samples with mean diameters of 3.7, 4.9, and 6.0 nm. Note that band-edge PL is clearly observed in all the samples. This makes it possible to study the PL dynamics in detail. QD-dispersed PVA film samples were prepared to investigate the PL characteristics at low temperatures as well as their temperature dependence.

The temperature dependence of the PL spectra was measured. Band-edge PL was observed as the main PL band at all temperatures ranging from 10 K to RT (not shown). The PL intensity at RT is ~60% of that at 20 K. Therefore, the influence of thermal quenching is small. These results suggest that the prepared CdS QDs are suitable for detailed studies of the temperature dependence of PL dynamics.

Figure 2 shows the decay profiles of band-edge PL at 10, 100, and 200 K for CdS QDs with mean diameters of 3.7, 4.9, and 6.0 nm, respectively. In this temperature range, the PL-decay profiles become longer as the temperature increases; the temperature dependence of the 3.7-nm QDs is smaller than that of the other two larger QDs. The decay profile at 200 K has a slow decay time of the order of hundreds of nanoseconds. This long decay-time component suggests a contribution by the optically passive dark-exciton state to the PL processes [9]. Usually, as the temperature increases, the non-radiative-decay rate increases, and the PL intensity and PL decay time decrease; however, the observed PL-decay profiles became longer with increasing temperature. In Ref. [9], we proposed a three-state model consisting of a ground state and two excited states (a lower-lying bound-exciton state and a higher-lying dark- exciton state), to explain the anomalous temperature dependence of PL dynamics. With increasing temperature, the thermal population of the higher-lying dark-exciton state becomes larger, and the PL-decay time becomes longer.

The observed PL-decay profiles exhibit multi-exponential decay. To quantitatively discuss the PL-decay profiles, we used a combination of monoexponential function and stretched-exponential function [9]: $A_1 \exp(-t/\tau_1) + A_2 \exp[-(t/\tau_2)^\beta]$. As discussed in Ref. [9], the decay components, τ_1 and τ_2, correspond to the decay times of the bright-exciton state and that characterized by the three-state model, respectively. As an example, the fitting results of the PL-decay profiles in 4.9-nm CdS QDs are shown

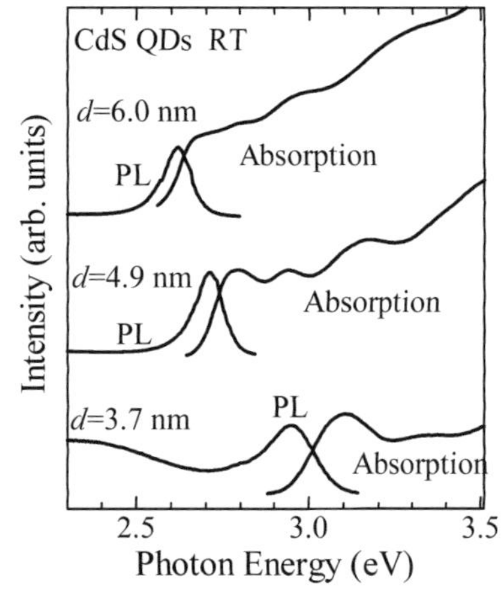

Figure 1. Absorption and PL spectra of CdS QD solution samples with mean diameter of 3.7, 4.9, and 6.0 nm.

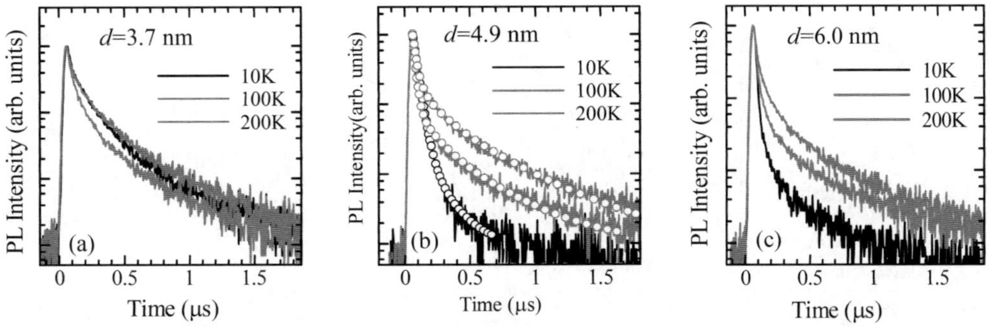

Figure 2. Decay profiles of band-edge PL at 10, 100, and 200 K of CdS QDs with mean diameter of 3.7, 4.9, and 6.0 nm.

by the open circles in figure 2(b). The fitting results and the experimental results demonstrate agreement.

The temperature dependence of the decay times of τ_1 and τ_2 obtained from the analysis of the PL-decay profiles in each sample are shown in figures 3(a)-3(c). In CdS QDs with diameters of 4.9 and 6.0 nm, τ_2 becomes longer with increasing temperatures up to ~180 K and becomes shorter in the temperature region higher than 200 K. The temperature dependence of τ_2 in the temperature range up to 180 K qualitatively corresponds to the temperature dependence of the decay time characterized by the three-state model [9]. The decay time of τ_1 becomes longer with increasing temperature in the higher-temperature region. Conversely, for smaller CdS QDs with diameter of 3.7 nm, both τ_1 and τ_2 are nearly independent of temperature.

The value of Δ_{ST} in 6.0-nm CdS QDs is ~9 meV [10], which is smaller than the thermal energy at RT. As a result, τ_1, which reflects the decay time of the bright-exciton state, becomes longer with increasing temperature in the high-temperature region owing to the thermal-energy-assisted mixing of bright- and dark-exciton states, which was not considered in smaller-sized CdS QDs having a large Δ_{ST} [9]. Because Δ_{ST} in 3.7-nm CdS QDs is ~50 meV [10], there is no influence of the mixing of bright- and dark-exciton states on them. Therefore, τ_1 and τ_2 are considered to be constant regardless of the temperature.

Finally, we quantitatively discuss the temperature dependence of the PL-decay time. In Ref. [11], a simple approach to evaluate the distribution of the decay time and the statistical average PL-decay time $\langle\tau\rangle$ from experimentally observed stretched-exponential PL decays was presented. Figure 3(d) shows the temperature dependence of $<\tau_2>$ in 4.9-nm CdS QDs. Based on the three-state model, the temperature dependence of the decay time is expressed by the following equation [9]:

$$\frac{1}{\tau}=\frac{\dfrac{1}{\tau_{\mathrm{Bx}}}+\dfrac{g_{\mathrm{Dx}}}{g_{\mathrm{Bx}}}\dfrac{1}{\tau_{\mathrm{Dx}}}\exp\left(-\dfrac{\Delta E}{k_B T}\right)}{1+\dfrac{g_{\mathrm{Dx}}}{g_{\mathrm{Bx}}}\exp\left(-\dfrac{\Delta E}{k_B T}\right)}. \quad (1)$$

Here, $1/\tau_{\mathrm{Dx}}$ ($1/\tau_{\mathrm{Bx}}$) denotes the radiative-decay rate, g_{Dx} (g_{Bx}) denotes the density of states of the dark-exciton (bound-exciton) state, and ΔE corresponds to the energy difference between the dark- and bound-exciton states.

The broken curve in figure 3(d) represents the result calculated using equation (1) with $\tau_{\mathrm{Dx}} = 1600$ ns, $\tau_{\mathrm{Bx}} = 25$ ns, $\Delta E = 8$ meV, and $g_{\mathrm{Dx}}/g_{\mathrm{Bx}} = 90$. This result demonstrates agreement with $<\tau_2>$ becoming longer as the temperature increases up to ~180 K. Conversely, at temperatures higher than 180 K, the experimental result shows a decreasing behavior, while the calculated result continues to increase gradually. This discrepancy appears to be due to the influence of the non-radiative recombination process, i.e., thermal-quenching at temperatures higher than 180K. To consider the influence of the non-radiative recombination process, we assumed the following thermally active non-radiative process:

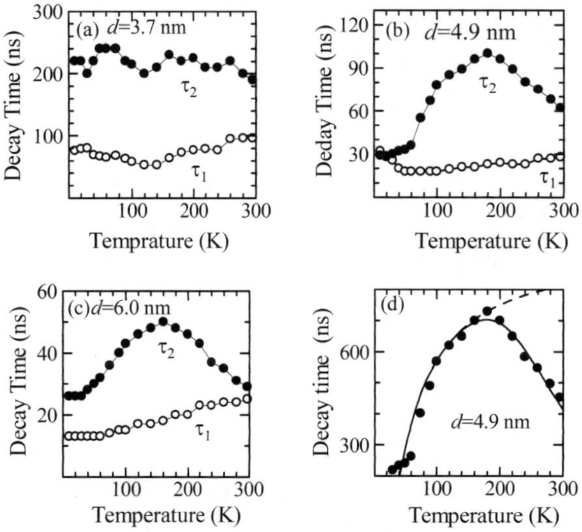

Figure 3. (a)-(c) Temperature dependence of the decay time of τ_1 and τ_2 in CdS QDs with mean diameter of 3.7, 4.9, and 6.0 nm. (d) Temperature dependence of $<\tau_2>$ in CdS QDs with mean diameter of 4.9 nm.

$1/\tau_{nr}(T) = 1/\tau_{nr}(0)\cdot \exp(-E_a/k_BT)$, where E_a represents the thermal-activation energy for the non-radiative process. The solid curve in the figure shows the result calculated using the parameters of $\tau_{nr}(0) = 8$ ns and $E_a = 120$ meV. This result quantitatively explains the temperature dependence of $\langle\tau_2\rangle$. Therefore, we can explain the PL-decay time becoming shorter in the high-temperature region (higher than 180 K) by considering the thermally activated non-radiative recombination process.

4. Conclusions

The size dependence of the PL dynamics in CdS QDs prepared by the colloid method was investigated. In all the samples, band-edge PL was clearly observed. The PL intensity at RT is ~60% of that at 10 K, demonstrating that the non-radiative recombination process is remarkably suppressed. The PL-decay profiles show a peculiar behavior different from ordinary semiconductors, in which the decay time increases with temperature. The temperature dependence of the PL-decay time can be quantitatively explained using a three-state model assuming dark- and bound-exciton states.

References

References

[1] Nirmal M, Norris D J, Kuno M and Bawendi M 1995 *Phys. Rev. Lett.* **75** 3728
[2] Takeoka S, Fujii M and Hayashi S 2000 *Phys. Rev.* B **62** 16820
[3] Efros A L, Rosen M, Kuno M, Nirmal M, Norris D J and Bawendi M 1996 *Phys. Rev.* B **54** 4843
[4] Blokland J H, Claessen V I, Wijnen F J P, Groeneveld E, Donegá C M, Vanmaekelbergh D, Meijerink A, Maan J C and Christianen P C M 2011 *Phys. Rev.* B **83** 035304
[5] Efros A L, Rosen M, Kuno M, Nirmal M, Norris D J and Bawendi M 1996 *Phys. Rev.* B **54** 4843
[6] Donegá C M, Bode M and Meijerink A 2006 *Phys. Rev.* B **74** 085320
[7] Crooker S A, Barrick T, Hollingsworth J A, Klimov V I 2003 *Appl. Phys. Lett.* **82** 2793
[8] Yu Z, Li J, O'Connor D B, Wang L W and Barbara P F 2003 *J. Phys. Chem.* B **107** 5670
[9] Kim D, Mishima T, Tomihira K and Nakayama M 2008 *J. Phys. Chem.* C **112** 10668
[10] Demchenko D O and Wang L W 2006 *Phys. Rev.* B **73** 155326
[11] Mauckner G, Thonke K, Baier T, Walter T and Sauer R 1994 *J. Appl. Phys.* **75** 4167

IOP Conf. Series: Journal of Physics: Conf. Series **1220** (2019) 012030 doi:10.1088/1742-6596/1220/1/012030

Photoluminescence polarization characteristics of self-trapped excitons in an undoped β-Ga_2O_3 single crystal

Suguru Yamaoka, Yusuke Mikuni, and Masaaki Nakayama

Department of Applied Physics, Graduate School of Engineering, Osaka City University, Sugimoto, Sumiyoshi-ku, Osaka 558-8585, Japan

nakayama@a-phys.eng.osaka-cu.ac.jp

Abstract. We have investigated the photoluminescence (PL) polarization characteristics of the self-trapped exciton (STE) in an undoped β-Ga_2O_3 single crystal at 77 K under three-photon excitation. From analysis of the polarization characteristics, we found that the STE PL is polarized almost parallel to the *a*-axis. The STE-PL polarization corresponds to the orientation of the self-trapped hole which was investigated in previous works using an electron-paramagnetic-resonance experiment and a first-principles calculation.

1. Introduction

Since β-Ga_2O_3 is one of wide-gap group-III semiconductor oxides, the band gap of which is ~4.7 eV [1], much attention has been paid to high-power field-effect transistors [2] and optoelectronic devices in a deep ultraviolet region [3-5]. The crystal structure of β-Ga_2O_3 is monoclinic with the lattice parameters of a=1.223 nm, b=0.304 nm, c=0.580 nm, and the angle between the *a*- and *c*-axes, β=103.7° [6]. Figure 1 shows the unit cell of the crystal structure of β-Ga_2O_3. The unit cell consists of two types of gallium-atom site, Ga(I) and Ga(II), and three types of oxygen-atom site, O(I), O(II), and O(III). This crystal structure will be considered in the discussion of this paper. In our previous work [7], it was confirmed from analysis of the temperature dependence of the Urbach tail that a broad photoluminescence (PL) band, the peak energy of which is ~3.35 eV, is attributed to the self-trapped exciton (STE). In alkali halides, which are typical materials for the investigation of the STE, it was reported that the STE PL has a polarization nature [8,9]; however, the polarization characteristics in β-Ga_2O_3 have not been revealed until now.

In the present work, we have investigated the PL polarization characteristics of the STE in an undoped β-Ga_2O_3 single crystal at 77 K under three-photon excitation. The PL polarization was clearly observed. It was found that the PL band of the STE is polarized almost parallel to the *a*-axis. We discuss the PL polarization characteristics of the STE, comparing with the orientation of the self-trapped hole (STH) which was investigated by an electron paramagnetic resonance (EPR) experiment [10] and a first-principles calculation [11].

2. Experimental details

The sample used was a (010)-oriented undoped β-Ga_2O_3 single crystal with a thickness of 0.60 mm grown by an edge-defined film-fed growth method at Tamura Corporation. In the PL measurements, the excitation light source was a mode-locked Ti:sapphire laser with a pulse duration of 110 fs and a repetition rate of 76 MHz. The energy of the fundamental laser light was 1.72 eV for the three-photon

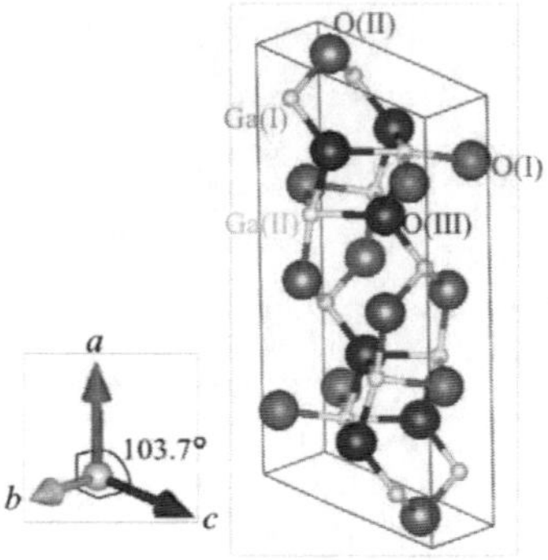

Figure 1. Unit cell of the crystal structure of β-Ga_2O_3.

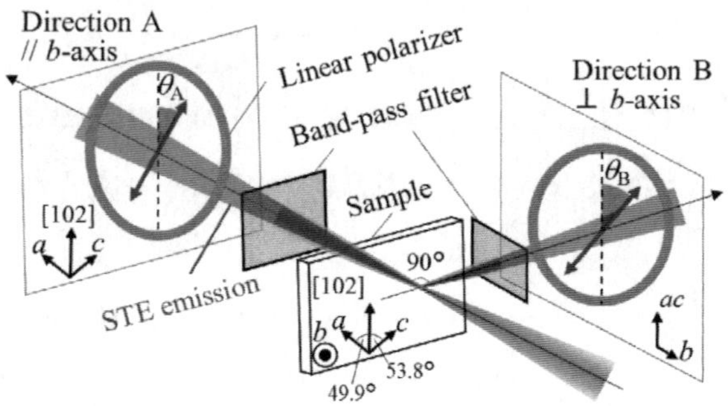

Figure 2. Schematic diagram of the experimental setup for the measurement of the PL polarization of the STE.

excitation to avoid influence of defects and impurities around the surface on the PL properties because the sample is transparent to the excitation light. In other words, the three-photon excitation enabled us to observe the intrinsic properties of the STE. The excitation fluence was ~1.4 mJ/cm^2 in the PL polarization measurement. Figure 2 shows the schematic diagram of the experimental setup for the measurement of the PL polarization of the STE. The incident angle of the excitation light was normal to the sample surface along the *b*-axis [010] and the PL from the sample was detected at the parallel and perpendicular directions to the excitation light, which are labelled the direction A and B in Fig. 2, respectively. In the measurement at the direction A (B), we obtained the PL polarization in the *ac*-plane (the plane perpendicular to the *ac*-plane). Thus, the PL polarization was determined in three dimensions. A band-pass filter was used to cut off the excitation laser light. The band-pass filter distorted the PL spectrum because of its absorption property; however, the spectrum distortion did not affect the PL polarization characteristics owing to the fact that the STE PL is a single band. The PL spectra were detected with a cooled charge-coupled device attached to a single monochromator with a resolution of 1.5 nm. All the optical measurements were performed at 77 K because only a liquid nitrogen cryostat was available in our experimental equipment to detect the PL both in the direction A and in the direction B.

3. Results and discussion

Figure 3 shows the PL spectrum at 77 K of the β-Ga_2O_3 single crystal in a back scattering configuration without the band-pass filter. A broad PL band peaking at 3.37 eV is attributed to the STE [7]. Because the STH in β-Ga_2O_3 is thermally stable in the temperature region lower than 90 K [10], the STE, which consists of the STH and an electron, is also stable at 77 K. The inset in Fig. 3 shows the integrated PL intensity as a function of excitation fluence. It is evident from the fitted line that the integrated PL intensity exhibits the cubic dependence on the excitation fluence, which demonstrates the occurrence of the three-photon excitation.

Figure 4(a) shows the polarization-angle dependence of the integrated PL intensity of the STE band detected at the direction A shown in Fig. 2 at 77 K, where the open circles indicate the experimental results. The inset shows the PL spectra of the STE at various polarizer angles for the reference of the polarization property. The polarization characteristics of the STE exhibit a typical cosine law; therefore, the following equation was used to analyze the experimental result:

$$I_{PL}(\theta) = I_0 \cos^2(\theta - \theta_{pol}) + I_{offset}, \quad (1)$$

where θ and θ_{pol} indicate the polarizer angle and polarization angle of the STE PL, respectively, and I_{offset} is the non-polarized PL intensity. The solid curve indicates the fitted result. We found that the polarization angle of the STE PL at the direction A is $\theta_{pol,A}$=138°. The degree of polarization (DOP), which is defined as DOP=$(I_{max}-I_{min})/(I_{max}+I_{min})=I_0/(I_0+2I_{offset})$, was evaluated to be 0.61, where I_{max} and

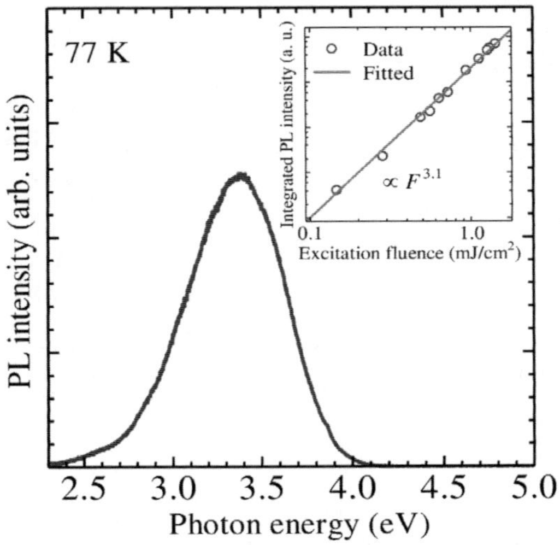

Figure 3. PL spectrum at 77 K of the β-Ga_2O_3 single crystal. The inset shows the integrated PL intensity as a function of excitation fluence (F). The solid line indicates the fitted result: $I_{PL} \propto F^{3.1}$.

I_{min} are the maximum and minimum PL intensities, respectively. The polarization angle $\theta_{pol,A}$ is almost parallel to the a-axis corresponding to θ =180°-49.9°=130.1° at the direction A as shown in Fig. 2. Thus, the STE PL is clearly polarized. The non-polarized PL component corresponding to I_{offset} seems to be due to the random orientation originating from the hopping motion of the STEs. Figure 4(b) shows the polarization-angle dependence of the integrated PL intensity of the STE band detected at the direction B shown in Fig. 2 at 77 K, where the open circles and solid curve indicate the experimental results and fitted result using Eq. (1). The inset shows the PL spectra of the STE at various polarizer angles. The polarization angle and DOP of the STE PL at the direction B were estimated to be $\theta_{pol,B}$=10° and 0.53, respectively. The polarization angle of $\theta_{pol,B}$=10° corresponds to that the PL polarization of the STE is inclined by 10° relative to the ac-plane. These results indicate that the polarization in three dimensions of the STE PL is almost parallel to the a-axis.

Hereafter, we discuss the PL polarization characteristics of the STE, comparing with the orientation of the STH that is the core of the STE. The hole in β-Ga_2O_3 is on a single O^- site with its $2p^5$ configuration ($2p_x^2 2p_y^2 2p_z^1$) [10]. Kananen *et al.* found from the EPR experiment for β-Ga_2O_3 that the STH is localized in a nonbonding O $2p$ orbital which is approximately along the a-axis [10]. This result indicates that the STH is localized at the O(I) in Fig. 1 because only the nonbonding O(I) $2p$ orbital is oriented nearly along the a-axis. This STH nature is consistent with the theoretical study of

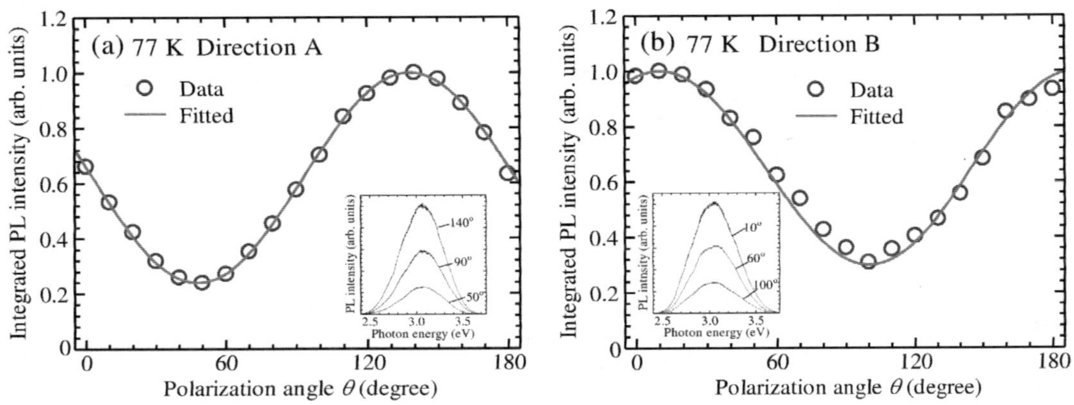

Figure 4. Polarization-angle dependence of the integrated PL intensity of the STE band at 77 K detected at the (a) direction A and (b) direction B, which are defined in Fig. 2, where the open circles and solid curve indicate the experimental results and fitted result using Eq. (1). The inset shows the PL spectra of the STE at various polarizer angles.

IOP Conf. Series: Journal of Physics: Conf. Series **1220** (2019) 012030 doi:10.1088/1742-6596/1220/1/012030

the STH in β-Ga_2O_3 based on hybrid functional calculations [11]. Thus, the STE-PL polarization almost along the a-axis corresponds to the orientation of the STH which was reported in Refs. [10,11]. In alkali halides, the polarization of the singlet (triplet) STE is parallel (perpendicular) to the orientation of the STH [8,9]. Consequently, the fact that the STE-PL polarization is parallel to the orientation of the STH suggests that the STE has a singlet nature. In our previous work [12], we confirmed that the STE-PL decay profile exhibits a fast component with a decay time of ~80 ns corresponding to the singlet characteristics.

4. Conclusion

We clearly observed the PL polarization characteristics of the STE in an undoped β-Ga_2O_3 single crystal at 77 K under three-photon excitation. Analyzing the PL polarization characteristics of the STE, we confirmed that the STE PL is polarized almost parallel to the a-axis, which is consistent with the STH orientation investigated by the EPR experiment and the first-principles calculation. This fact suggests that the STE has a singlet nature.

References

[1] Onuma T, Saito S, Sasaki K, Masui T, Yamaguchi T, Honda T and Higashiwaki M 2015 *Jpn. J. Appl. Phys.* **54** 112601
[2] Higashiwaki M, Sasaki K, Kuramata A, Masui T and Yamakoshi S 2012 *Appl. Phys. Lett.* **100** 013504
[3] Ueda N, Hosono H, Waseda R and Kawazoe H 1997 *Appl. Phys. Lett.* **70** 3561
[4] Muhammed M M, Roldan M A, Yamashita Y, Sahonta S -L, Ajia I A, Iizuka K, Kuramata A, Humphreys C J and Roqan I S 2016 *Sci. Rep.* **6** 29747
[5] Guo D, Wu Z, Li P, An Y, Liu H, Guo X, Yan H, Wang G, Sun C, Li L and Tang W 2014 *Opt. Mater. Express* **4** 1067
[6] Geller S 1960 *J. Chem. Phys.* **33** 676
[7] Yamaoka S and Nakayama M 2016 *Phys. Status Solidi C* **13** 93
[8] Pellaux J P, Iida T, von der Weid J P and Aegerter M A 1980 *J. Phys. C: Solid State Phys.* **13** 1009
[9] Purdy A E, Murray R B, Song K S and Stoneham A M 1977 *Phys. Rev. B* **15** 2170
[10] Kananen B E, Giles N C, Halliburton L E, Foundos G K, Chang K B and Stevens K T 2017 *J. Appl. Phys.* **122** 215703
[11] Varley J B, Janotti A, Franchini C and Van de Walle C G 2012 *Phys. Rev. B* **85** 081109(R)
[12] Yamaoka S, Furukawa Y and Nakayama M 2017 *Phys. Rev. B* **95** 094304

IOP Conf. Series: Journal of Physics: Conf. Series **1220** (2019) 012031 doi:10.1088/1742-6596/1220/1/012031

Energy relaxation from STE to In^+ centers in $NaI:In^+$ crystals

S Watanabe, T Kawai* and K Mizoguchi

Department of Physical Science, Osaka Prefecture University, 1-1 Gakuen-cho, Naka-ku, Sakai-shi, Osaka 599-8531, Japan

E-mail:buri@p.s.osakafu-u.ac.jp

Abstract. The energy relaxation from host crystals to impurity centers has been investigated by measuring the absorption and excitation spectra and the temperature dependence of the luminescence spectrum of $NaI:In^+$. In the absorption spectrum, the A, B, and C bands due to the In^+ centers are confirmed at 3.8, 4.0, and 4.5 eV, respectively. The excitation spectrum for the A_T luminescence due to the In^+ centers responds to the energy position of the A, B, and C bands and also the energy region above the exciton transition of NaI. The fact suggests the existence of the energy relaxation from host crystals to the In^+ centers. The total intensity of the A_T and A_X luminescence due to the In^+ centers exhibits different temperature dependence between the excitations at the C band and host crystals. Under excitation at the C band, the total intensity of the A_T and A_X luminescence monotonously declines with increasing temperature from 10 to 300 K. On the other hand, under excitation at host crystals, the total intensity of the A_T and A_X luminescence exhibits an increase with increasing temperature from 10 to 50 K, and a decrease with increasing temperature from 50 to 100 K. We discuss the temperature dependence of the total intensity of the luminescence under excitation at host crystals in relation to free excitons and self-trapped excitons.

1. Introduction

For impurity-activated scintillator materials, energy transfers through the energy relaxation from host crystals to impurity centers play a crucial role in the scintillation process. The energy transfer mechanisms have been extensively investigated for Tl^+-activated CsI and NaI crystals, which are widely used as scintillator materials. In $CsI:Tl^+$ and $NaI:Tl^+$, free excitons (FE), self-trapped excitons (STE) self-trapped holes (STH), etc have an important part to play in the energy transfer processes [1-5].

Though the Tl^+ ion is well-known activated ions, In^+ ions are also potential candidates for activated ions, because the In^+ ions have the outermost $5s^2$ electron configuration in the ground state and belong to a family of Tl^+-type ions [6,7]. Thus, when the In^+ ions are doped into NaI crystals, it is expected that the In^+ centers in the NaI crystals exhibit high scintillation efficiency like $NaI:Tl^+$. However, the optical studies of $NaI:In^+$ have not been reported to our knowledge. In this study, we have investigated the luminescence properties of $NaI:In^+$ crystals under excitation in the energy region below and above the fundamental absorption edge of NaI host crystals, in order to obtain an information about the energy transfer processes from host crystals to impurity centers through the FE and STE states.

2. Experimental methods

$NaI:In^+$ crystals were grown by the Bridgman method using NaI beads (ALDRICH, 99.999%) and InI beads (ALDRICH, 99.999 %). The optical measurements up to the near vacuum ultraviolet region were

IOP Conf. Series: Journal of Physics: Conf. Series **1220** (2019) 012031 doi:10.1088/1742-6596/1220/1/012031

performed at the BL-3B line of UVSOR (Institute for Molecular Science, Okazaki, Japan). Details of this beam line have been reported in Ref. [8]. For optical measurements, the crystals were cleaved into the size about $5 \times 5 \times 1$ mm^3 from the ingots in a glove box under N_2 atmosphere. The samples were mounted on a cold stage of a liquid helium-flow type cryostat and were set in a vacuum chamber.

3. Experimental results and discussion

Figure 1 shows the absorption and luminescence spectra at low temperatures. The absorption bands due to the In^+ centers are observed at 3.8, 4.0, and 4.5 eV, and are called A, B, and C, respectively. The A, B, and C absorption bands are attributed to intra-ionic transitions from the $^1A_{1g}$ ground state to the $^1T_{1u}$, $^3T_{2u}$+^3E_u, and $^3T_{1u}$ excited states in the In^+ center, respectively [6,7]. The concentration of the In^+ center in our NaI:In^+ crystal was estimated to be about 2.0×10^{17} cm^{-3} from the absorption intensity. When the NaI:In^+ crystal is excited by the light corresponding to the C absorption band, the luminescence bands called A_T and A_X appear at 2.8 and 2.1 eV, respectively, which are attributed to the radiative transition from two kinds of minima on the adiabatic potential energy surfaces (APESs) of the $^3T_{1u}$ and $^3A_{1u}$ excited-states [9,10].

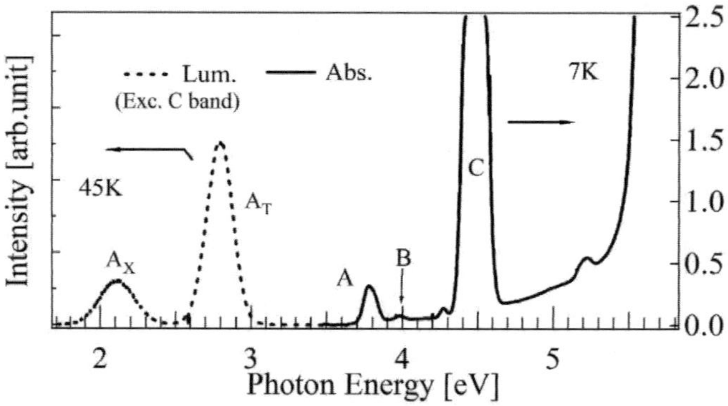

Figure 1. Absorption (solid) and luminescence (broken) spectra of NaI:In^+ crystals at 7 and 45 K, respectively. The peak intensity of the C absorption band at 4.5 eV is off the scale.

Figure 2 shows the excitation spectrum for the A_T luminescence at 2.8 eV and reflection spectrum at 7 K. The excitation spectrum for the A_T luminescence band exhibits the remarkable responses at the energy positions corresponding to the A, B, and C absorption bands. It should be noted that the A_T luminescence band is effectively excited at the energy region above the exciton transition of NaI host crystals, whose energy can be estimated to be about 5.6 eV from the reflection spectrum. As shown in the insert of Figure 2, the A_T and A_X luminescence bands due to the In^+ center can be seen in addition to the intrinsic luminescence due to STE at 4.15 eV under excitation at 6.2 eV, which corresponds to the energy above the exciton transition of NaI host crystals. The fact indicates the existence of the energy transfer from host lattice to the In^+ centers through the FE and STE states at low temperature.

Figure 3 shows the temperature evolution of the luminescence spectra of the NaI:In^+ crystals under excitation at 6.2 eV. At the temperature region below 30 K, the A_T luminescence band due to the In^+ center and the STE luminescence band are observed at 2.8 and 4.2 eV, respectively. As the temperature increases from 30 to 110 K, the A_T luminescence band decreases and the A_X luminescence band increases with compensation for the decrease of the A_X luminescence intensity. The A_X luminescence band and the STE luminescence band decrease with increasing temperature from 60 K. The STE luminescence band disappears at around 110 K, while the A_X luminescence band exhibits a broadening of the band-shape and a slight decline of the intensity. In order to discuss the energy transfer to the In^+ centers, we focus the total integrated luminescence intensity of the A_T and A_X bands due to the In^+ center.

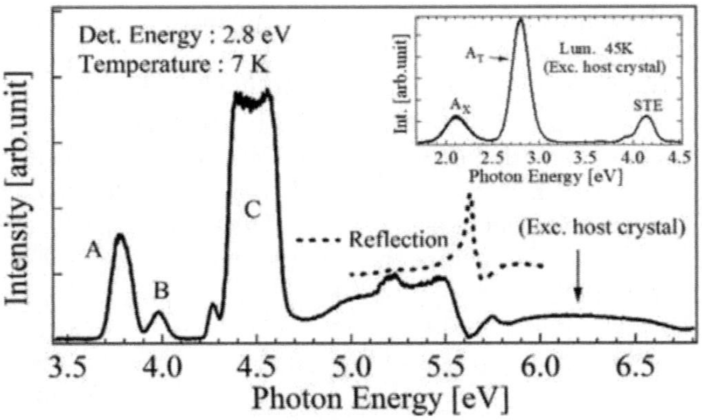

Figure 2. Excitation spectrum (solid) for the A_T luminescence at 7 K. For reference, the reflection spectrum of NaI:In^+ is shown by a dashed-dotted line. The insert is the luminescence spectra at 45 K under excitation at host crystals, whose energy is shown by downward arrow in Figure 2.

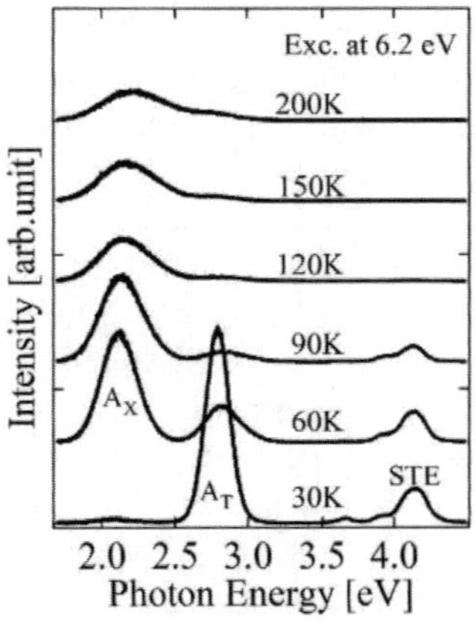

Figure 3. Luminescence spectra of NaI:In^+ crystals observed at various temperatures under excitation at 6.2 eV.

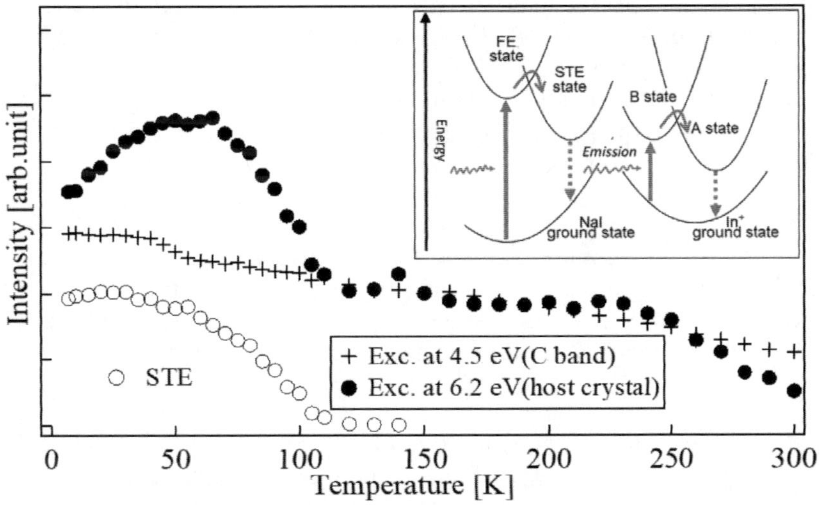

Figure 4. Temperature dependence of the total intensity of the A_T and A_X luminescence under the excitation at the C band (crosses) and the host crystals (filled circles). Temperature dependence of the total intensity is normalized at 150 K. Temperature dependence of the luminescence intensity of STE is shown by open circles. The insert is the schematic diagram of the STE luminescence and reabsorption by the In^+ centers.

The total intensity of the A_T and A_X luminescence under excitations at 4.5 and 6.2 eV and the STE luminescence intensity are plotted as a function of the temperature in Figure 4. The total intensity of the A_T and A_X luminescence under excitation at 4.5 eV exhibits a monotonous decline with increasing temperature from 10 to 300 K. The monotonous decline will come from the non-radiative transition from the $^3T_{1u}$ APES. In the temperature range above 110 K, the total intensity of the A_T and A_X luminescence under excitation at 6.2 eV also exhibits a monotonous decline with increasing temperature.

We pay attention to the difference of the temperature dependence between the excitations at 4.5 and 6.2 eV in the temperature region 10 to 110 K. According to the previous paper [8], STEs in NaI crystals migrate through lattices by a thermally activated hopping motion above the temperature of about 70 K. Then, the STEs can encounter the impurity ions during the migration and transfer their energies to the impurity ions. As a result, it is reported that the luminescence related to the impurity centers in NaI crystals is enhanced. In our NaI:In$^+$ crystal, however, the total intensity of the A_T and A_X luminescence decreases with increasing temperature from 70 K in a similar way to the STE luminescence intensity. The difference between the temperature dependence of the In$^+$ center luminescence in our NaI:In$^+$ crystal and that reported in the previous paper might come from that of the impurity concentration. Since the concentration of the impurity ions in the previous paper is much larger than that in our NaI:In$^+$ crystals, the low concentration of the In$^+$ centers might cause the earlier reaching of the STEs to the surface and/or other impurity ions, where STEs recombine nonradiatively, before encountering the In$^+$ centers. Then, another energy transfer processes might be dominant.

As discussed in our previous paper [11], the similar temperature dependence of the intrinsic STE and impurity luminescence indicates the potential of an emission reabsorption mechanism for the energy transfer from the host lattice to impurity centers. Therefore, we consider that the temperature dependence of the total intensity of the A_T and A_X luminescence in our NaI:In$^+$ crystal comes from the energy transfer from the host lattice to impurity centers through reabsorption processes of the FE and STE luminescence. In the temperature region below 60 K, the total intensity of the A_T and A_X luminescence increases with increasing temperature. In this temperature range, the STE luminescence intensity is almost constant, while the quantum yield of the FE-related luminescence, which appears at the energy region around 5.6 eV, increases gradually as the temperature rises from 10 to 40 K [1]. Since the A_T luminescence is also excited at 5.6 eV, as can be seen in the excitation spectrum of Figure 2, the increase of the FE-related luminescence might induce that of the total intensity of the A_T and A_X luminescence. In the NaI:In$^+$ crystal, the lower energy tail of the broad STE luminescence overlaps with the B absorption band of the In$^+$ center. Reflecting the overlap between the STE luminescence and the B band, a slight dent structure can been confirmed at 3.98 eV in the STE luminescence band, which has a Gaussian band-shape intrinsically. Thus, the decrease of the total intensity of the A_T and A_X luminescence with increasing temperature from 70 K will come from that of the STE luminescence.

4. Conclusion

The temperature change of the luminescence spectra of NaI crystals doped with In$^+$ centers were investigated under the excitation at the C band and host crystals. From the comparison between the temperature changes of the total luminescence intensity due to the In$^+$ centers, we concluded the energy relaxation process from host crystals to impurity centers through emission reabsorption.

Acknowledgment

The optical measurements were supported by the Use-of-UVSOR Facility Program of the Institute for Molecular Science, Japan. This work was partially supported by JSPS KAKENHI Grand Number JP16H04003.

References

[1] Nagata S, Fujiwara K and Nishimura H 1991 *J. Lumin.* **47** 147.
[2] Kubota S, Shiraishi F and Takami Y 1998 *J.Phys. Soc. Jpn.* **68** 291.
[3] Ratner M, Ratner A and Hryn'ova T 2002 *Nucl. Instrum. Methods Phys. Res.* A **486** 463.
[4] Gektin A, Gridin S, Vasyukov S, Vasil'ev A, Belsky A and Shiran N 2015 *Phys. Status Solidi* B **252** 380.
[5] Yakovlev V, Trefilova L, Alekseev V and Krasnov V 2016 *Funct. Mater.* **23** 540.
[6] Ranfagni A, Mugnai D, Bacci M, Viliani G and Fontana M P 1983 *Advances in Physics* **32** 823.
[7] Jacobs P W M 1991 *J. Phys. Chem. Solids* **52** 35.
[8] Fukui K, Ikematsu R, Imoto Y, Kitaura M, Nakagawa K, Ejima T, Nakamura E, Sakai M, Hasumoto M and Kimura S 2014 *J. Synchrotron Rad.* **21** 452-455.
[9] Fukuda A 1970 *Phys. Rev.* B **1** 4161.
[10] Mugnai D, Pazzi G P, Fabeni P and Ranfagni A 2014 *J. Lumin.* **156** 137.
[11] Nagata S, Kawai T and Hirai T 2013 *Opt. Mater.* **35** 1257.

IOP Conf. Series: Journal of Physics: Conf. Series **1220** (2019) 012032 doi:10.1088/1742-6596/1220/1/012032

Nonlinear polarization optical response to entangled state of biexciton

T Tsuji, G Oohata*, and K Mizoguchi

Department of Physical Science, Graduate School of Science, Osaka Prefecture University,1-1 Gakuen-cho, Naka-ku, Sakai 599-8531, Japan

E-mail: oohata@p.s.osakafu-u.ac.jp

Abstract. We have investigated the nonlinear optical responses of biexcitons which show the polarization entanglement by using pump-probe spectroscopy. Measuring all polarization combinations of pump-probe signals and precise analysis using the idea of quantum tomography enable us to extract the specific property of the entanglement. The obtained results indicate the two kinds of the third-order nonlinear response, namely the photo-induced absorption and the optical Kerr effect. From the analysis with the optical Kerr effect signals, the polarization correlations of the optical responses are clearly derived.

1. Introduction

Quantum entanglement is an important physical phenomenon on the field of the quantum information science. Recently, it is reported that the quantum entanglement plays important roles also in various physical phenomena including biological systems [1,2]. Correspondingly, efficient and selective detection methods for the quantum entanglement are highly desired for the demonstration and further applications of the entanglement in the biological systems. Most of conventional detection methods for the quantum entanglement are implemented by direct measurement of the emitted entangled photons [3,4]. An example of them is measurements of photon pairs emitted in the optical process called Resonance Hyper Parametric Scattering (RHPS) in CuCl, which is the two-photon emission by the decay of the biexciton [5]. In the RHPS process, because the emitted photon-pairs are closely related to the quantum state of the biexciton, we can investigate the quantum entanglement of the biexciton by coincidence counting of the photon-pairs. However, since extremely efficient photon counting techniques are required for the direct detection of the photon-pairs, it is difficult to apply the detection methods to studies for the entanglement properties of various materials.

In this paper, we report on the observation of the entanglement of the biexciton in a CuCl single crystal. Based on the concept of the quantum tomography which is an analytical method to estimate a quantum state from the results of all projection measurements [6,7], pump-probe measurements for biexcitons were implemented in all polarization combinations. In addition, the obtained spectra by the pump-probe measurements were decomposed into the elements of optical Kerr effect (OKE) and photo-induced absorption (PIA) by using a simple theoretical model. Then we discuss the polarization correlation of the OKE and demonstrate the entanglement of the biexciton.

2. Theoretical background

The optical responses related to the biexciton have the third-order nonlinearity. Therefore, it is

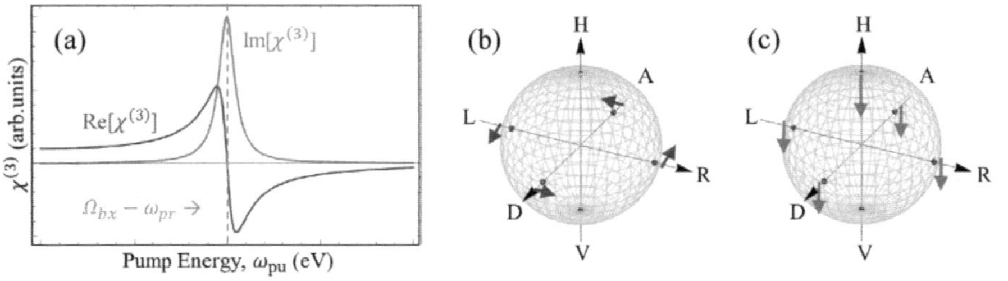

Figure 1. (a)Typical profiles of the real and imaginary part of the $\chi^{(3)}$ function shown in equation (1). The dashed line means the resonance energy of the biexciton corresponding to the energy of the probe beam, namely $\omega_{pu} = \Omega_{bx} - \omega_{pr}$. (b), (c)The change of the Bloch sphere in the case that the OKE and PIA are applied only to the H-polarization component. The OKE rotates each point on the H-V axis, and the PIA moves each point to the V polarization pole.

predicted that the pump energy dependence of the pump-probe signals for the biexcitons correspond to the profile of the third-order nonlinear susceptibility $\chi^{(3)}$. In general, the $\chi^{(3)}$ function can be expressed as

$$\chi^{(3)} \propto \left[\Omega_{bx} - (\omega_{pu} + \omega_{pr}) + i\Gamma\right]^{-1}, \quad (1)$$

where ω_{pu} and ω_{pr} are the energies of the pump and probe beam respectively, and Ω_{bx}(=6.372 eV in case of a CuCl) and Γ are the resonance energy and the damping constant of the biexciton respectively [8]. The profiles of the real and imaginary parts of the $\chi^{(3)}$ are shown in figure 1(a), that means the change of the refractive index and the absorption coefficient respectively which are caused by the pump beams. These optical processes are referred to as the OKE and the PIA respectively.

Here, let us consider how these optical responses affect the polarizations. For the simplicity, we assume the case that the polarization of the initial state is the diagonal polarization, that is, $|D\rangle \propto |H\rangle + |V\rangle$, where $|H\rangle$ and $|V\rangle$ are the state of horizontal and vertical polarizations respectively and the third-order nonlinear responses are applied only to the H polarization component. Considering the contribution of the OKE and PIA, the initial state $|D\rangle$ changes to $|D'\rangle_{OKE} \propto e^{-i\Delta\theta}|H\rangle + |V\rangle$ and $|D'\rangle_{PIA} \propto \sqrt{1-\Delta\alpha}|H\rangle + |V\rangle$, where $\Delta\theta$ is the variation of the phase difference between H and V components which originate from the variation of the refractive index and $\Delta\alpha$ is the variation of the absorption coefficient of H component. Then, the initial state $|D\rangle$ is slightly modulated to the state of right-circular polarization ($|R\rangle \propto |H\rangle + i|V\rangle$) by the OKE, and to the state $|V\rangle$ by the PIA. Similarly, we consider the change of the 6 polarizations: H, V, D, anti-diagonal A, R, and left-circular L polarizations. The polarization change caused by the OKE and PIA can be expressed as the change of a Bloch sphere for the 6 polarizations shown in Fig. 1 (b), (c) respectively. It can be seen that the OKE corresponds to the rotation of the sphere on the particular axis and the PIA does to the deformation in one direction of the sphere.

3. Experimental setup and results

In the experiment, we used a CuCl single crystal prepared by vapor deposition. The second harmonic

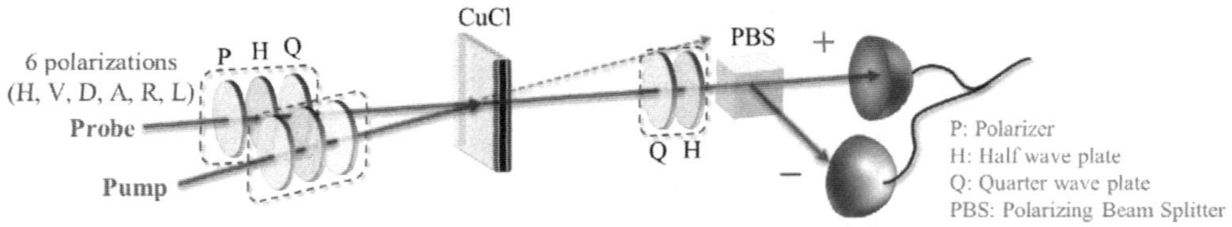

Figure 2. The schematic diagram of the experimental setup. The polarizations of the pump and probe beam are set to 6 polarizations: H, V, D, A, R, and L polarizations. The intensity differences of the transmitted probe beam with H-V, D-A, and R-L bases are measured.

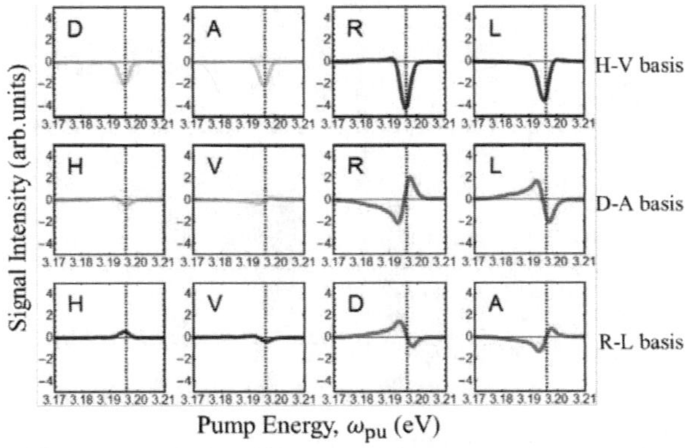

Figure 3. The pump energy dependence of the intensity differences obtained in respective three bases in the case of H-polarization pump beam. The letters on the upper left side mean the polarization of the probe beam. The dashed line at 3.196 eV means the resonance energy of the biexciton corresponding to the fixed energy of the probe beam, namely $\omega_{pu} = \Omega_{bx} - \omega_{pr}$.

wave of a mode-locked Ti:Sapphire laser was used as an excitation pulse source and its energy was controlled by 4f optical systems with diffraction gratings. The energy of a probe beam, ω_{pr} was fixed at 3.176 eV and the energy of the pump beam, ω_{pu} was changed from 3.154 to 3.217 eV. The schematic diagram of the experimental setup is shown in figure 2. Each beam was set to 6 polarizations: H, V, D, A, R, and L polarizations by using polarizers and waveplates. Then, the beams were focused on the sample at 10 K. The transmitted probe beam was split into two beams with three bases of polarizations: H-V, D-A, and R-L bases by using waveplates and a polarizing beam splitter. The intensity differences between the two beams in the respective bases were obtained by the pump-probe signals using a lock-in amplifier.

In the case that the polarization of the pump beam is H, the intensity difference observed in the respective polarization bases are plotted as a function of pump energy in figure 3. Each spectrum shows typical three structures at the resonance energy of the biexciton corresponding to the fixed energy of the probe beam: the peak (dip) profile, differential profile of Lorentz function, and nearly no signal. These optical responses correspond to the profiles of the real part and the imaginary part of the third-order nonlinear susceptibility $\chi^{(3)}$ shown in figure 1(a). Therefore, these results suggest that the OKE and the PIA occur at the resonance energy of the biexciton.

4. Discussion

We discuss each spectrum in detail using a simple theoretical model. As previously mentioned and experimentally confirmed, optical responses of the biexciton correspond to the profile of the third-order nonlinear susceptibility $\chi^{(3)}$. Therefore, we decomposed the results into the real part (differential shape) and imaginary part (peak (dip) shape) of the $\chi^{(3)}$. In figure 4, we show the pump energy dependence of the intensity differences and the fitting result in D-A basis in the case of the polarization combination of the H-pump and the R-probe beams. From the fitting with the theoretical model, the component of the real part of the $\chi^{(3)}$ is relatively large, indicating that OKE is dominant. The values of the OKE and the PIA can be estimated from the coefficients of each part of the $\chi^{(3)}$ obtained in the comparison between the experimental results and the theoretical model. In the following discussion, we focus only on the OKE signals.

As previously mentioned in section 2, the probe polarization change can be expressed as the change in the Bloch sphere for the 6 polarizations. Therefore, from the experimental results it is possible to consider the values of the OKE on each rotation axis independently. Then, the normalized relative values of the OKE on each axis are obtained from 12 experimental results for each pump polarization. The results are shown in figure 5 as a bar chart. For each polarization of the pump beam, it is found that the value on the particular rotation axis is clearly large which normalized to ± 1, whereas the values on the other rotation axes are almost zero. Therefore, these results demonstrate that the rotation axes and values of the OKE have obvious correlations with the polarization combinations of the pump and probe beam.

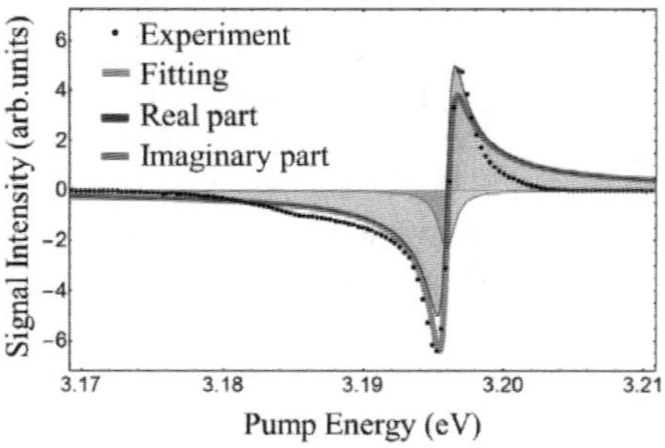

Figure 4. The pump energy dependence of the intensity difference and the fitting result in D-A basis in the case of the polarizations of the pump and probe beam are H and R. The pump energy dependence of the signal is decomposed into the real and imaginary part of general $\chi^{(3)}$ function expressed as equation (1).

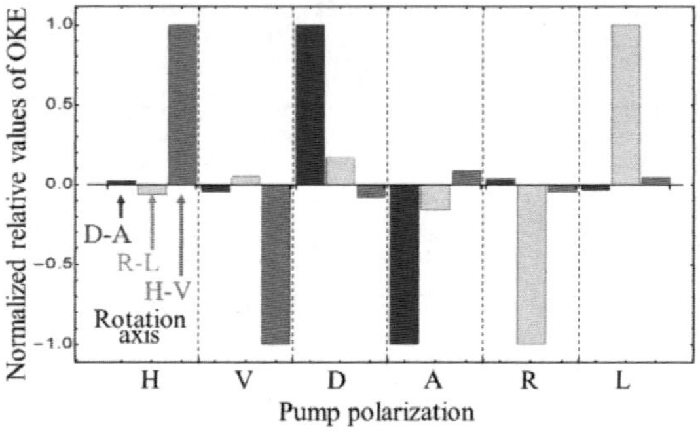

Figure 5. The normalized relative values of the OKE on the 3 axes (H-V, D-A, and R-L) obtained from the experimental results. The values on the 3 axes correspond to the bar on the left, center, and right side respectively.

Finally, we discuss the rotation axes and the values of the OKE related to the entangled state of the biexciton. Considering the spin state of the biexciton, $|\psi\rangle_{\mathrm{bx}} = \frac{1}{\sqrt{2}}(|\uparrow\rangle|\downarrow\rangle + |\downarrow\rangle|\uparrow\rangle)$, the wave function of the photons which interact with the biexcitons is expressed in terms of polarizations as

$$|\psi\rangle_{\mathrm{ph}} = \frac{1}{\sqrt{2}}(|H\rangle|H\rangle + |V\rangle|V\rangle) = \frac{1}{\sqrt{2}}(|D\rangle|D\rangle + |A\rangle|A\rangle) = \frac{1}{\sqrt{2}}(|R\rangle|L\rangle + |L\rangle|R\rangle). \tag{2}$$

This equation determines the combinations of the pump and probe polarizations that can interact with the biexciton. We can find that the relation of equation (2) is analogous to the experimental results shown in figure 5. Consequently, it is clear that the obtained polarization correlations of the optical responses express the spin state of the biexciton.

5. Conclusion

We have observed the OKE and the PIA due to the biexciton state in 72 combinations of polarization. From the obtained results, we demonstrate that the nonlinear optical response related to the quantum entanglement of the biexciton appears in the pump energy dependence of the pump-probe signals by the analysis based on the quantum tomography.

Acknowledgements
This work was supported by JSPS KAKENHI Grant Number JP16H04003 and JP26610088.

References

[1] Sarovar M, Ishizaki A, Fleming G R and Whaley K B 2010 *Nat. Phys.* **6** 462
[2] Gauger E M, Rieper E, Morton J J L, Benjamin S C and Vedral V 2011 *Phys. Rev. Lett.* **106** 040503
[3] Freedman S J and Clauser J F 1972 *Phys. Rev. Lett.* **28** 938
[4] Kwiat P G, Mattle K, Weinfurter H, Zeilinger A, Sergienko A V and Shih Y 1995 *Phys. Rev. Lett.* **75** 4337
[5] Edamatsu K, Oohata G, Shimizu R and Itoh T 2004 *Nature* **431** 167
[6] James D F V, Kwiat P G, Munro W J and White A G 2001 *Phys. Rev. A* **64** 052312
[7] Chuang I L and Nielsen M A 1997 *J. Mod. Opt.* **44** 2455
[8] Ivanov A L, Haug H and Keldysh L V 1998 *Phys. Rep.* **296** 237

IOP Conf. Series: Journal of Physics: Conf. Series **1220** (2019) 012033 doi:10.1088/1742-6596/1220/1/012033

Switching of radiation pressure acting on microparticles by using T-type photochromic reactions

K Tanaka, K Setoura, S Ito, H Miyasaka
Division of Frontier Materials Science and Center for Promotion of Advanced Interdisciplinary Research, Graduate School of Engineering Science, Osaka University, 1-3 Machikaneyama-cho, Toyonaka, 560-8531, Japan

sito@chem.es.osaka-u.ac.jp, miyasaka@chem.es.osaka-u.ac.jp

Abstract. Manipulation of small particles can be achieved by using the radiation pressure of a laser beam. The strength and direction of the radiation pressure depends on the dielectric constant of target materials. It can therefore be expected to induce mechanical motions of small materials in optical trap by changing their optical property by chemical reactions. Along this line, we recently demonstrated switching of radiation pressure acting on single optically trapped microparticles by using photochromic molecules (Refs. 3, 4). To explore more flexible mesoscopic motions synchronizing photochemical reactions, we have investigated in the present study the micromechanical motion of optically trapped particles containing a T-type photochromic molecule, naphthopyran (NP), with multiple thermal back-reaction rates. The single microparticle with NP was optically trapped with CW 532-nm laser and the photochromic reaction was induced by UV irradiation, resulting in the modification of the optomechanical responses of the particle under laser trapping.

1. Introduction

Noncontact and noninvasive trapping of small particles can be achieved by using the radiation force of a laser beam. [1, 2] The radiation force acting on a small particle under photo-irradiation can be categorized into three terms: scattering, absorption, and gradient forces. The gradient force directs along the spatial gradient of light field, while the scattering and absorption forces push a particle toward the direction of light propagation. An off-resonant laser beam with the target particles is generally used in optical tweezing. Although photo-damage due the absorption of trapping beam by target material can be avoided under off-resonant photoirradiation condition, the radiation force under such condition is in general not so strong. On the other hand, resonant absorption can induce strong absorption force. It is therefore expected to manipulate small particles at lower light intensity under resonant condition. For instance, reversible motions of micro-objects was demonstrated by switching absorption force through the reversible change of the colour of the target objects through organic photochromic reactions.[3, 4]

In the present study, with an aim to demonstrate a variety of micro-mechanical motions regulated by the rate of chemical reaction, we prepared microparticles containing a T-type (thermally reversible) photochromic molecule, naphthopyran (NP). Upon UV exposure, NP undergoes photo-isomerization to transoid-cis (CT) and transoid-trans (TT) forms with different lifetimes. The numerical ratio between CT and TT forms depends on UV irradiation condition; it can be expected to modify the

IOP Conf. Series: Journal of Physics: Conf. Series **1220** (2019) 012033 doi:10.1088/1742-6596/1220/1/012033

mechanical motion of the microparticles due to the back-reaction dynamics by controlling UV irradiation condition.

2. Experimental Section

2.1. Preparation of sample

As specimens, we prepared micrometer-sized (ca. 6-8 μm in diameter) polymer particles with the T-type photochromic compounds, NP. As polymer microparticles, we employed micro particles of cross-linked polyacrylic acid ester (AFX) and cross-linked polybutyl methacrylate (BMX) provided by Sekisui Plastics. These polymer microparticles were used as received. The particles were immersed in acetonitrile solutions of the photochromic molecule. The stock solution was kept in a dark area in an experimental room for 24-48 hours, resulted in the inclusion of the photochromic molecules into the polymer particles with acetonitrile molecules by swelling. After that, we put a few drops of the colloidal solution onto a well-cleaned glass plate and dried it at atmospheric condition to remove the solvent. Then the polymer particles on the glass plate were dispersed into ultrapure water in a small sample chamber consisting of two coverslips and a 0.3-mm thick silicon-rubber spacer.

2.2. Optical setup

For all optical trapping experiments, we used an inverted optical microscope (IX71, Olympus) and a CW fiber laser with three-colour output at 1064, 532, and 355 nm which correspond to its fundamental wavelength, the second harmonic, and the third harmonic (3-band CW laser, Oxide). The second (532 nm) and third (355 nm) harmonics were used respectively for optical trapping of

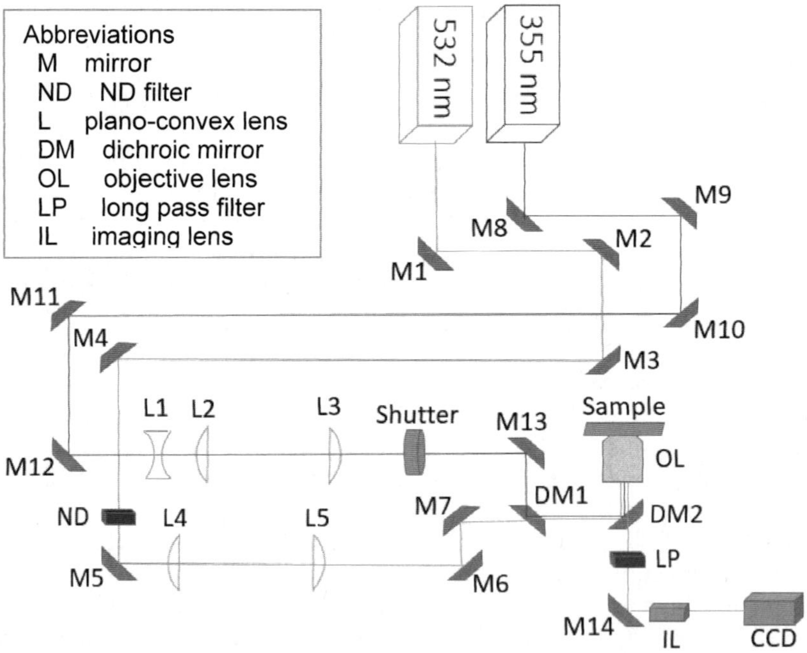

Fig. 1. Experimental setup for inducing and monitoring of photochromic reactions in an optically trapped microparticle.

the single microparticle and inducing the photochromic reaction of the embedded molecules. These two beams were focused by using a microscope objective (UPlanFLN, 100x NA = 1.30, Olympus). The micro-motion of optically trapped single microparticles due to the photochromic reaction was observed by the same objective lens and recorded using a monochromatic CCD camera (Infinity 3-1 URM, Lumenera).

2.3. Estimation of the Z-Position of trapped microparticles

One of the microparticles was optically trapped with the CW 532-nm laser and the photochromic reaction of the guest molecule was induced by UV (355 nm) irradiation, resulting in the modification of the opto-mechanical responses of the particle under laser trapping. This modification induced displacement of the particle along the optical axis (Z-axis). The Z-displacement was detected in the following manner. Before experiment, we obtained the relationship between the diameter of outer circumferential scattering ring of the polymer particle in optical transmission image and the absolute Z-position of the particle controlled by using a 3D piezo scanning stage. Using a calibration curve thus obtained we could deduce the Z-position of a trapped microparticle.

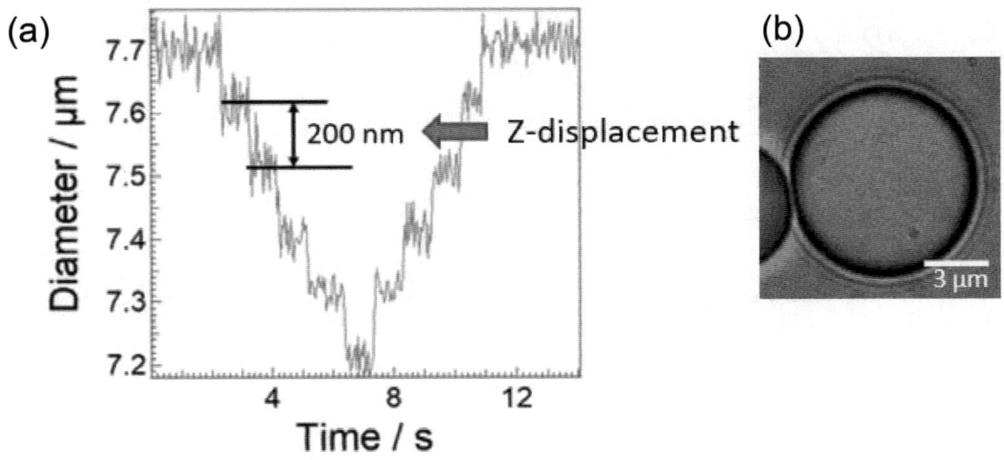

Fig. 2. (a) Calibration curve showing the relationship between the diameter of scattering ring and absolute Z-position of the polymer particle and (b) optical transmission micrograph of the single polymer particle.

3. Result

Figure 3 shows a time-trajectory of the Z-position of a trapped polymer microparticle including NP. UV irradiation at 3 s induced the upward motion of the trapped particle by 1.5 µm along the Z-axis; this motion is attributable to the generation of absorption force of the visible light due to the spectral change of the included NP molecules. After stopping the UV irradiation at 5 s, the Z-position decreased and the particle returned to the initial position before UV irradiation. This decay curve is divided into two components. The time constant of the fast decay component is 1.2 s and that of the slow component is 55 s. We consider that NP underwent isomerization to TC (transoid-cis) form and TT (transoid-trans) form by the UV irradiation. We presume that the fast decay component is caused by the short-lived TC form, while the slow one by the long-lived TT form. The time profile of the backward motion after stopping UV irradiation depended on the reaction kinetics of guest molecules and photoirradiation condition.

Thus, we have successfully demonstrated mesoscopic mechanical motions of optically trapped polymer microparticles synchronizing photochromic reactions. We foresee this approach enables a variety of mesoscopic photochemomechanical motion by combining laser manipulation and a diversity of chemical reactions.

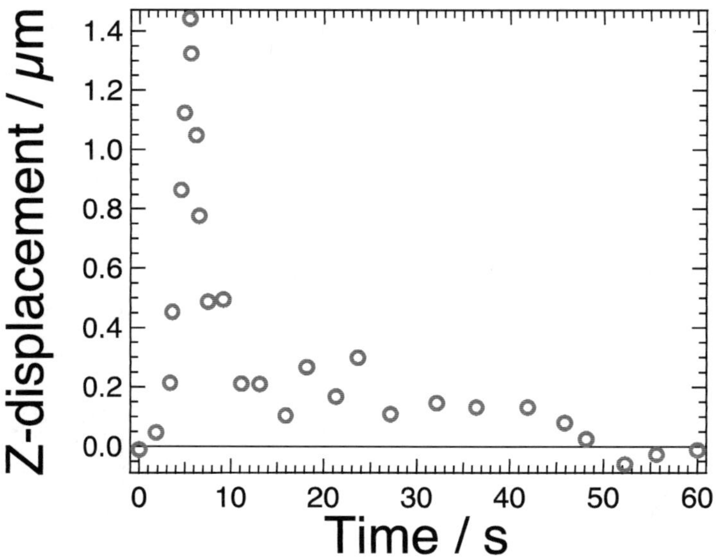

Fig. 3. Time-trajectory of the Z-position of a trapped polymer particle including NP caused by UV irradiation.

References

[1] A. Ashkin, *Phys. Rev. Lett.,* **24**, 156 (1970).
[2] A. Ashkin, J. M. Dziedzic, J. E. Bjorkholm, and S. Chu, *Opt. Lett.,* **11**, 288 (1986).
[3] K. Setoura, A. M. Memon, S. Ito, Y. Inagaki, K. Mutoh, J. Abe, and H. Miyasaka, *J. Phys. Chem. C*, in press.
[4] S. Ito, M. Mitsuishi, K. Setoura, M. Tamura, T. Iida, M. Morimoto, M. Irie, and H. Miyasaka, *J. Phys. Chem. Lett.,* **9**, 2659 (2018).

IOP Conf. Series: Journal of Physics: Conf. Series **1220** (2019) 012034 doi:10.1088/1742-6596/1220/1/012034

Formation of a single poly(*N*,*N*-diethylacrylamide) micro-droplet in water by coupling of photothermal effects and an optical force

M Matsumoto[1], T Asoh[2], T Shoji[1], T Nishiyama[3], Hideo Horibe[3], Yukiteru Katsumoto[4] and Yasuyuki Tsuboi[1]*

[1] Graduate School of Science, Osaka City University, 3-3-138 Sugimoto, Sumiyoshi-ku, Osaka, 558-8585, Japan
[2] The OCU Advanced Research Institute for Natural Science and Technology, Osaka City University, 3-3-138 Sugimoto, Sumiyoshi-ku, Osaka, 558-8585, Japan
[3] Graduate School of Engineering, Osaka City University, 3-3-138 Sugimoto, Sumiyoshi-ku, Osaka, 558-8585, Japan
[4] Faculty of Science, Fukuoka University, 8-19-1 Nanakuma, Jonan-ku, Fukuoka, 814-0180, Japan

*twoboys@sci.osaka-cu.ac.jp

Abstract. Poly(*N*-isopropylacrylamide) (PNIPAM) exhibits phase separation with lower critical solution temperature (LCST). In the 1990s, Masuhara and co-workers reported the first demonstration of optical trapping of PNIPAM forming a micrometer-sized polymer droplet. Since then, this technique has attracted much attention to create a molecular assembly in a micro-space. In the present study, we targeted poly(*N*,*N*-diethylacrylamide) (PDEA), which has an analogous chemical structure to PNIPAM. We demonstrated that optical tweezers formed the unique micro-morphologies of a phase separated PDEA droplet. Fluorescence microscopic images and Raman spectra of the PDEA droplet showed that a lot of smaller-sized water-rich micro-domains were inhomogeneously formed in the droplet. Such unique phase separation behavior was never observed in steady-state heating of an aqueous PDEA solution above its LCST. Our results indicate that a novel micro-structure can be formed by coupling of an optical gradient force and a local temperature elevation.

1. Introduction

Poly(*N*-isopropylacrylamide) (PNIPAM), which is a representative thermoresponsive polymer, is homogeneously dissolved in water at room temperature, taking hydrated random-coiled structures. The PNIPAM chains turn into globular structure accompanied by dehydration of the polymer chains upon heating above the lower critical solution temperature (LCST). Subsequently, these dehydrated globules aggregate with each other due to a hydrophobic interaction, resulting in micrometer-sized polymer assemblies or polymer-rich droplets. Such phase separation behavior of PNIPAM has been investigated over 50 years since the first demonstration by Heskins and Guillet [1,2], because PNIPAM has LCST near physiological conditions (c.a. 30 °C) thus rendering this water-soluble polymer suitable for biotechnological applications such as drug delivery system and artificial actuator. Moreover, PNIPAM has also attracted much attention to fundamental aspects such as understandings of protein folding and

DNA packing. However, little is known about their physical properties of the polymer-rich droplets such as their polymer concentration (C_{poly}), because it is hard to analyze individual droplet due to their rapid thermal fluctuation.

To address this issue, we focused on optical tweezers that is a powerful tool to trap micro-particles at the focal point of a tightly focused laser beam. In the 1990s, Masuhara and co-workers demonstrated that optical tweezers can trap PNIPAM chains dissolved in solution and form PNIPAM micro-droplet [3]. Inspired by these studies, we previously developed a method that combines optical tweezers with micro-Raman spectroscopy and determined C_{poly} of the optically trapped droplet of PNIPAM (Figure 1) [4]. Moreover, we recently found that the molecular weight of PNIPAM chains had a strong effect on C_{poly} [5].

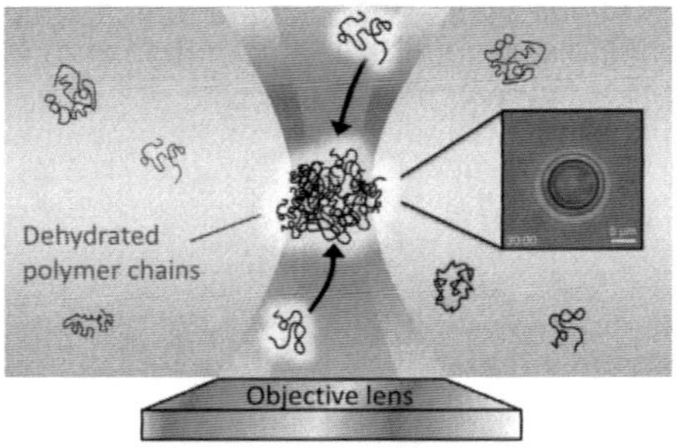

Figure 1. An illustration of optical tweezers combined with micro-Raman spectroscopy.

In the present study, we targeted poly(*N,N*-diethylacrylamide) (PDEA), which is another LCST-type thermoresponsive polymer. Although PDEA has an analogous chemical structure to PNIPAM, and both polymers have similar LCSTs in water, their phase separation behavior differs from each other [6]. We aimed at demonstrating optical trapping of PDEA and determining C_{poly} of the optically trapped PDEA-assembly.

2. Experimental Section

2.1. Sample preparation and characterization

PDEA was synthesized by reversible addition-fragmentation chain transfer polymerization and was characterized in the same manner as our previous report [6]. The weight-averaged molecular weight (M_w) and the polydispersity (M_w/M_n) of the synthesized PDEA were evaluated using size-exclusion chromatography (SEC) to be 66000 and 1.4, respectively. The LCST of the aqueous PDEA solution was determined to be 29 °C.

2.2. Optical setup

We have already explained the details of the optical tweezers combined with a confocal micro-Raman spectroscopy in our previous report [4], and briefly explained here. All the experiments of optical tweezers were performed using an inverted optical microscope equipped with an objective lens (×100, N.A. = 1.3), CMOS camera, and a thermostatic stage. For optical trapping, a continuous-wave (CW) near-infrared (NIR) laser beam (λ = 1064 nm) was focused into the sample solution. As a Raman excitation light source, a CW 532 nm laser beam was coaxially introduced with the NIR laser beam. The confocal system was arranged with a pinhole. Raman signals were detected with a cooled CCD.

IOP Conf. Series: Journal of Physics: Conf. Series **1220** (2019) 012034 doi:10.1088/1742-6596/1220/1/012034

3. Results and Discussion

3.1. Optical trapping of PDEA chains in water
Focusing the NIR laser beam into the dilute PDEA solution at room temperature, a single polymer-rich droplet was formed at the focal point (Figure 2a). The size of the particle became larger and reached a steady-state after several minutes. After switching off the NIR laser beam, the droplet immediately disappeared. The formation of the micro-droplet was due to a photothermal effect and an optical force [7].

To gain further insight into the micro-structures inside the droplet, we observed it under the fluorescence microscope. In this experiment, we used aqueous PDEA solution containing rhodamine B. Figure 2b shows a representative fluorescence microscopy image. There were a lot of water-rich domains fluctuating in the PDEA droplet. This inhomogeneous structure was never observed in steady-state heating of an aqueous PDEA solution above LCST. These results indicate that optical tweezers will open a new channel for the creation of ordered polymer structures based on the joint action of a photothermal effect and an optical force.

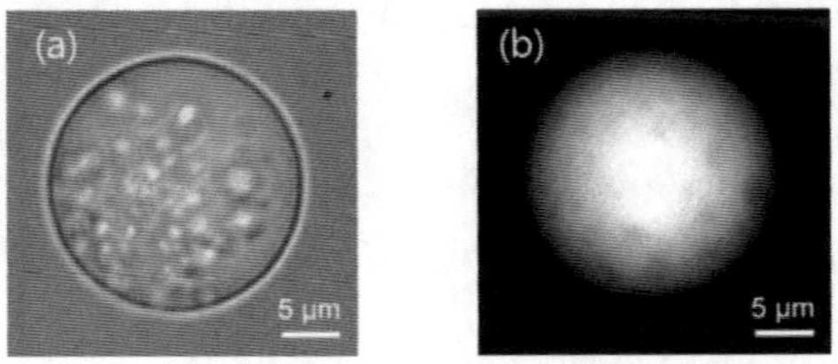

Figure 2. Bright-field (A) and fluorescence (B) optical micrographs of a single droplet formed by a focused near-infrared laser beam of 0.18 W into an aqueous PDEA solution (2.0 wt%). The droplets in the images are at steady state (10 min after the laser was turned on).

3.2. Determination of the polymer concentration of the PDEA droplet
To determine C_{poly}, we obtained the Raman spectra of the aqueous solution (without NIR laser irradiation) and the droplet of PDEA below the LCST (Figure 3). In the Raman spectra, two Raman bands appear: the C-H stretching vibrational mode of the polymers in 2800–3050 cm^{-1} range and the O-H stretching vibrational mode of the water in 3100–3700 cm^{-1} range. By the NIR laser irradiation, the intensity of the C-H band relatively increased, while that of the O-H band decreased because of the formation of the polymer droplet. Based on the comparison of integrated intensity ratio between the C-H band and the O-H band, we determined C_{poly} to be 40 wt%. The C_{poly} of PDEA was much lower than that of PNIPAM (60 wt%) in our previous report [4]. This means that in spite of the fact that PDEA (LCST = 29 °C) was slightly more hydrophobic than PNIPAM (LCST = 32 °C), the PDEA-droplet contained a substantial amount of water.

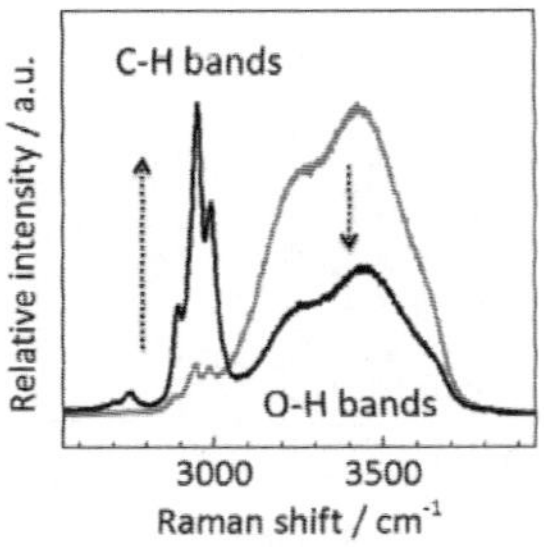

Figure 3. Raman spectra of an aqueous PDEA solution (2.0 wt%) below the LCST (Gray line) and of a PDEA droplet formed by a near-infrared laser beam of 0.16 W focused into the solution (Black line).

4. Conclusion

In this study, a single PDEA droplet was formed using optical tweezers. We found that the PDEA droplet had an inhomogeneous structure with a lot of water-rich domains fluctuating inside the droplet. Moreover, we determined C_{poly} of the PDEA droplet based on micro-Raman spectra. We concluded that a novel micro-structure was formed in the PDEA solution by coupling of an optical gradient force and a local temperature elevation.

Acknowledgments

This work was supported by JSPS KAKENHI Grant Numbers JP16K17922, JP17K04974, and JP16H06507 in Scientific Research on Innovative Areas "Nano-Material Manipulation and Structural Order Control with Optical Forces".

References

[1] Heskins M and Guillet J E 1968 Solution Properties of Poly(*N*-isopropylacrylamide) *J. Macromol. Sci. Part A - Chem.* **2** 1441–55

[2] Halperin A, Kröger M and Winnik F M 2015 Poly(*N* -isopropylacrylamide) Phase Diagrams: Fifty Years of Research *Angew. Chemie Int. Ed.* **54** 15342–67

[3] Ishikawa M, Misawa H, Kitamura N and Masuhara H 1993 Poly(*N*-isopropylacrylamide) Microparticle Formation in Water by Infrared Laser-Induced Photo-Thermal Phase Transition. *Chem. Lett.* 481–4

[4] Shoji T, Nohara R, Kitamura N and Tsuboi Y 2015 A method for an approximate determination of a polymer-rich-domain concentration in phase-separated poly(*N*-isopropylacrylamide) aqueous solution by means of confocal Raman microspectroscopy combined with optical tweezers *Anal. Chim. Acta* **854** 118–21

[5] Asoh T-A, Nishiyama T, Horibe H, Tsuboi Y, Fujiwara K, Shoji T and Matsumoto M 2018 Raman microspectroscopic study on an optically formed poly(*N*-isopropylacrylamide) rich microparticle: molecular weight dependence of a polymer concentration in the particle *Optical Manipulation Conference* ed T Omatsu (SPIE) p 63

[6] Matsumoto M, Wakabayashi R, Tada T, Asoh T-A, Shoji T, Kitamura N and Tsuboi Y 2016 Rapid Phase Separation in Aqueous Solution of Temperature-Sensitive Poly(*N,N*-diethylacrylamide) *Macromol. Chem. Phys.* **217** 2576–83

[7] Hofkens J, Hotta J, Sasaki K, Masuhara H and Iwai K 1997 Molecular assembling by the radiation pressure of a focused laser beam: Poly (*N*-isopropylacrylamide) in aqueous solution *Langmuir* **13** 414–9

Dielectric screening effect on exciton resonance energy in monolayer WS_2 on SiO_2/Si substrate

Yuto Kajino*, Masaya Arai, Kenichi Oto and Yasuhiro Yamada*

Department of Physics, Chiba University, Inage, Chiba 263-8522, Japan

* corresponding author e-mail: kajino@chiba-u.jp, yasuyamada@chiba-u.jp

Abstract. We studied the dielectric screening effects by a SiO_2/Si substrate on the excitonic properties of monolayer WS_2 by means of microscopic reflectance and photoluminescence spectroscopy. Through the observation of high-order exciton resonances from 1s to 5s at 10 K, we estimated the band gap and exciton binding energy. Our theoretical calculation using the screened Keldysh potential well reproduced the high-order Rydberg energies. In addition, we discussed the deviation of oscillator strength obtained from experimental results in comparison with theoretical calculation.

1. Introduction

Transition metal dichalcogenides (TMDC) MX_2 (M = Mo, W; X = S, Se) are atomically thin materials having honey-comb lattice structure similar to graphene. It has been revealed that the monolayer MX_2 is the direct-gap semiconductor with the bandgap energy of 1.8 ~ 2.5 eV [1]. Recently the MX_2 monolayers have attracted great attention as a stage of new physics research involving the complicated strong light-matter interaction [2], large exciton binding energy [3], and spin-valley physics [4,5]. In addition, using weak interlayer van der Waals interaction, the flexible design of artificial stacked structures has been actively studied [6].

Due to the strong two-dimensional (2D) nature, the physical properties of monolayer materials are predominantly influenced by the surrounding environment. In particular, the dielectric screening effect by surrounding materials has recently emerged as a critical issue in monolayer semiconductors, which will open a new way of controlling optical properties [7,8]. Because the electric force lines between photoexcited electrons and holes extend outside of the monolayer, the surrounding dielectric material screens the electron-hole Coulomb force, which can give rise to an enhancement of exciton binding energy. Since the larger portion of electric force lines extends outside of monolayer for the longer distance between electrons and holes (i.e., exciton Bohr radius), the dielectric screening is the more significantly observed for the higher-order exciton Rydberg states. This leads to the deviation of high-order exciton resonance energy from that predicted from simple 2D hydrogenic model. In this study, we experimentally observed 1s to 5s exciton resonances in monolayer WS_2 at low temperature and estimated the exciton binding energy. Furthermore, from the comparison with theoretical calculation using a screened Keldysh potential [9,10], we evaluated the impact of the dielectric screening effects on the excitonic properties in monolayer WS_2.

2. Fabrication and experimental setup

Monolayer WS_2 was fabricated from a bulk crystal on Si substrate with 280-nm SiO_2 surface layer (SiO_2/Si) by mechanical exfoliation [11]. The number of layers was identified by the optical microscope image shown in figure 1. Microscopic reflectance and photoluminescence (PL) spectroscopy were performed. The excitation light was focused to approximately 1 μm spot on the monolayer WS_2 by an objective lens, and the excitation power density was set to 1 W/cm^2. We used supercontinuum white-light laser (repetition rate of 80 MHz and pulse duration of ~ 1 ps) as an excitation light source. The sample was placed in an optical cryostat, and measurements were performed at room temperature and 10 K. The reflectance contrast and PL spectra under 2.32-eV excitation were measured. Note that the reflectance contrast is defined by $\Delta R/R = \left(R_{1LWS_2} - R_{substrate}\right)/R_{1LWS_2}$, where R_{1LWS_2} and $R_{substrate}$ are the reflectance of monolayer WS_2 and substrate, respectively. It has been shown that the reflectance contrast of ultrathin film on transparent substrate is approximately proportional to the absorption coefficient [12].

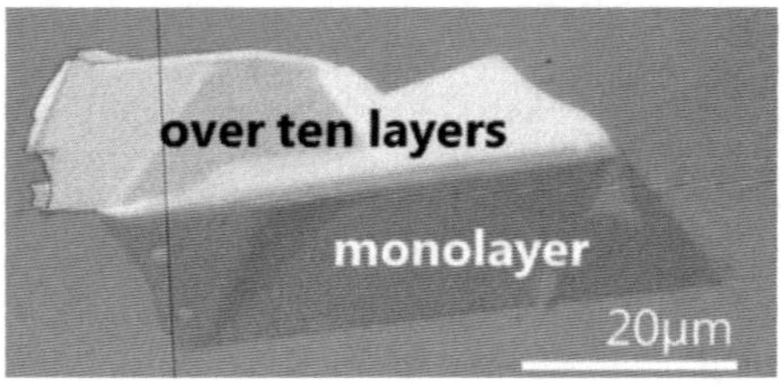

Figure 1. An optical microscope image of monolayer WS_2 on SiO_2/Si substrate.

3. Results and discussion

Figure 2 shows the reflectance contrast (blue curve) and PL (red curve) spectra at (a) room temperature and (b) 10 K. Reflectance contrast spectrum exhibits a few negative peaks with broad background caused by interference induced by the SiO_2 layer between the monolayer and Si. At room temperature, 1s A exciton peak (2.01 eV) was observed both in the reflectance contrast and PL spectra as shown in figure 2(a). B exciton also appears at around 2.4 eV in the reflectance spectrum, which corresponds to the high-energy band caused by the spin-orbit splitting of the valence bands [13]. At 10 K, two reflectance contrast peaks, corresponding to A (2.09 eV) and B (2.5 eV) excitons, were observed as shown in figure 2(b). The PL peak energy of the A exciton agrees well with the reflectance contrast peak energy. We confirmed that the Stokes shift is negligibly small. A small peak just below the A exciton peak has been identified as a negative trion [14]. This suggests that the monolayer WS_2 is slightly doped. The exciton and trion peak energies are consistent with the previous reports [14]. The low-energy broad PL band has been assigned to bound exciton [15].

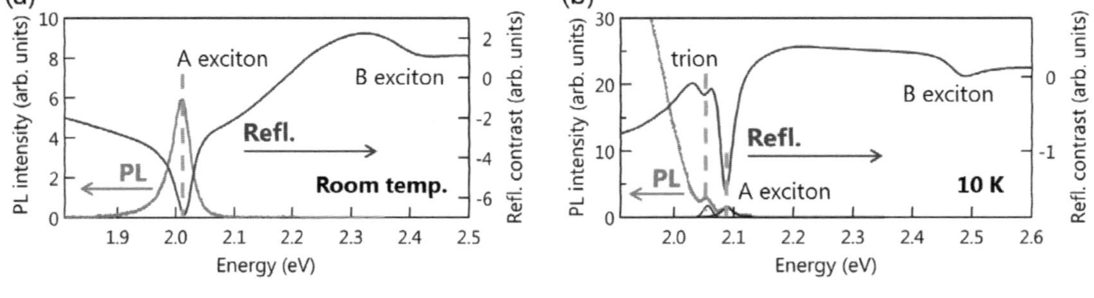

Figure 2. Reflectance contrast (blue curve) and PL (red curve) spectra of monolayer WS_2 at (a) room temperature and (b) 10K.

In order to focus on the higher-order exciton resonance, the 1st derivative of the reflectance contrast spectrum at 10 K is shown in figure 3(a). Above the 1s A exciton peak, we observed several peaks, which we assigned to *n*s Rydberg excitons (*n* = 1, 2, 3, …). The peak energy was plotted as a function of quantum number n in figure 3(b) (black circle). In a simple 2D hydrogenic model, the exciton binding energy for quantum number n, $E_{b,n}$, is given by the following equation;

$$E_{b,n} = \frac{\mu e^4}{(4\pi\varepsilon)^2 2\hbar^2 (n-1/2)^2}, \tag{1}$$

where μ and ε are the reduced mass and dielectric constant, respectively. However, because of the dielectric screening effect mentioned above, the quantum-number dependence of the exciton resonance energy deviates from the simple 2D hydrogenic model.

As a rough estimation, we first conduct curve fitting using equation (1) for n = 3, 4, and 5 (red curve in figure 3(b)) under an assumption of constant effective dielectric constant for large n, and estimated the band gap energy E_g = 2.36 (± 0.08) eV and exciton Rydberg energy $E_{b,1}$ = 275 (± 80) meV.

Next, we take the dielectric screening into consideration. We performed the numerical calculation of exciton binding energy using the screened Keldysh potential given by

$$V_{eh}(r) = -\frac{\pi e^2}{2r_0}\left[H_0\left(\frac{(1+\varepsilon_s)r}{2r_0}\right) - Y_0\left(\frac{(1+\varepsilon_s)r}{2r_0}\right)\right], \tag{2}$$

where H_0 and Y_0 are Struve and Bessel functions [9,10]. The screening length: r_0 is related to the 2D polarizability of the monolayer WS_2. ε_s is the dielectric constant of SiO_2. We adopted the SiO_2 dielectric constant ε_s = 1.98 at ~ 100 GHz. This frequency is close to that corresponds to the estimated exciton binding energy. Using equation (2), we calculated the exciton resonance energy as shown in figure 3(b). Here we determined r_0 to best reproduce the experimental results. We obtained E_g = 2.36 eV and $E_{b,1}$ = 270 meV when r_0 = 7 nm. The screening length estimated here is slightly larger than that obtained from density functional theory under random phase approximation (r_0 = 4 nm) [16]. The good accordance of experimental data and theoretical calculation supports that the Coulomb interaction between the photoexcited electrons and holes in monolayer WS_2 changes due to the exciton Bohr radius and surrounding dielectric constant.

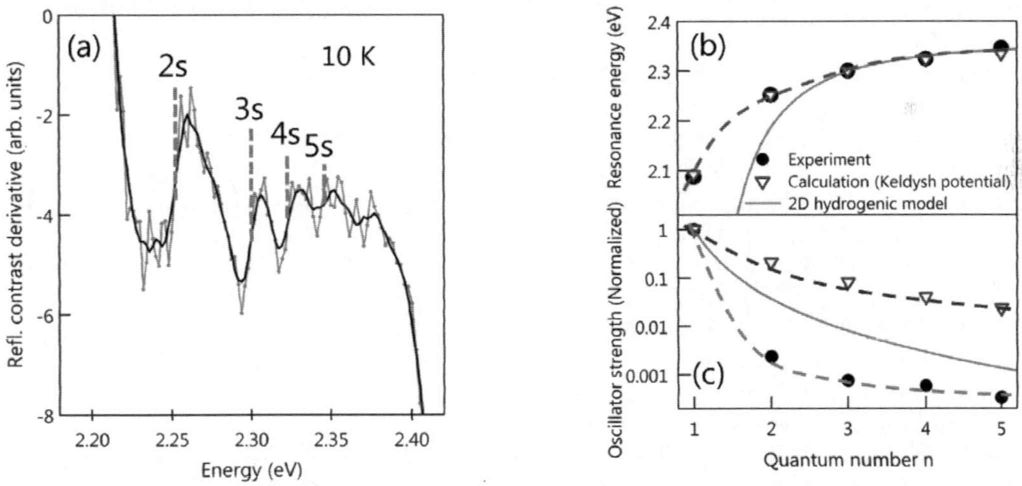

Figure 3. (a) The1st derivative of the reflectance contrast spectrum of the monolayer WS_2 (red curve). The smoothed curve is also displayed for reference (black curve). The high-order Rydberg states are labelled by their quantum number n. (b) Resonance energies and (c) oscillator strengths for the exciton states as a function of the quantum number n obtained by the experiment (black circle) and theoretical calculation (blue triangle). To estimate the band gap energy, fitting (red curve) is performed using the hydrogenic model for n = 3, 4 and 5 exciton states. Broken curves are for eye guiding.

IOP Conf. Series: Journal of Physics: Conf. Series **1220** (2019) 012035 doi:10.1088/1742-6596/1220/1/012035

Finally, we discuss the oscillator strengths obtained from the experimental results in comparison with that from theoretical calculation. As shown in figure 3(c), the oscillator strengths of the higher-order excitons obtained from theoretical calculation were larger than that from simple 2D hydrogenic model. However the experimental results are rather smaller than theoretical calculation. We consider intense excitation effect as the origin of this reduced oscillator strength. It has been reported in a Cu_2O bulk crystal that the oscillator strengths of high-order Rydberg excitons deviate from the simple hydrogenic model under intense photoexcitation due to nonlinear exciton-exciton interaction: the higher-order excitons have the larger interaction cross-section because the Bohr radius increases with quantum number [17]. In our experiments, we used pulsed laser as an excitation light source and focused to ~ 1 μm. Thus, the initial exciton density can be so high to give rise to the nonlinear exciton-exciton interaction. Further, exciton confinement in 2D materials might enhance exciton-exciton interaction compared with bulk crystals. Similarly, exciton-electron (hole) interaction may cause the observed reduction in the oscillator strength.

4. Conclusion

In this study, the exciton resonance energy of monolayer WS_2 was evaluated by reflectance and PL spectroscopy at room temperature and 10 K. By the observation of high-order Rydberg excitons using microscopic spectroscopy at 10 K, we estimated the band gap and exciton binding energy. We compared the experimental results with theoretical calculation and confirmed the dielectric screening effects dominate the excitonic properties in monolayer WS_2. The oscillator strength is not well reproduced by our calculation, probably due to nonlinear exciton-exciton interaction. Further studies are needed for the thorough understanding of the dielectric screening effects. For this purpose, use of high-quality monolayer TMDC and measurements under weak continuous-wave excitation are essential.

References

[1] Mak K F, Lee C, Hone J, Shan J and Heinz T F 2010 *Phys. Rev. Lett.* **105** 136805
[2] Splendiani A, Sun L, Zhang Y, Li T, Kim J, Chim C-Y, Galli G and Wang F 2010 *Nano Lett.* **10** 1271–5
[3] Ramasubramaniam A 2012 *Phys. Rev. B* **86** 115409
[4] Jones A M *et al.* 2013 *Nat. Nanotech.* **8** 634–8
[5] Mak K F, He K, Shan J and Heinz T F 2012 *Nat. Nanotech.* **7** 494–8
[6] Geim A K and Grigorieva I V 2013 *Nature* **499** 419–25
[7] Chernikov A, Berkelbach T C, Hill H M, Rigosi A, Li Y, Aslan O B, Reichman D R, Hybertsen M S and Heinz T F 2014 *Phys. Rev. Lett.* **113** 076802
[8] Raja A *et al.* 2017 *Nat. Commun.* **8** 15251
[9] Keldysh L V. 1979 *JETP Lett.* **29** 658–60
[10] Cudazzo P, Tokatly I V. and Rubio A 2011 *Phys. Rev. B* **84** 085406
[11] Novoselov K S, Geim A K, Morozov S V, Jiang D, Zhang Y, Dubonos S V, Grigorieva I V and Firsov A A 2004 *Science* **306** 666–9
[12] McIntyre J D E and Aspnes D E 1971 *Surface Science* **24** 417–34
[13] Zhao W, Ghorannevis Z, Chu L, Toh M, Kloc C, Tan P-H and Eda G 2013 *ACS Nano* **7** 791–7
[14] Plechinger G, Nagler P, Kraus J, Paradiso N, Strunk C, Schüller C and Korn T 2015 *Phys. Status Solidi RRL* **9** 457–61
[15] Shang J, Shen X, Cong C, Peimyoo N, Cao B, Eginligil M and Yu T 2015 *ACS Nano* **9** 647–55
[16] Berkelbach T C, Hybertsen M S and Reichman D R 2013 *Phys. Rev. B* **88** 045318
[17] Kazimierczuk T, Fröhlich D, Scheel S, Stolz H and Bayer M 2014 *Nature* **514** 343–7

Raman imaging studies on perforated MoS_2 films prepared by RF sputtering method

N Hasuike[1], S Yamauchi[1], K Nishio[1], S Kamoi[2], Woo Sik Yoo[3] and K Kisoda[4]

[1] Department of Electronics, Kyoto Institute of Technology, Matsugasaki, Sakyo-ku, Kyoto, 606-8585, Japan
[2] Kyoto Prefectural Technology Center, Chudoji Minami-cho, Shimogyo-ku, Kyoto, 600-8813, Japan
[3] WaferMasters, Inc., Dublin, CA 94568, United States of America
[4] Department of Physics, Wakayama University, Sakaedani, Wakayama, 640-8510, Japan

Corresponding author e-mail: hasuike@kit.ac.jp

Abstract. We investigated the spatial distribution of the crystallinity for sputtered-MoS_2 films by Raman imaging technique. MoS_2 lattice could not be structured in as-sputtered sample prepared at room temperature. The crystallinity could be improved by thermal annealing at 630 °C in vacuum condition. However, annealed MoS_2 films had two kinds of circular perforated areas on their top surface. In one area, MoS_2 film was evaporated and the substrate was exposed. In another area, MoS_2 films formed a hillock shape and showed tensile strain. In addition, the crystallinity was deteriorated in hillock formed area, which is due the generation of sulfur vacancies by thermal annealing.

1. Introduction

Scientific researches on layered materials such as graphene and transition metal dichalcogenides (TMDs) have led to a new understanding of fundamental properties of nanomaterials and suggested the possibilities for the realization of high performance electronic devices. Molybdenum disulfide (MoS_2), which is one of TMDs, has unique and tunable electronic properties, and some electronic devices using single or bilayer MoS_2 micro-flakes fabricated by mechanical exfoliation have been reported [1-4]. However, it is necessary to establish the fabricating method of a few layer MoS_2 films with large area for the integration of devices. A few layer MoS_2 films could be grown on a substrate by chemical vapor deposition so far, but they also had micro-flake shapes [5,6]. In this study, we fabricated MoS_2 films by RF sputtering method and investigated their crystallinity by Raman imaging technique.

2. Experimental

MoS_2 thin film samples were prepared on SiO_2/Si substrate at room temperature for 30 minutes by RF sputtering method. The film thickness was estimated at 70 nm by using stylus-type surface profile measurement. After sputtering processes, the samples were annealed at 300 °C and 620 °C for 2 hours in vacuum (~10^{-6} Torr). Raman imaging measurement was carried out by a confocal microscope using 532 nm laser. The incident laser light was expanded linearly by using the cylindrical lens, and it was focused on the sample surface. And then, scattered light was recorded using CCD detector with 400×1340 pixels.

IOP Conf. Series: Journal of Physics: Conf. Series **1220** (2019) 012036 doi:10.1088/1742-6596/1220/1/012036

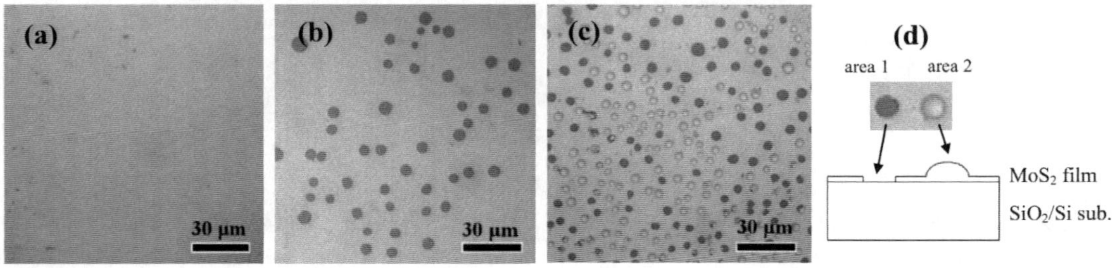

Figure 1. Optical microphotographs of as-sputtered MoS_2 (a) and annealed MoS_2 at 300 °C (b), 620 °C (c). The schematic diagram of the sample surface morphology after annealing is shown in (d).

3. Results and discussion

3.1. Surface morphology

Figure 1(a), (b) and (c) show the optical microphotographs of the samples. As-sputtered MoS_2 film had flat surface with no noticeable unique structure as shown in fig.1(a). On the other hand, annealed MoS_2 films showed two kinds of circular structures on their top surface with the size of micro-meter. They could be distinguished as film evaporated area (area 1) and hillock formed area (area 2). The schematic diagram of the surface morphology is shown in the fig.1(d). In addition, the density of these areas increased as increase in annealing temperature.

3.2. Scanning auger electron microscopy

We observed Auger electron spectroscopy (AES) to investigate the chemical composition of the sample annealed at 620 °C. Figure 2(a) shows scanning secondary electron microscope (SEM) image of the sample, and figure 2(b), (c), (d) and (e) show AES images obtained for molybdenum, sulfur, oxygen and silicon, respectively. AES images revealed that the MoS_2 thin films were evaporated and SiO_2/Si substrate was exposed in area 1, because signals come from molybdenum and sulfur were hardly detected but signals from oxygen and silicon were strongly observed in area 1. AES measurement can detect auger electrons escaping from the top surface, and the escaping depth is roughly estimated at several nm from the surface. Therefore, AES images indicate the chemical composition of the top surface. On the other hand, AES images taken from the flat area and area 2 showed the same features, which indicates that they have almost same chemical composition at their top surface.

3.3. Raman imaging measurement

Raman imaging measurement was carried out to investigate the spatial distribution of the crystallinity in MoS_2 thin films. Hexagonal MoS_2 belongs to the space group D_{6h} *($P6_3/mmc$)*, and a group theoretical analysis gives 3 Raman-active modes of E_{1g} (287 cm^{-1}), E^1_{2g} (383 cm^{-1}) and A_{1g} (409 cm^{-1}) at the Γ-point in the Brillouin zone [7]. Figure 3(b), (d) and (f) show CCD detector images of Raman signals for the

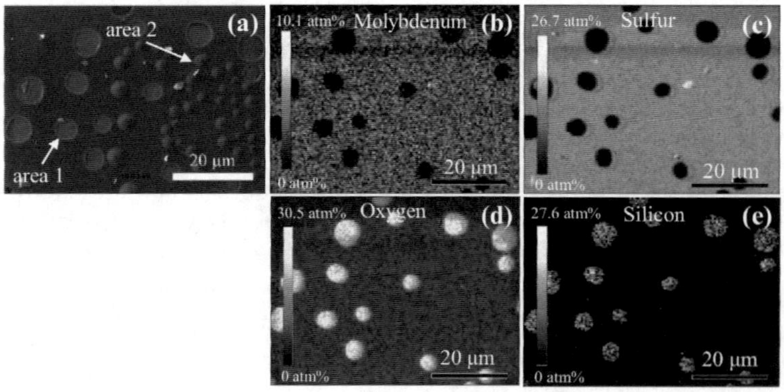

Figure 2. SEM image of the annealed sample at 620 °C (a), and AES images of molybdenum (b), sulfur (c), oxygen (d) and silicon (e), respectively.

samples. In the images, the vertical and bottom horizontal axes correspond to those of the CCD detector pixels, respectively. Raman shift and the signal intensity are denoted on the upper horizontal axis and the contrast using grayscale. The incident laser expanded linearly focused on the surface along the line shown in the optical micrographs of figure 3 (a), (c) and (e), respectively. As shown in figure 3 (b), the CCD image of as-sputtered MoS_2 showed only the Raman signal derived from SiO_2/Si substrate at 520 cm^{-1}, and the signal was uniform in all probing area. This indicates that MoS_2 lattice could not be structured in as-sputtered sample. On a side note, the broad and weak signal was observed slightly at the range from 900 to 1000 cm^{-1}, which is not Raman signal from the sample but the background signal from the equipment. In addition, in the sample annealed at 300 °C, Raman signals derived from MoS_2 was not observed yet as shown in figure 3 (d). However, Si signals observed at 300, 520 and 980 cm^{-1} taken from area 1 showed higher signal intensity compared with the flat area [8], because SiO_2/Si substrate was exposed in area 1 as revealed by AES measurement. These features are well reflected in the CCD image of the sample annealed at 620 °C as shown in figure 3 (f), but two additional Raman signals

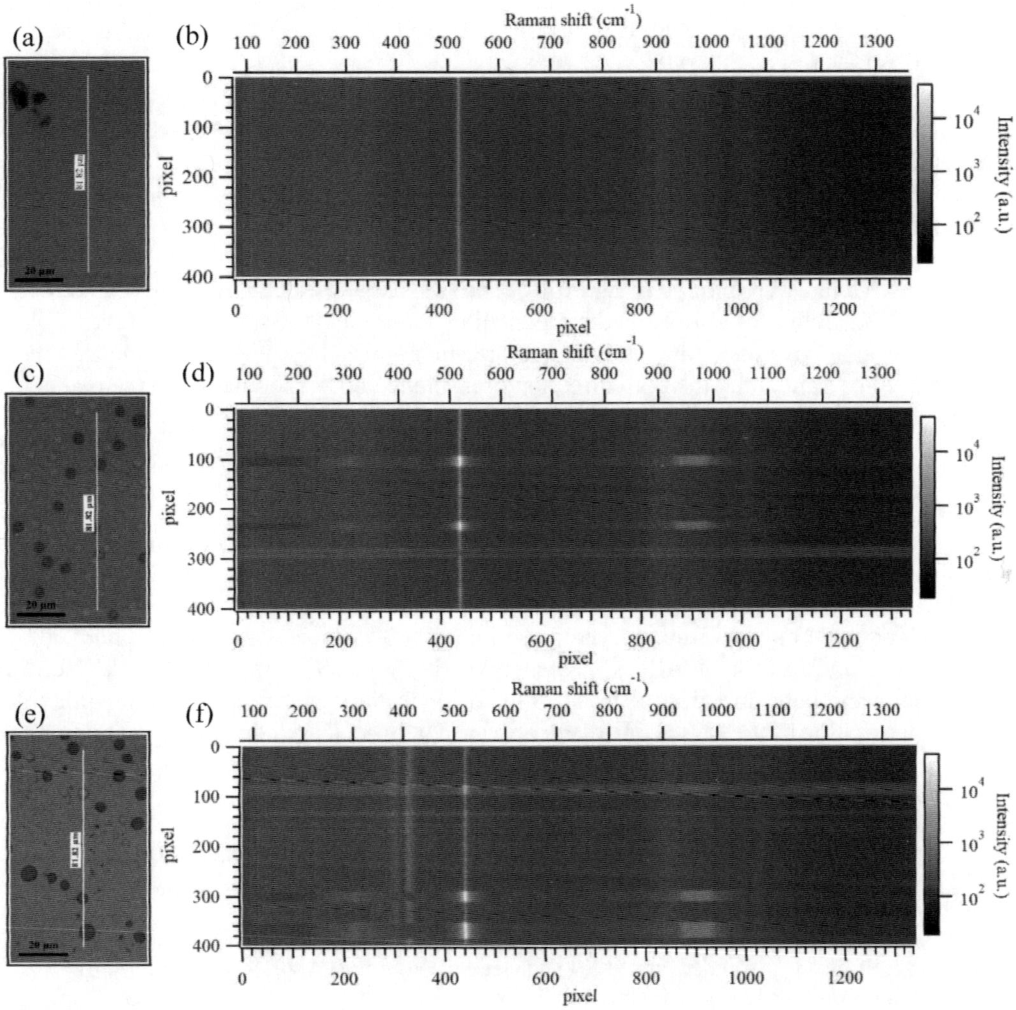

Figure 3. Optical microphotographs and CCD images of Raman signals for as-sputtered sample (a)(b), and the samples annealed at 300 °C (b)(c), 620 °C (c)(d), respectively. The incident laser was expanded linearly and focused on the surface along the line shown in the optical micrographs.

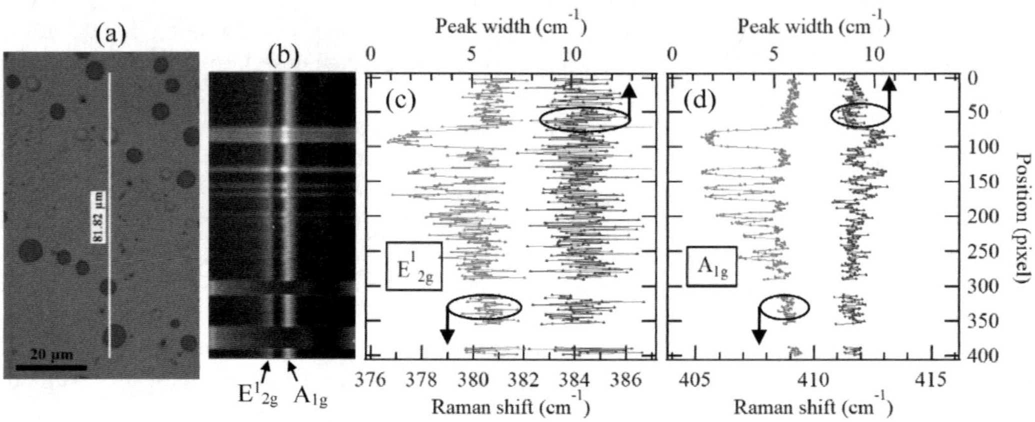

Figure 4. Optical microphotograph (a) and the close-up of CCD image around E_{2g} and A_{1g} modes (b) for the sample annealed at 620 °C. The peak widths and the frequencies of E_{2g} and A_{1g} were plotted in (c) and (d), respectively.

were observed at 380 cm^{-1} and 410 cm^{-1} at all probing places except area 1. These signals correspond to E_{2g} and A_{1g} modes of MoS_2 phonon, indicating that MoS_2 lattice was formed by thermal annealing at 620 °C. Here, we consider about the area 2. Although Si signals taken from area 2 showed higher signal intensity than that of the flat area, Si substrate was not exposed as revealed by AES measurement. These results suggested that MoS_2 film formed a hillock shape with hollowed inside in area 2. In addition, Raman signals shifted to lower frequency side and both peaks were broadened slightly compared with the flat area. These values were plotted in figure 4(c) and (d) so that the detailed values can be compared. The plots showed that both peak shifts and the broadening occurred at the same time in area 2. And then, they gradually became advance from the edge toward the center. The peak shift indicates that MoS_2 lattice had the tensile strain along the both direction of in-plane and perpendicular to the plane in area 2. And, the broadening indicates the deterioration of the crystallinity, which was caused by the generation of sulfur vacancies in MoS_2 lattice by thermal annealing due to its high vapour pressure.

4. Conclusion

In this study, we investigated the spatial distribution of the crystallinity for as-sputtered and annealed MoS_2 thin films by using Raman imaging technique. Although the lattice ordering was not structured in as-sputtered MoS_2 sample, the crystallinity was improved by thermal annealing. However, annealed MoS_2 films showed two kinds of circular structures on their top surface, which were generated by evaporation of the films and forming a hillock shape, respectively. MoS_2 film with hillock shape showed tensile strain along the both direction of in two-dimensional plane and perpendicular to the plane. In addition, the deterioration of the crystallinity was observed in hillock area due to the generation of sulfur vacancies.

References

[1] Li H, Zhang Q, Ray Yap C C, Tay B K, Edwin T H T, Olivier A and Baillargeat D 2012 *Adv. Funct. Mater.* **22** 1385
[2] Radisavljevic B, Whitwick M B and Kis A 2011 *ACS NANO* **5** 9934
[3] Bae H, Kim C K and Choi Y K 2017 *AIP ADVANCES* **7** 075304
[4] Buscema M, Steele G A, van der Zant H S J and Catellanos-Gomez A 2014 *Nano Res.* **7** 561
[5] Xu E Z, Liu H M, Park K, Li Z, Losovyj Y, Starr M, Werbianskyj M, Fertig H A and Zhang S X 2017 *Nanoscale* **9** 3576
[6] Özden A, Sar H, Yeltik A, Madenoğlu B, Sevik C, Ay F and Perkgöz K 2016 *Phys. Stat. Sol. RRL* **10** 792
[7] Wieting T J and Verble J L 1971 *Phys. Rev. B* **3** 4286
[8] Temple P A and Hathaway C E 1973 *Phys. Rev. B* **7** 3685

Potential effect in Balmer light emission near surface irradiated with highly charged ions

N Nishida[1], Y Hori[1], A Yamauchi[1], M Sakurai[1], D Kato[2] and H A Sakaue[2]

[1] Department of Physics, Kobe University, 1-1 Rokkodai, Nada-ku, Kobe, 657-8501, Japan
[2] National Institute for Fusion Science, 322-6 Oroshi-cho, Toki, 509-5292, Japan

E-mail: 157s119s@stu.kobe-u.ac.jp

Abstract. The interaction between highly charged ions (HCIs) and the solid surface causes various phenomena due to high potential energy of HCIs. Luminescence is also included in them. In order to investigate the origin of the light emission from the surface irradiated with HCIs, we conducted spectroscopic measurements and then we investigated characteristics of Balmer light emission because potential effect of HCI was remarkable in Balmer light emitted from hydrogen atoms desorbed from the surface. In the present study, the Balmer light intensity was revealed to increase rapidly with the charge state of HCIs.

1. Introduction

The highly charged ion (HCI) possesses large potential energy increasing with its charge state. When a HCI is incident on the solid surface, emission of photons in the range of visible to X-ray, hundreds of secondary electrons, sputtering of secondary ions and modification of surface structure in nanometer scale are caused. The effect of potential energy concentrates on only a few atomic layers of the topmost surface [1,2].

In the previous experiments, measurement of luminescence of Er_2O_3 irradiated with HCIs had been performed. The emission intensity from the region of 607–680 nm strongly depended on charge state of HCIs. In addition, the emitted light was not dependent on kinetic energy. The luminescence caused by the potential effect of HCIs. The luminescence from Er_2O_3 irradiated with HCIs nonlinearly increased depending on the charge state of HCIs. The light emission of particles sputtered from the surface by HCIs is dominant [3,4]. In the present study, spectroscopic measurements and secondary ion mass spectrometry (SIMS) were performed in order to specify atoms participating in the luminescence. In addition, the characteristics of the Balmer light emission was investigated since the light emission from the adsorbed hydrogen atoms is remarkable at spectroscopic measurements. We investigated the dependence of Balmer light emission on charge state and temporal change of the emission intensity.

2. Experimental

HCIs were produced by using an electron beam ion source (EBIS) installed at Kobe University, Japan [5-7]. The Kobe EBIS comprises an electron gun, drift tubes, an electron collector and a superconducting magnet. HCIs are created by successive ionization processes in the collision of electrons focused by the strong magnetic field (3 T) with ions. The current density of electrons at drift tube region is in the order of 1000 A/cm^2. The potential of drift tubes (the acceleration voltage to

IOP Conf. Series: Journal of Physics: Conf. Series **1220** (2019) 012037 doi:10.1088/1742-6596/1220/1/012037

HCIs) is 3 kV. The ultimate pressure of the Kobe EBIS is in the range of 10^{-8} Pa. A sector magnet is used to select charge state of HCIs extracted from the ion source. HCIs with specific charge state enter the irradiation chamber maintained at ultrahigh vacuum of about 10^{-7} Pa. Primary ion current is monitored by a Faraday cup and the ion beam shape and position are monitored by a micro-channel plate (MCP). Incident ion current to the sample during irradiation is also monitored.

In the present measurement, the light emission caused by the interaction between HCIs and samples was measured with a liquid nitrogen cooled CCD. In order to keep kinetic energies of HCIs incident on the surface constant (16 keV) for each measurement, the retarding voltage was applied to the sample. The diameter of HCI beam was about 3 mm for each experiment. Spectroscopic measurements were conducted using a polychromator. Wavelength distribution for the range of 420–670 nm and spatial distribution of horizontal direction were measured from the two-dimensional CCD image. Er_2O_3, Si, highly oriented pyrolytic graphite (HOPG) and Cu were used as samples. We used commercially available Si wafer for Si sample. Because no cleaning or passivation procedure was done, the native oxide film was formed on the Si surface. These samples were stored in the atmosphere before each measurement, and were irradiated with HCIs (Ar^{q+} and Ne^{q+}). For HOPG, the surface was cleaned before introduction to the irradiation chamber by pealing-off procedure. In spectroscopic measurements, the exposure time was 2 hours. SIMS was performed using a quadruple mass analyser (QMS). In the Balmer light measurement, the distribution of light emission in two-dimensional space was obtained using a filter instead of a polychromator. The emission intensity in the 656 ± 5 nm region (H_α) was evaluated and the exposure time was 10 minutes.

3. Results and discussions

Figure 1 shows the spectrum of Si irradiated with Ne^{6+}. The emission intensity is calculated as a summation of count at the pixels corresponding to each wavelength. Since many noises appeared in the spectrum due to cosmic rays during measurement, data points which are remarkably stronger than that of adjacent pixels are replaced with average values of data on both sides in order to remove noises. In figure 1, emitted light lines are observed around 486 nm, 590 nm and 656 nm. These lines also appeared in the spectra of Er_2O_3 irradiated with Ar^{11+} [4]. From these results, it is suggested that these peaks are not derived from irradiated ions or samples since three emitted light lines appear in common.

The analysis of ions produced by irradiation with HCIs was performed using a QMS to identify the origin of the light emission. Figure 2 shows a secondary ion mass spectrum of Si irradiated with Ar^{11+}. A very strong signal of H with a mass number of 1 and a signal of Na with a mass number of 23 are observed in the spectrum in addition to Si as a sample and Ar as a primary ion. It should be noted that the intensity of mass number of 1 (H^+) is overwhelmingly more than that of 2 (H_2^+) in the spectrum. This result is completely different from the quadruple mass spectrum of the residual gas [8]. These results indicate that hydrogen atoms contained in molecules on the surface are ionized and desorbed as an interaction of molecules with HCIs. From these results, it seems that emitted light lines observed by

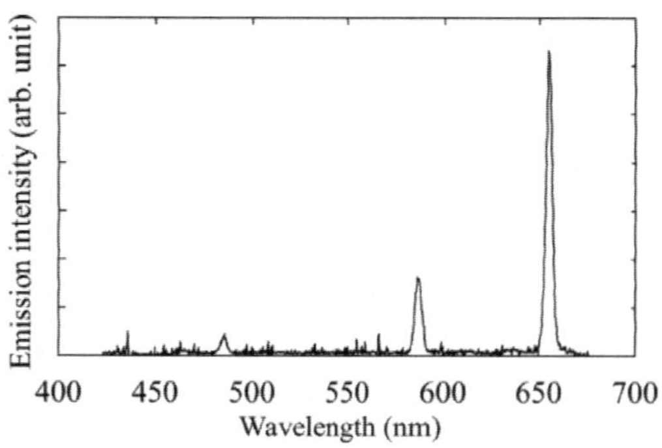

Figure 1. The emission spectrum of Si irradiated with Ne^{6+}.

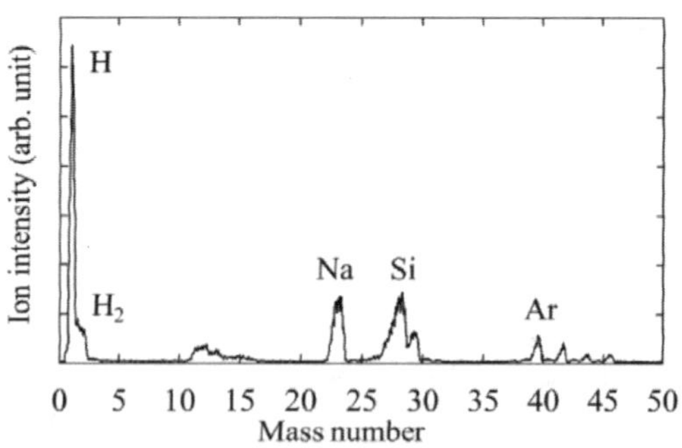

Figure 2. The secondary ion mass spectrum of Si irradiated with Ar^{11+}.

IOP Conf. Series: Journal of Physics: Conf. Series **1220** (2019) 012037 doi:10.1088/1742-6596/1220/1/012037

spectroscopic measurements come from H and Na, i.e. Balmer series (H_α and H_β) and the resonance line of Na (D line). It is considered that H is originated from either water or hydrogen adsorbed or occluded on the sample surface. The peak at 590 nm did not appear in the emission spectrum of HOPG, therefore the origin of Na is contamination.

To investigate the nature of Balmer light produced by the injection of HCIs, the emission intensities at the 656 ± 5 nm region (H_α) were evaluated for various experimental conditions. The emission intensity was obtained from the summation of data points at light emitting part subtracting background in the CCD image, and normalized by incident ion current divided by the charge state of HCIs. Finally, the intensity was converted to the emission intensity per single ion. Figure 3 shows the dependence of emission intensities per single HCI on the charge state of Ar^{q+} (q = 6, 8, 11, 14, 16). The intensities increase with charge state. It is considered that the transfer of electrons caused by large potential energy leads to the excitation of more hydrogen atoms. Figure 4 and figure 5 present temporal changes of emission intensities from Si and HOPG surfaces per the injection of single HCI. Here, the time is converted to the number of incident ions since the number of incident ions per unit time is different for each measurement. The intensities tend to decrease by continuous irradiation with HCIs. It is suggested that the number of hydrogen atoms on the surface decreased by the irradiation.

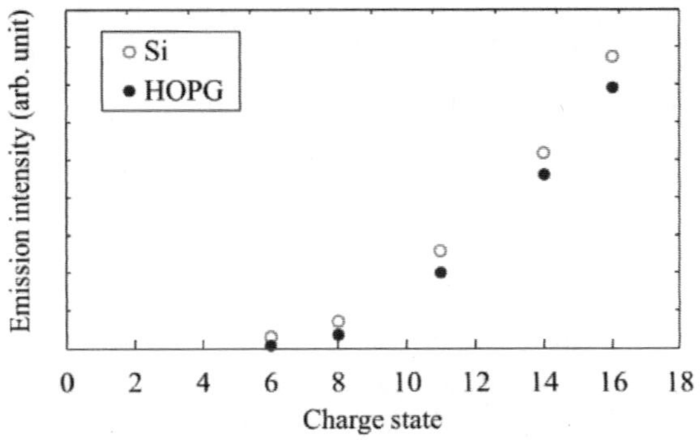

Figure 3. Dependence of emission intensities per single ion on charge state of Ar^{q+} (q = 6, 8, 11, 14, 16).

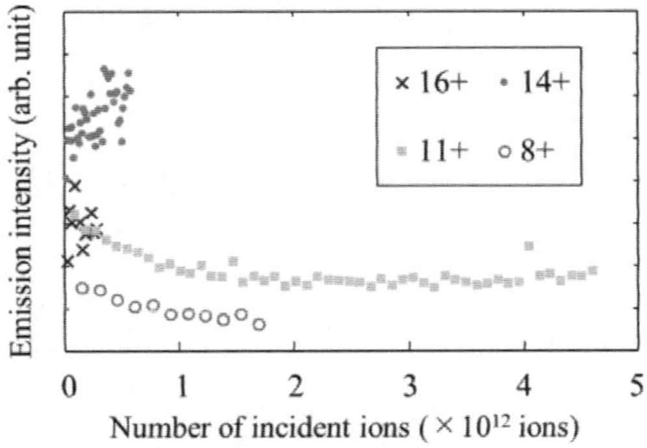

Figure 4. Temporal changes of emission intensities of Si irradiated with Ar^{q+}.

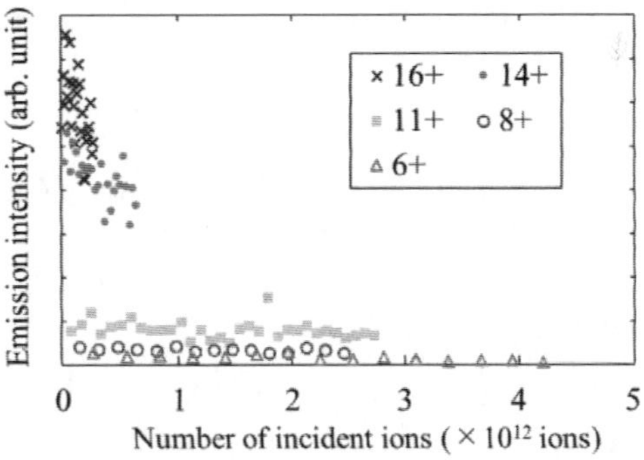

Figure 5. Temporal changes of emission intensities of HOPG irradiated with Ar^{q+}.

It is suggested that water or hydrogen are desorbed and decreased from the surface by HCI injection, since the SIMS spectrum indicates that atomic hydrogen ions are produced by irradiation with HCIs. Comparing temporal changes of Si and HOPG, decrease of emission from HOPG is slower than that from Si. This is probably because the Si sample is stored in the atmosphere, while the HOPG surface was prepared a new surface for every measurement, so that exposure time of HOPG to air is shorter than that of Si. Therefore, it is considered that the origin of emission is both hydrogen attached to the sample in the irradiation chamber and water adsorbed in the atmosphere. However, the number of molecules present on the surface is unknown in these measurements due to the difference of exposure time to air. Figure 6 shows temporal changes of the emission from HOPG irradiated with Ar^{11+} for various exposure time to air after pealing-off procedure. Because the emission intensity of the sample exposed to air for 7 days is stronger than that of the sample exposed for 1 day, it is considered that water in air was adsorbed to the HOPG surface.

Water molecules seem to mostly contribute to the Balmer light emission, however, there remains a possibility of contribution from adsorbed hydrogen molecules which come from residual gas. It is essential to control the condition of the surface in order to identify the origin of Balmer light. As future prospects, in order to reduce the influence of water, the sample surface will be cleaned using an ion gun in vacuum before measurement.

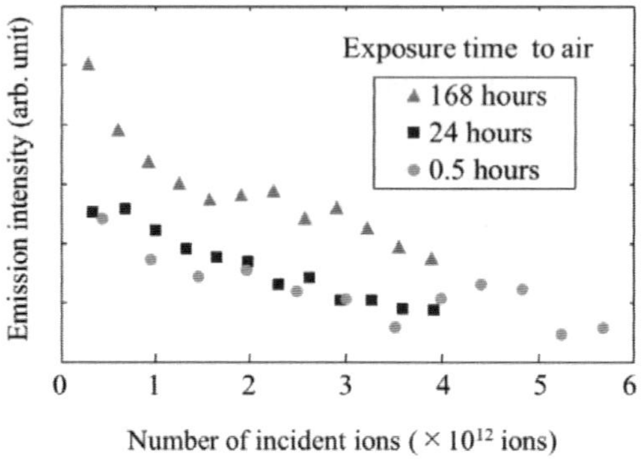

Figure 6. Temporal changes of emission intensities of HOPG irradiated with Ar^{11+}. HOPG surfaces were exposed to air for 0.5 hours, 24 hours (1 day) and 168 hours (7 days).

4. Conclusions

The emission light from the surface irradiated with HCIs was observed. The origin of luminescence was identified as H as the main light emission species from the spectroscopic measurement and SIMS. The emission intensity of Balmer light increases with the charge state of HCIs and decreases as the irradiation continues.

Acknowledgement

This work is performed with the support and under the auspices of the NIFS Collaboration Research Program (NIFS18KOCF001).

References

[1] Tona M, Takahashi S, Watanabe H, Yoshiyasu N, Nakamura N, Yamada C, Sakurai M, Terui T, Mashiko A and Ohtani S 2007 *Surf. Sci.* **601** 723
[2] Aumayr F and Winter H P 2005 *Nucl. Instrum. Methods. Phys. Res.* B **233** 111
[3] Sakurai M, Miyamoto T, Sasaki K, Koto D and Sakaue H A 2015 *J. Vac. Soc. Jpn.* **50** 147
[4] Sakurai M, Sasaki K, Miyamoto T, Kato D and Sakaue H A 2016 *e-J. Surf. Sci. Nanotech.* **14** 1
[5] Sakurai M, Nakajama F, Fukumoto T, Nakamura N, Ohtani S and Mashiko S 2005 *Shinku* **48** 317
[6] Sakurai M, Sakaue H A, Watanabe H, Nakamura N, Ohtani S, Kawase Y, Mitsumori K, Terui T and Mashiko S 2007 *Shinku* **50** 390
[7] Sakurai M, Tona M, Watanabe H, Nakamura N, Ohtani S, Terui T, Mashiko S and Sakaeu H A 2009 *J. Phys. Conf. Ser.* **163** 012115
[8] Inayoshi S S 2007 *Shinku* **50** 228

Intermediate band formation by InGaAs/GaAs multistacked quantum dots without strain compensation

Keishiro Goshima[1], Keisuke Inukai[1], Norio Tsuda[1], and Takeyoshi Sugaya[2]

[1] Aichi Institute of Technology, 1247 Yachigusa Yakusa, Toyota, Aichi, Japan
[2] Advanced Industrial Science and Technology, 1-1-1 Umezono Tsukuba, Ibaraki, Japan

Abstract. Quantum dots (QDs) are widely used for enhancing the performance of optical devices. Intermediate-band solar cells that use multistacked QDs can surpass the conversion efficiency of conventional silicon cells. We performed the optical characterization of $In_{0.4}Ga_{0.6}As$/GaAs multistacked QDs without strain compensation. We used the theoretical and experimental techniques prescribed for intermediate-band solar cells, i.e., photoluminescence (PL) spectroscopy and two-color excitations spectroscopy. The interdot spacings were not uniform and were found to be 15 nm and 7 nm. The results verify the formation of intermediate bands by the multistacked QDs. Using the theoretical studies and experimental results, we performed an in-depth study on the mechanism underlying the formation of intermediate bands by the multistacked QDs and the effect of different interdot spacings.

1. Introduction

Recently, self-assembled quantum dots (QDs) have been shown to be possible candidates for solar cells, semiconductor lasers, and quantum cryptography mediums. To realize these optoelectronic devices, it is necessary to increase the optical gain in QDs. Multiple-stacking in QDs is one possible solution for this[1]. Moreover, multistacked QDs not only exhibit a higher optical gain but can also realize an intermediate band by controlling the barrier thickness between the QDs. For example, intermediate solar cells achieved by creating the intermediate band between the valence band and the conduction band using quantum mechanical coupling. Photo carriers can arise with the cascade process from the conduction band to the valence band via the intermediate band, and they are used to lower the energy photon below the band gap energy. Thus, the ultimate conversion efficiency at sun intensity can increase above 60%[2,3]. The intermediate band structure is realized by a superlattice structure grown by multistacked QDs. The intermediate band in the QD array is caused by the electronic coupling of states in the neighboring QDs. When the interdot spacing between the QDs of growth detection is less than the de Broglie length, the QD states eventually delocalize over the QDs arrays, such as the coupling effect, and new energy bands with the existing carriers are formed [4]. To realize intermediate solar cells, it is necessary to obtain multistacked QDs structures with small interdot spacings. Meanwhile, growth techniques and self-assembled QDs with multistacked growth have been studied. The crystal quality of QDs degrades owing to issues such as crystal defects and transitions in crystals resulting from the internal strain. Therefore, strain compensation techniques have been introduced to fabricate multistacked QDs with good crystal quality [1,5].

However, the disadvantage of this technique is that it cannot fabricate small interdot spacing structures. Multistacked QDs with an intermediate band require a small interdot spacing because they use the coupling effect owing to the quantum mechanical interactions.

We studied the fabrication of InGaAs/GaAs QDs because InGaAs has a slight lattice mismatch with the GaAs substrate. We successfully fabricated small interdot multistacked InGaAs QD structures without using a strain compensation technique by the intermittent deposition of InGaAs layers using an As_2 source [6,7]. In previous studies, we observed the electronic band structure and properties of multistacked QDs with different interdot spacings [8,9].In this study, we investigated the intermediate-band (sub-band) formation and the cascade excitation process in multistacked QDs with various barrier thicknesses by theoretical analysis and optical measurements.

2. Sample and Experimental Procedure

The structure was fabricated by the intermittent deposition of $In_{0.4}Ga_{0.6}As$ layers, using an As_2 source. Consequently, the crystal quality was not degraded. We grew ten-stack $In_{0.4}Ga_{0.6}As$ QD layers on a Si-doped GaAs(001) substrate using an As_2 source by molecular beam epitaxy. Figure 1 shows the schematic of the multistack QD structure. We grew ten-stack $In_{0.4}Ga_{0.6}As$ QD structures with a GaAs barrier layer of the thicknesses 10 nm and 20 nm, corresponding to the interdot spacing of 7 nm and 15 nm, respectively. The height, width, and shape of a dot were 5 nm, 20 nm, and pyramidal, respectively. Further details of our growth process can be found in the literature [6,7]. Photoluminescence (PL) was investigated using a spectroscopy system. Laser light was focused on the QD samples through objective lens. Pulse laser with the pulse width of 150 ps and the reputation rate of 50 MHz was used as an excitation source in the PL measurements. The QD sample was mounted on a helium cryostat, and the temperature was tuned to 10 K. The PL signals were detected by a monochrometer, whose spectrum resolution was 0.1 nm, and a silicon-avalanche photodiode (Si-APD) detector was used. We used two color lasers: color 1 laser was a pulse laser with a wavelength of 780 nm, and color 2 laser had CW operation with a wavelength of 1300 nm. The energy of color 2 laser tuned to the ground state of the QDs.

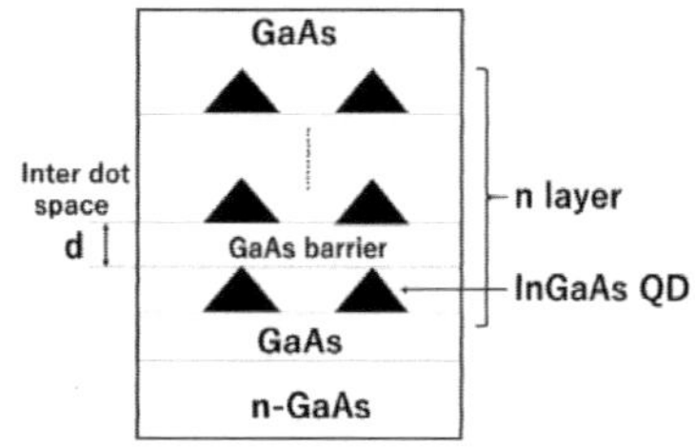

Figure 1: Schematic of multistacked QD structure.

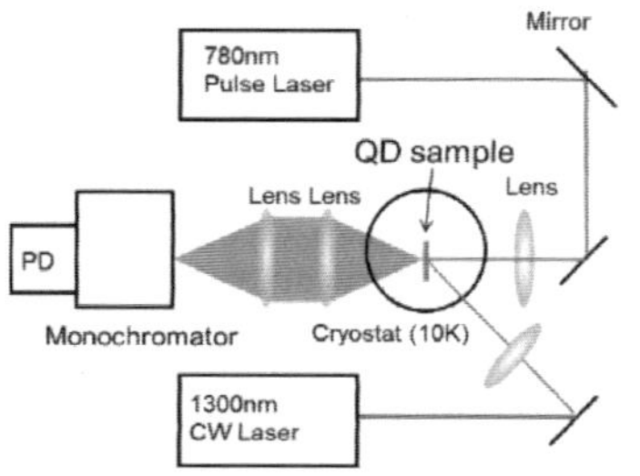

Figure 2: Photoluminescence and two-color excitation setup.

3. Results and Discussion

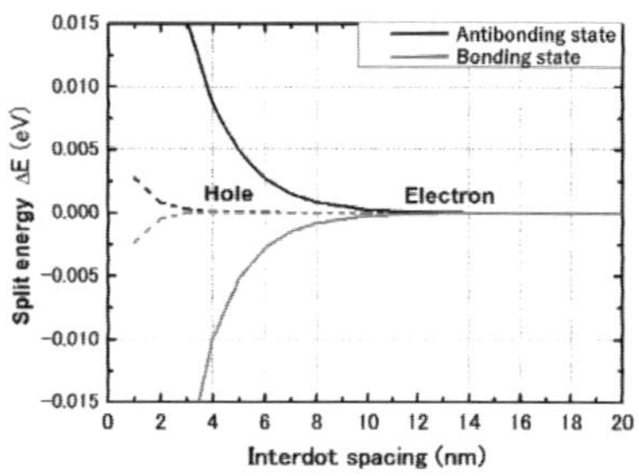

Figure 3: This Split energy as a function of interdot spacing

3.1 Numerical Circulation

Figure 3 shows the numerical calculations of the electric structures according to barrier thickness. To compute the electronic states of the QD embedded in the GaAs surrounding, we solved the Schrödinger equation in the effective mass approximation using three-dimensional finite element modeling using COMSOL Multiphysics [10]. We calculated that the bonding and anti-bonding states for the electron were formed by a barrier with a thickness less than 10 nm. The lesser the barrier thickness, the more the energy deference between the bonding and the anti-bonding states were spread. In another case, the bonding and anti-bonding states for the hole were formed within 4 nm. These different results were caused by the effective mass between the electron and hole. We expected perfect intermediate-band (sub-band) states to be formed by a barrier thickness of within 4 nm.

IOP Conf. Series: Journal of Physics: Conf. Series **1220** (2019) 012038 doi:10.1088/1742-6596/1220/1/012038

3.2 Photoluminescence

We observed the PL at 10 K. Figure 4 and 5 show the PL spectra for the interdot spacings of d = 15 nm and d = 7 nm, respectively.
For d = 15 nm, the primary peak and FWHM of the PL spectrum were 1.25 eV and 53 meV, respectively. From the calculation results, the wavefunction was not coupled, and the intermediate band was not created because of the large interdot spacing. Therefore, the spread of the PL spectra was caused by the size deviation of the individual QDs.

Meanwhile, for d = 7 nm, three peak components are shown in the PL spectrum. The primary peak and two subpeaks were 1.29 eV, 1.317 eV, and 1.253 eV, in which the peaks were identified by fitting a Gaussian curve. From the calculation results, the electron wavefunction was coupled, and the intermediate band was created because of the small interdot spacing. We now discuss the primary peak of 1.29 eV, and the subpeak of 1.317 eV. It was assumed that the primary peak and subpeak energies were caused by the transition regime from the bonding and anti-bonding states of the electron, respectively. The observed energy difference between the primary and subpeaks was 15 meV. The FWHM of d = 7 is 65 meV, which is larger than d = 15 nm. The difference of FWHM between two samples was 12 meV, which is expected the energy difference between the bonding and anti-bonding states with d = 7 nm. This result may in fact show a coupling effect about d = 10 nm or less. The peak of 1.253 eV is caused by the enlargement of QDs in the stacking direction because of the lower energy side of the primary peak [11].

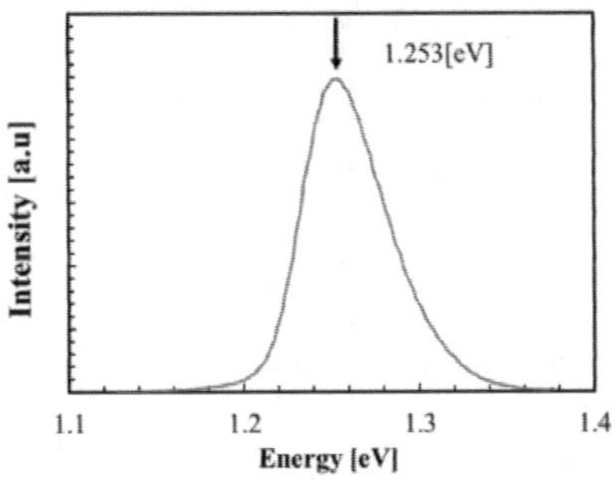

Figure 4: PL spectrum of the d = 15

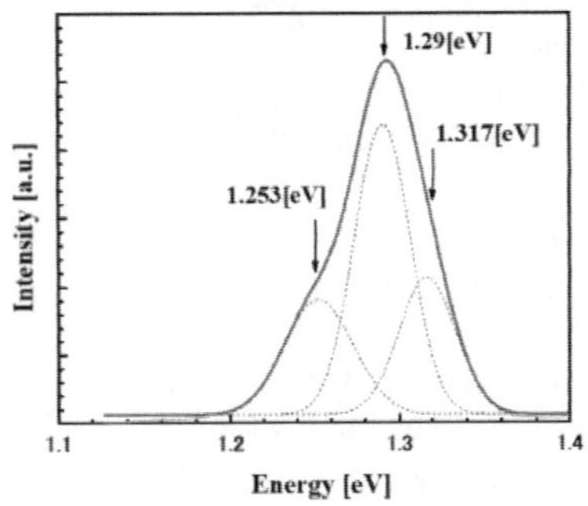

Figure 5: PL spectrum of the d = 7 nm sample at 10 K. Dashed line shows the three components used for fitting.

3.3 Two-color excitation photoluminescence

To citify the intermediate band, we performed the carrier dynamics using two-color excitation. Figures 7 and 8 show the photoluminescence spectra with ground state in the multistacked QDs. The interdot spacing of these sample were d = 15 and d = 7 nm, respectively. We used two color lasers: color 1 laser was a pulse laser with a wavelength of 780 nm, and color 2 laser had a CW operation with a wavelength of 1300 nm. One-color (only color 1) and two-color (color 1+2) excitations are shown with black and red lines, respectively.

In the case of d = 15 nm, the PL intensities did not change between one-color and two-color excitations. However, in the case of d = 7 nm, we observed that the PL intensities were lower for the two-color excitation. Figure 2 shows the band diagram of the multistacked QDs with an intermediate band. In the one-color excitation, the carriers were initially excited directly to the conduction band by the first laser. Subsequently, the excited carriers relaxed to the ground state in the QDs. Finally, the carriers recombined with the PL to the hole ground state. In the two-color excitation, the second laser excitation energy was tuned below the ground-state energies. The photocarriers excited by the first laser relaxed to the intermediate state in the QDs. However, these relaxed carriers were excited again from the intermediate band to the conduction band by the second laser, as follows in Fig.8,9. Therefore, we assumed that the recombination from the ground state decreased by the cascade process into the carriers via the intermediate band. In addition, the peak and FWHM of the PL spectra at d = 7 nm did not change under one-color and two-color excitations. We assumed that the carriers in intermediate band were uniformly re-excited by the second excitation.

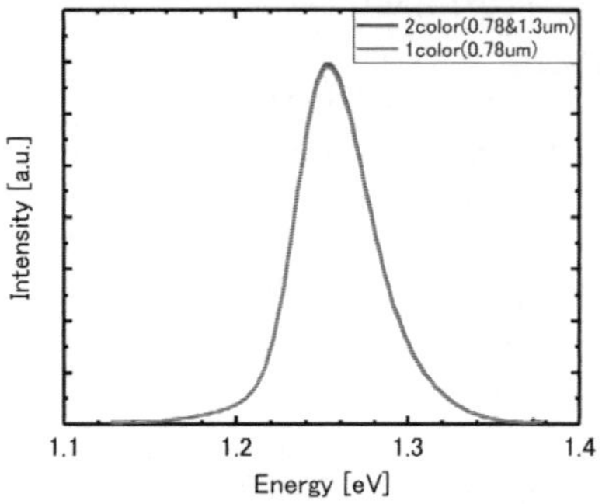

Figure 6: Photoluminescence spectra for d = 15 nm with the one-color (only color 1) and two-color (color 1+2) excitations.

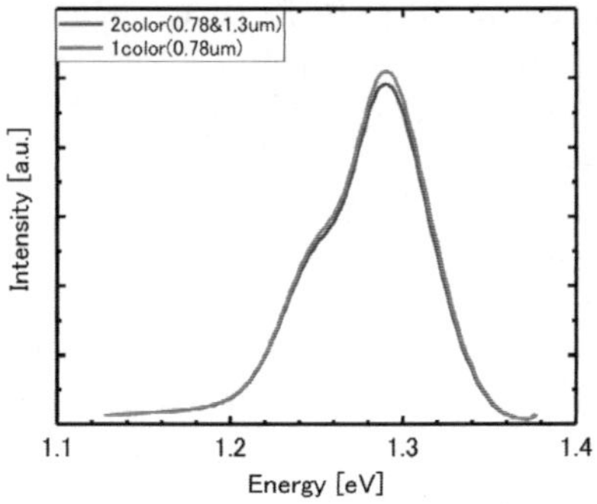

Figure 7: Photoluminescence spectra for d = 7 nm with one-color (only color 1) and two-color (color 1+2) excitations.

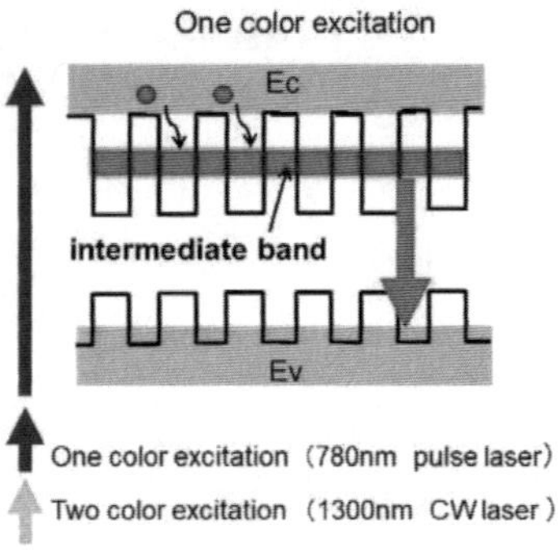

Figure 8: Band diagram of multistacked QDs with an intermediate band. Carrier dynamics with one-color excitation.

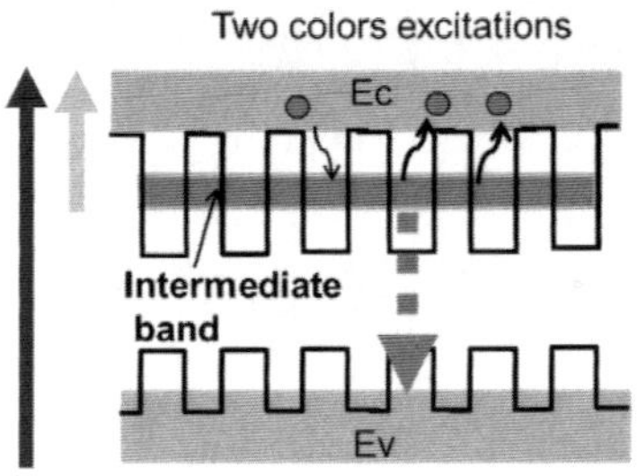

Figure 9: Band diagram of multistacked QDs with an intermediate band. Carrier dynamics with two-color excitation.

4. Summary

We investigated the electronic structures, intermediate formation, bandwidth, and carrier dynamics for different interdot spacings of multistacked QDs using optical methods. From the PL spectra method, we observed that the shape of the spectra was different for interdot spacings of d = 15 nm and d = 7 nm. We assumed that the broadening of the spectra for d = 7 nm was caused by the anti-bonding and bonding states of the electron, and the enlarged QDs in the stacking direction. For the one-color and two-color excitation methods, we observed different PL intensities for different interdot spacings. The intensity of the PL peak decreased for d = 7 nm when the two-color excitation was used. This behavior was caused by the carrier dynamics that reactivated the process via the intermediate state.

References

[1] Akahane K, Ohtani N, Okada Y and Kawabe M 2002 *J. Cryst. Growth* **245** 31-36
[2] Luque A and Marti A 1997 *Phy. Rev. Lett.* **78**(26) 5014-5017
[3] Antolion E, Marti A, Farmer C D, et al. 2010 *J. Appl. Phys.* **108** 064513
[4] Tomic S 2010 *Phys. Rev. B.* **82** 195321
[5] Jang Y D, Badcook T J, Mowbray D J, Skolnick M S, et al. 2008 *Appl. Phys. Lett.* 92 251905
[6] Sugaya T et al. 2011 *Solar Ener. Mater. Solar Cells* **95** 2920-2923
[7] Sugaya T et al. 2010 *Jpn. J. Appl. Phys.* **49** 030211
[8] Goshima K, Tsuda N and Sugaya T 2018 *Jpn. J. Appl. Phys.* **57** 06HE08
[9] Goshima K, Tsuda N and Sugaya T 2016 *Photovoltaic Science and Engineering Conference.*1-1-1e
[10] Melnik R and Willatzen M 2004 *Nanotechnol.* **15** 1-8
[11] Wasilewski Z R et al. 1999 *J. Cry. Grow.* **201-202** 1131-1135

IOP Conf. Series: Journal of Physics: Conf. Series **1220** (2019) 012039 doi:10.1088/1742-6596/1220/1/012039

Non-destructive dispersion of quantum dots into gases

M. Kumakura[1], D. Koide[1], T. Shimomura[1], Y. Baba[1], T. Takiyama[2], T. Kameyama[2], T. Torimoto[2], T. Moriyasu[1]

[1]Graduate School of Engineering, University of Fukui, Bunkyo, Fukui, ,910-8507, Japan
[2]Graduate School of Engineering, Nagoya University, Chikusa, Nagoya, 464-8601, Japan

E-mail: mtstk_kumakura@apphy.u-fukui.ac.jp

Abstract. We report on our experimental results for the nondestructive dispersion of CdSe/ZnS core-shell quantum dots and ZnS-$AgInS_2$ solid solution quantum dots into buffer gases by using a nebulizer. By monitoring scattering light from droplets and fluorescence from quantum dots, we discuss the evaporation and loss of the droplet and the fluorescence quenching of the quantum dot after the evaporation.

1. Introduction

A semiconductor quantum dot (QD) is one of nanoparticles attracting a great deal of interest in many fields.[1] Since its electronic structure and spectral property are strongly influenced with their size and environment[2], more precise size control of the QD is considered to be one of key factors involved in its future applications for efficient opt-electronic devices. The aim of our study is to carry out the size-separation of the QDs in gases, which have low viscosity and high thermal diffusivity, by optical means. In liquid helium, the optical size-separation has been successfully demonstrated for CuCl QDs created by laser ablation.[3] Since a typical viscosity value for a gas at normal temperature is close to the one for the liquid helium, a similar manipulation would be expected to be possible in a buffer gas. The use of the buffer gas enables us to control temperature more widely and reduce the influence of convection and foam in addition to relative ease of an experimental setup. For realizing similar optical separation for a variety of chemically synthesized QDs at normal temperature, we are developing an experimental approach for their non-destructive dispersion into buffer gases.

In this paper, we report our recent experimental results for the dispersion and isolation of CdSe/ZnS core-shell QDs[4] and $(Ag\ In)_x\ Zn_{2(1-x)}\ S_2$ solid solution (ZAIS) QDs[5] into buffer gases of dry nitrogen by using droplets generated with nebulizers.

2. Experimental setup

To non-destructively introduce quantum dots into buffer gases, we constructed the experimental apparatus using a mesh-type nebulizer (Omron NE-U) with a PZT oscillator (Fig.1). The CdSe/ZnS QDs and ZAIS QDs were dissolved, respectively, in diethyl ether and hexane (the typical QD number

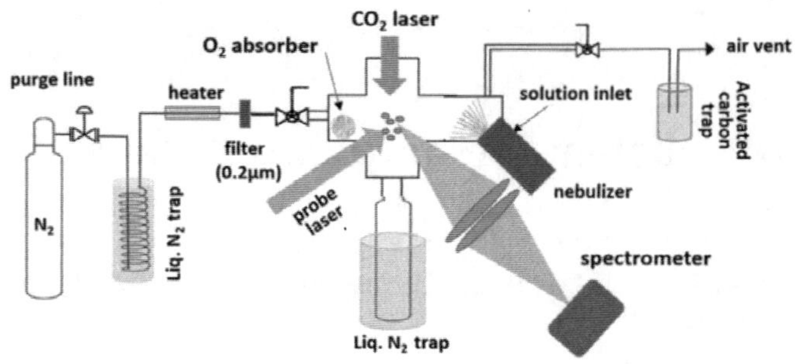

Figure 1. Experimental apparatus for the non-destructive dispersion of quantum dots into a dry nitrogen buffer gas.

density of the solution was approximately 10^{13}-10^{14}/ml), and each solution (0.5 ml) was injected into a gas cell (~ 100 ml) by using the nebulizer. The CdSe/ZnS QDs used in the experiment are commercial products from Sigma-Aldrich Co.. Their diameter is approximately 6.3 nm, and the peak wavelength of the fluorescence spectrum is approximately 630 nm. The ZAIS QDs are chemically synthesized and characterized in Nagoya University, and their diameter is nearly 5 nm. In this experiment, we used two types of ZAIS QDs with the ratio $x = 0.3$ and 0.7, whose absorption and emission spectra are different from each other; the peak wavelengths of their fluorescence spectra are 570 nm and 630 nm, respectively (532nm excitation).[5] Nominal diameter of the generated droplet is around 5 μm, and the typical number of QDs included in a single droplet is estimated to be 100-1000. To isolate QDs in the buffer gas, we enhanced the evaporation of the solvent from the droplet by two means; the heating of the gas cell (~320K) and the cold trapping of solvent vapor with liquid nitrogen.

3. Results and discussions

The generation and evaporation of the droplet were monitored by observing the scattering intensity of the probe laser beam (628 nm, 10mW). The observed time dependence of the scattering intensity is shown in Fig.2. From this figure, it is found that almost all the visible droplets were lost in nearly 1 min. Possible causes of this loss are considered to be the falling by gravity, the sticking on the cell wall through the diffusion, and the evaporation. To prevent the former two loss processes, we applied the CO_2 laser heating (~10 W/ 20 mm^2) to the droplets. As a result, it was confirmed that the evaporation was successfully enhanced and almost all the droplets evaporated less than 30 s as see in Fig.2.

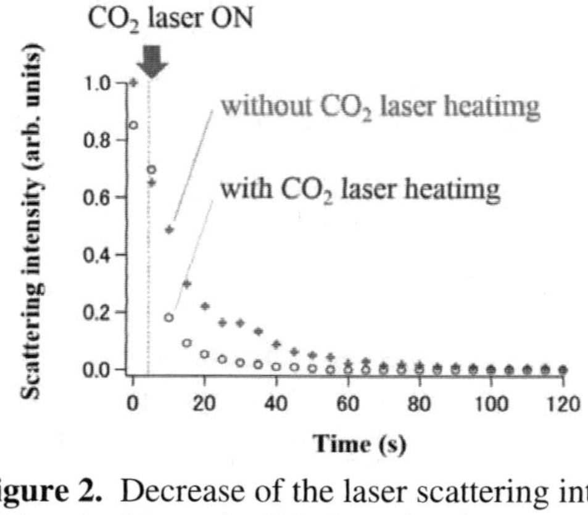

Figure 2. Decrease of the laser scattering intensity with and without the CO_2 laser heating.

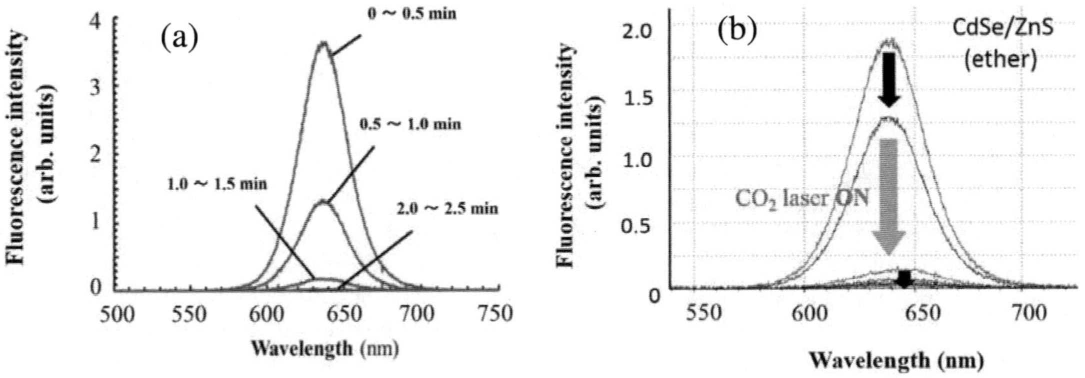

Figure 3. Decrease of the fluorescence intensity from the CdSe/ZnS QD; (a) without and (b) with the CO_2 laser heating.

The fluorescence from the QDs was monitored with a CCD monochrometer. The wavelength of the excitation laser is 532 nm, and the intensity is typically 1.4 W/ 7 mm^2. Figure 3(a) shows the time dependence of the fluorescence spectrum observed in the case of the CdSe/ZnS QD. Just after the droplet generation relatively high fluorescence was observed due to the QDs in the droplets. However, the fluorescence intensity gradually decreased in approximately 1 min. This decay rate is similar to the one observed in the scattering intensity, and fluorescence from the isolated QD was not observed apparently. The observed fluorescence decay is considered to be caused from the loss of the QDs due to the falling and sticking of the droplets as seen in the scattering measurement.

To observe the fluorescence from the isolated QD, we enhanced the evaporation of the droplet by the CO_2 laser heating during the fluorescence measurement. Figure 3(b) shows the change of the fluorescence intensity after the CO_2 laser heating. As seen in this figure the fluorescence intensity sharply decreases as the droplets evaporate. This result suggests the occurrence of the strong fluorescence quenching due to the evaporation of the droplet.

Since the exciton in the CdSe is weakly binding, it is well known that the change of the QD surface considerably affects the emission efficiency.[6] In contrast, in the case of the ZAIS QD, the emission quantum efficiency is expected to be less sensitive to its surface change, since the photon emission is

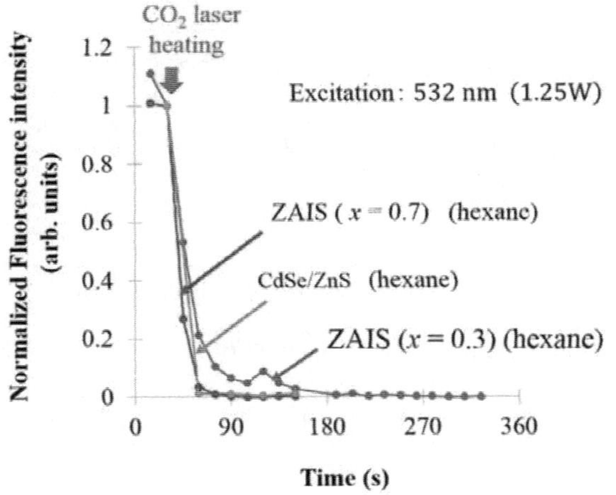

Figure 4. Decrease of the fluorescence intensity from the QDs after the CO_2 laser heating.

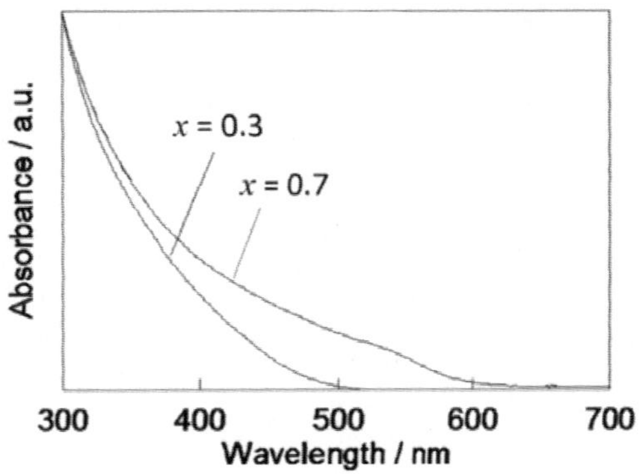

Figure 5. Absorption spectra of ZAIS QDs

caused by the defect localized in the QD. However, the fluorescence observed from the ZAIS QD (x = 0.7) showed the same strong quenching to the CdSe/ZnS QD (Fig.4). On the other hand, in the case of the ZAIS QD (x = 0.3), which was optically excited under the condition of the red-side far-wing excitation (Fig. 5), the fluorescence decay rate was apparently lower. This fact indicates that the thermal relaxation after the optical excitation should play an important role in the fluorescence quenching. The slight recovery of the fluorescence intensity around 120 s is due to the fluctuation of the QD number density and photo detector noise.

Finally, to reduce the influence of impurity contamination on the emission efficiency, we also tried to decrease the size of the droplet. We newly constructed an experimental apparatus with a cluster generator by using a metal mesh rotor (Corona Co. CNR-400B). The typical cluster size is in the range of 10 to 100 nm in the case of H_2O clusters. From the scattering measurement it was confirmed that the almost all the visible clusters evaporates in approximately 30 s, which indicates that the generated clusters are smaller than the previous droplets. However, the similar fluorescence quenching was observed for both ZAIS QDs (x = 0.3 and 0.7) even after 30 s from the generation of the clusters.

4. Conclusion

The CdSe/ZnS QDs and ZAIS QDs (x = 0.3 and 0.7) were nondestructively dispersed in a dry nitrogen buffer gas by using micro-droplets and nano-clusters. Evaporation behavior of the droplet and clusters were monitored by measuring scattering and fluorescence. The enhancement of the evaporation by the CO_2 laser heating was demonstrated successfully. From the fluorescence measurement, it was found that the fluorescence from the QD was strongly quenched after the evaporation of the droplets and clusters. The difference in the fluorescence decay rate between QDs suggests that the thermal relaxation after the optical excitation is one of possible causes of the fluorescence quenching.

5. Acknowledgement

This work was supported by JSPS KAKENHI Grant Number JP16H06505 in Scientific Research on Innovative Areas "Nano-Material Optical-Manipulation".

References

[1] Smith A M and Nie S 2010 *Acc. Chem. Res.* **43** 190
[2] Ibnaouf K H, Prasad S, Al Salhi M S, Hamdan A, Zaman M B, and El Mir L 2014 *J. Lumin* **149** 369
[3] Inaba K, Imaizumi K, Katayama K, Ichimiya M, Ashida M, Iida T, Ishihara H, and Itoh T 2006 *Phys. Status Solidi* B **243** 3829
[4] Dabbousi B O, Rodriguez-Viejo J, Mikulec F V, Heine J R, Mattoussi H, Ober R, Jensen K F, and Bawendi M G 1997 *J. Phys. Chem.* B **101** 9463
[5] Kameyama T, Takahashi T, Machida T, Kamiya Y, Yamamoto T, Kuwabata S, and Torimoto T 2015 *J. Phys. Chem.* C **119** 24740
[6] Thuy U T D, Liem N Q, Thanh D X, Protière M, and Reiss P 2007 *Appl. Phys. Lett.* **91** 241908

IOP Conf. Series: Journal of Physics: Conf. Series **1220** (2019) 012040 doi:10.1088/1742-6596/1220/1/012040

Proposed scheme of realization of artificial nano-heterostructures by optical force under double resonance

Yuto Yamada[1], **Masayuki Hoshina**[2], **Tomohiro Yokoyama**[1], **and Hajime Ishihara**[1,2]

[1] Division of Frontier Materials Science, Osaka University, 1-3 Machikaneyama-cho, Toyonaka, Osaka 560-8531, Japan

[2] Department of Physics and Electronics, Osaka Prefecture University, 1-1 Gakuen-cho, Naka-ku, Sakai, Osaka 599-8531, Japan

E-mail: yamada.y@opt.mp.es.osaka-u.ac.jp

Abstract. We theoretically study a new configuration control of nanoparticles based on an optical force, which enables to trap two kinds of nanoparticles simultaneously in two-dimensional hetero-arrangement. Our proposed scheme utilizes two resonance effects on a localized surface plasmon and an electronic transition of nanoparticles. These resonances change the optical force behavior significantly. We propose periodic rhomboid metallic nanostructures with two linear polarized laser beams, the directions of which polarization are parallel to the two diagonals of rhomboid. The polarization direction fractionates the trapped nanoparticles when the two laser frequencies correspond to the electronic transition energies of two kinds of particles, respectively. We evaluate the feasibility of our proposed scheme numerically. The map of the optical force shows a clear spatial separation of trapping position for the two kinds of particles. Our results demonstrate an essential advantage of the optical manipulation that enables creating novel nanostructures unrealizable by conventional nanofabrication techniques.

1. Introduction

Optical manipulation is a technique to manipulate minute objects by using an optical force. It has been applied for micrometer-scale objects since A. Ashkin and his collaborators have demonstrated the trapping of particles with a single laser beam in 1986 [1]. In recent years, attempts to manipulate nanometer-scale objects by the optical force have been advanced. Nanoparticles smaller than the wavelength of visible light exhibit significant quantum effects which are hardly observed in micrometer-size objects. Thus, a development of control technology of nanoparticle's motion and position within nanometer-scale resolution will open a new perspective to obtain unprecedented, indented, and useful functions by configuration and ordered structures of them. From this point of view, the optical manipulation has advantages for the operation over a large-area in a non-contact manner. However, the optical manipulation of nanoparticles is still a challenging task because of a weak exerted force, which is approximately proportional to the volume of the target within the Rayleigh scattering regime. As strategies for the nanoscale optical manipulation, the localized surface plasmon (LSP) resonance and the electronic resonance of nanoparticles are key ingredients [2,3]. The LSP is a collective oscillation of electrons on the surface of metallic structures. Its resonant frequency is sensitive to the size and shape of metallic structures. The LSP for a nanoscale metallic-gap results in an extremely

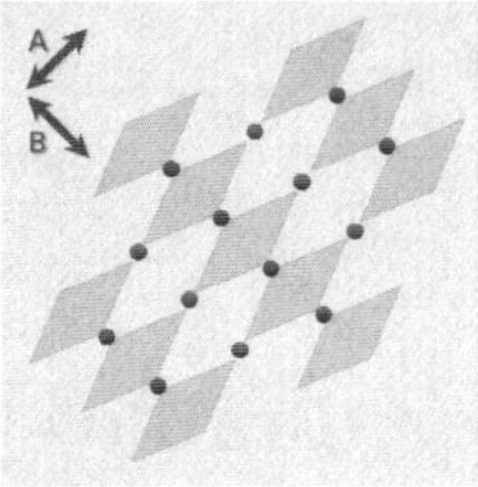

Figure 1. Schematic view of hetero-arrangement of two kinds of nanoparticles formed by the optical trapping in our proposed rhomboid nanostructure. Red and blue arrows indicate the polarization directions of incident electric field A and B, respectively. The frequencies of A and B correspond to the electronic transition energies of red particles a and blue particles b, respectively.

enhanced electric field at the gap (plasmonic hotspot). This strong and localized electric field attracts nanoparticles toward the hotspot strongly and achieves a nanometer-scale resolution. The optical force is also enlarged when the frequency of applied field is resonant with the electric transition of nanoparticles. The resonance has also an advantage for a selective manipulation of nanoparticles with quantized electronic levels.

In this work, we theoretically demonstrate a new function of optical manipulation based on an interplay of two resonance effects for the LSP and the electronic transition of nanoparticles. The schematic view of our proposed system is illustrated in Fig. 1. We consider a periodically arranged rhomboid metallic nanostructure and two linear polarized laser beams, labeled as A and B. We suppose that the incident fields are plane waves. The direction of polarization of A is parallel to one diagonal of rhomboid metal. That of B is along to the other diagonal. We investigate a simultaneous optical trapping of two kinds of nanoparticles by this system. The polarization direction of two beams selects trapping positions for nanoparticles. Then, the optical force distributes nanoparticle a and b as shown in Fig. 1. The LSPs induced by light A and B show different resonant frequencies due to the rhomboidal shape of metal. To see a clear distribution, we assume that the electronic transition energies of nanoparticle a and b are close to the resonant energies of LSP by light A and B, respectively. Such assumption can be satisfied by proper fabrication of metallic nanostructure. We demonstrate a numerical calculation based on the discrete dipole approximation (DDA) method [6] and discuss an absorption cross section of the metals and the optical force on the particles to evaluate the feasibility of our proposal.

2. Theoretical model

In the numerical demonstration, we evaluate a time-averaged optical force $\langle \boldsymbol{F} \rangle$ [4] given by a total electric field $\boldsymbol{E}(\boldsymbol{r},t) = \boldsymbol{E}(\boldsymbol{r})\mathrm{e}^{-\mathrm{i}\omega t}$,

$$\langle \boldsymbol{F} \rangle = \frac{1}{2}\mathrm{Re}\left[\int_V \mathrm{d}\boldsymbol{r}\,(\nabla \boldsymbol{E}^*(\boldsymbol{r}))\cdot \boldsymbol{P}(\boldsymbol{r})\right], \tag{1}$$

where the integration is taken for the whole of nanoparticle. $\boldsymbol{P}$ is the induced polarization. In addition to $\langle \boldsymbol{F} \rangle$, we obtain the absorption cross section σ_{abs} from $\boldsymbol{E}$ as $\sigma_{\mathrm{abs}} = 4\pi k \int_V \mathrm{d}\boldsymbol{r}\, |\boldsymbol{E}(\boldsymbol{r})/\boldsymbol{E}_0(\boldsymbol{r})|^2 \mathrm{Im}[\chi(\boldsymbol{r};\omega)]$, where k is the wave number of incident field [5]. The total electric field is obtained from the Maxwell's equation and a constitutive equation. The latter is $\boldsymbol{P}(\boldsymbol{r}) = \chi(\boldsymbol{r};\omega)\boldsymbol{E}(\boldsymbol{r})$ with a local susceptibility $\chi(\boldsymbol{r};\omega)$, A solution of the Maxwell's equation is given by a free-space Green's function $\boldsymbol{G}(\boldsymbol{r},\boldsymbol{r}';\omega)$ as $\boldsymbol{E}(\boldsymbol{r}) = \boldsymbol{E}_0(\boldsymbol{r}) + \int \mathrm{d}\boldsymbol{r}'\, \boldsymbol{G}(\boldsymbol{r},\boldsymbol{r}';\omega)\boldsymbol{P}(\boldsymbol{r}')$, where $\boldsymbol{E}_0$ is an incident electric field. Substitution of the constitutive one into this Maxwell's equation provides a self-consistent equation that determines the total electric field $\boldsymbol{E}(\boldsymbol{r})$.

For complicated structures, we applied the DDA method [6] to solve the self-consistent equation. The DDA discretizes a 3D space to cubic cells. Then, the self-consistent equation becomes a set of simultaneous linear equations for $\boldsymbol{E}$. The resolution of numerical calculation is restricted by the size of cells. Let us put the cell size as $1\,\mathrm{nm}^3$. For the DDA calculation, we apply the Drude model to the susceptibility of metal with the parameters of gold. The

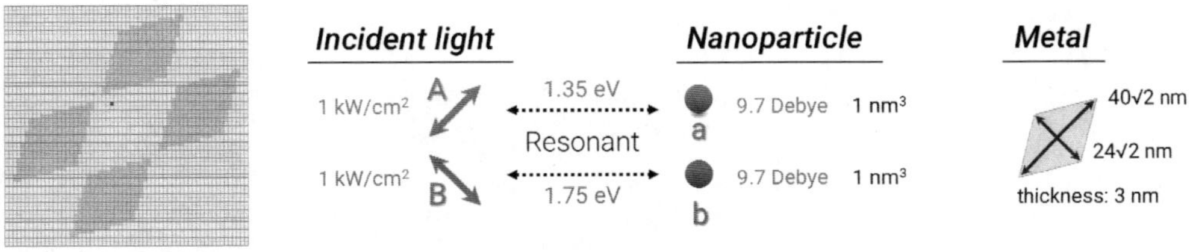

Figure 2. Rhomboid metallic nanostructures in the DDA method and parameters for the numerical simulation. Red cell in left figure is an example position of a nanoparticle.

nanoparticles are described by the Lorentz model using the parameters of typical fluorescent dye, where the resonance energies are 1.35 eV and 1.75 eV for a and b, respectively, and the damping constant is 1 meV. We apply a point dipole approximation to the nanoparticles with 9.7 D expressed by a single cell in the DDA. The parameters are summarized in Fig. 2. The incident fields are plane waves. As a medium, we assumed a vacuum.

3. Results and discussions

We perform the calculation when only four metallic nanostructures are present (see Fig. 2). Note that the length of all four gaps between the metals are equivalent with each other.

First, we investigate the absorption cross section of four metals in the absence of nanoparticles. Figure 3a shows the spectrum of $\sigma_{\rm abs}$ of four metals when either laser A or B is irradiated. The spectrum depends clearly on the direction of incident polarization. The difference of resonant energy is attributed to the difference of diagonal of rhomboid metal. This anisotropy agrees qualitatively with the LSP on prolate spheroidal metals [7]. Figure 3b shows the profiles of intensity of total electric field for an incident of light A at 1.35 eV and that of light B at 1.75 eV. Light A (B) excites long (short) diagonal's nanogaps with different frequencies, and does not induce strong electric field at the others.

Figure 4 exhibits a map of optical force exerted on a single particle a (red) or b (blue arrow) when both light A and B are irradiated simultaneously. Here, the energies of incident light A and B are slightly smaller than the resonace of particle a and b, respectively (light A: 1.3484 eV, B: 1.7484 eV) to obtain stronger gradient force reflecting a real part of Lorentz funciton. The intensity of incident electric fields is 1 kW/cm^2. We find that the optical force on particle a is strong in the vicinities of nano-gaps for long diagonals, whereas the force on b is for short diagonals. The difference of magnitudes is due to the difference of absorption cross section because of the shape anisotropy of metals. The clear separation of the forces distribution enables selective and simultaneous arrangement of two kinds of nanoparticles.

4. Conclusions

We have studied a simultaneous and selective optical trapping for two kinds of nanoparticles based on our proposed metallic nanostructure. There are three key ingredients: rhomboidal shape of metals, irradiation of two linear polarized lights in different directions, and interplay of the resonances on the LSP and nanoparticles. We have demonstrated a numerical simulation based on the DDA method and exhibited enough feasibility for the simultaneous trapping of two kind of particles to realize hetero-arrangement of them.

Our proposed structure would be extended for more than two kinds of nanoparticles and more complicated arrangement. Such variety of nanoparticle's arrangement leads to novel and high-performance functions by applying to a material surface.

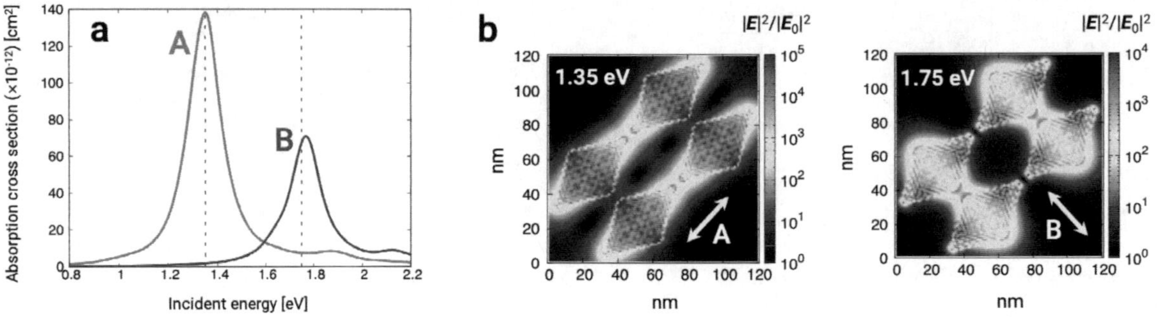

Figure 3. (a) Absorption cross section of the system comprising rhomboid metals shown in Fig. 2 when light A (red line) and light B (blue) are irradiated. The dashed lines indicate the energy 1.35 eV and 1.75 eV. (b) Profiles of intensity of the total electric field at 1.35 eV for light A (left panel) and at 1.75 eV for light B (right). The other parameters are mentioned in Fig. 2.

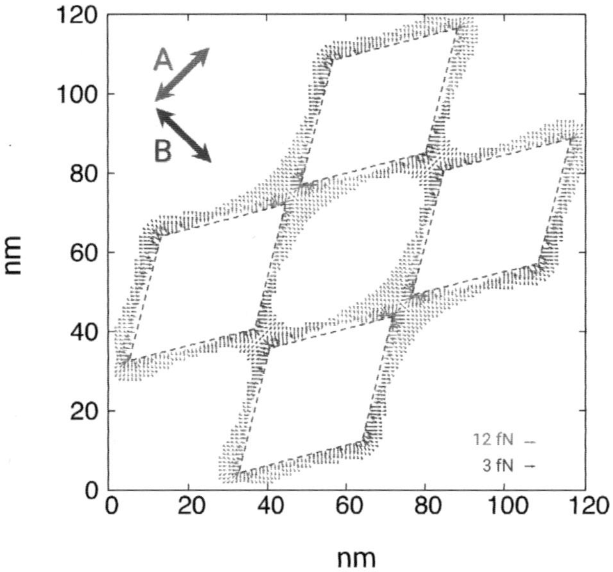

Figure 4. Map of the optical force exerted on a single nanoparticle under the irradiation both of light A and B. Red and blue arrows correspond to the particle a with resonance at 1.35 eV and the particle b at 1.75 eV, respectively. The length of arrows means the magnitude of the force. The gray dashed line indicates the edge of metallic nanostructures. The other parameters are mentioned in Fig. 2. The force weaker than 0.2 fN is not shown.

Acknowledgments

We acknowledge supports from JSPS KAKENHI Grant No. JP16H06504 in Scientific Research on Innovative Areas: "Nano-Material Optical-Manipulation".

Reference

[1] Ashkin A, Dziedzic J M, Bjorkholm J E, and Steven Chu 1986 *Opt. Lett.* **11** 288-290.
[2] Tsuboi Y, Shoji T, Kitamura T, Takase M, Murakoshi K, Mizumoto Y, and Ishihara H 2010 *J. Chem. Lett.* **1** 2327.
[3] Iida T and Ishihara H 2003 *Phys. Rev. Lett.* **90** 057403.
[4] Iida T and Ishihara H 2008 *Phys. Rev. B* **77** 245319.
[5] Goedecke G H and O'Brien S G 1988 *Appl. Opt.* **27** 2431.
[6] Purcell E M and Pennypacker C R 1973 *Astrophys. J.* **186** 705.
[7] Barber P W, Chang R K, and Massoudi H 1983 *Phys. Rev. Lett.* **50** 997.

Plasmonic optical trapping of pyrene-pendant polymer chains by controlling thermophoretic force

Kenta Ushiro, Tatsuya Shoji, Yasuyuki Tsuboi*

Graduate School of Science, Osaka City University, 3-3-138 Sugimoto, Sumiyoshi, Osaka, 558-8585, Japan

twoboys@sci.osaka-cu.ac.jp

Abstract. Plasmonic optical tweezers (POT) has a high potential for manipulation of nanomaterials due to an enhanced optical force. However, unfavorable thermal effects induced by plasmon excitation have frequently hindered the manipulation. For this issue, we have recently developed a novel non-plasmonic optical tweezers using a nanostructured silicon substrate (B-Si). We called it "Nano-Structured Semi-Conductor-Assisted Optical Tweezers (NASSCA-OT)". In the present study, we trapped pyrene-pendant polymer chains homogeneously dissolved in water for POT or NASSCA-OT. We used plasmonic gold nanopyramidal dimer arrays or B-Si in contact with the aqueous polymer solution. During plasmon excitation with a near-infrared laser light, any sign of optical trapping was never detected in fluorescence micro-spectroscopy. By contrast, trapping of the polymer chains was obviously observed for NASSCA-OT. Upon laser irradiation, pyrene excimer fluorescence was dramatically increased at the focal spot. These results indicate that NASSCA-OT is a powerful tool for manipulation of molecular materials.

1. Introduction

Nowadays, optical tweezers have undergone explosive developments in accordance with a great progress of lasers. So far, various types of small objects have been trapped under microscopes. In particular, optical tweezers are powerful tools in medical science. Egg cells, blood cells, bacteria, and so on have been manipulated by lab-made or commercially-available optical tweezers [1, 2]. These cells and bacteria have micron-sized dimension.

On the other hand, from a viewpoint of material science, nano-materials such as molecular clusters, DNA, proteins, quantum dots are fantastic targets for optical manipulation, since ordering, phase transition, crystallization, morphological changes, chemical reaction, self-organization etc. would be induced due to intermolecular interactions in micro-assemblies formed by optical trapping. Generally, optical force (radiation pressure) decreases with decreasing particle size. Therefore, it is rather difficult to trap such small nano-materials than micron-sized particles. For such nano-manipulation, a new physical principle enhancing optical forces should be involved into optical tweezers. A powerful candidate for such a novel optical tweezer is "plasmonic optical tweezers (POT)". However, POT has frequently suffered from thermal effects such as thermophoresis [3]. For addressing this issue, very recently we developed a novel heat-free optical tweezers; Nano-structured Semiconductor-Assisted optical tweezers (NASSCA-OT) [4]. This is based on silicon crystals with nano-structured surfaces (black silicon; B-Si). We demonstrated that NASSCA-OT has a stronger grip (stiffness) than POT.

IOP Conf. Series: Journal of Physics: Conf. Series **1220** (2019) 012041 doi:10.1088/1742-6596/1220/1/012041

In the present study, we examined NASSCA trapping of artificial polymer chains homogeneously dissolved in water. The polymer is labelled with pyrene. Trapping behavior was monitored using fluorescence micro-spectroscopy. We demonstrate that NASSCA-OT can optically manipulate these nanometer-sized macromolecules (Figure 1).

Figure 1. Schematic illustration of NASSCA optical trapping of pyrene-pendant water-soluble polymer chains

2. Experimental methods

2.1. Sample

We synthesized pyrene-labelled poly[*N*, *N*-dimethylacrylamide-*ram*-poly(ethylene glycol) methyl ether methacrylate] [poly(Py-DMA-PEG)] by free radical polymerization as a trapping target. This polymer is homogeneously dissolved in water. Pyrene is a representative fluorescence molecular probe. Fluorescence spectra of pyrene show two important peaks: on the one hand, a monomer fluorescence peak (375 ~ 450 nm), and on the other hand, an excimer emission band at around 490 nm emitted when two pyrene molecules are spatially close to each other. Number average molecular weight of the polymer chains used here was determined to be 1,500,000 by size exclusion chromatography. This polymer was dissolved in 0.1 mol/L NaOH aqueous solution as a sample solution.

2.2. Nanostructured substrates for optical trapping

As a plasmonic substrate, we fabricated gold nanopyramidal dimer arrays on a glass substrate by means of angular-resolved nanosphere lithography [5]. The plasmonic substrate has a broad extinction band around 800 nm that is ascribed to a gap-mode plasmonic resonance.

A B-Si substrate was fabricated by plasma-assisted reactive ion etching of a silicon wafer in SF_6/O_2 [6]. After the etching process nanoneedle structures formed on the silicon wafer. The optical reflection of B-Si is only a few percents over visible and near-IR spectral ranges due to the refractive index gradient around nanoneedles.

2.3. Optical setup

A cw near-infrared (NIR) laser beam (λ= 808 nm) for optical trapping was coaxially introduced into an oil-immersion objective lens (×100, NA = 1.30) on a fluorescence inverted microscope with an ultraviolet (UV) laser beam (λ= 377 nm). Fluorescence of pyrene was excited by the UV laser and Hg-lump for fluorescence microspectroscopy and fluorescence observation, respectively. For monitoring the optical trapping behaviour, we selectively obtained fluorescence spectra from the vicinity of the laser irradiation area on the substrate by using confocal optical setup.

3. Results and Discussion

First, we examined POT of the polymer. However, any sign of the trapping was never detected when light intensity was $I = 100$ kW/cm^2. Presumably, plasmon-enhanced optical force (attractive force) and thermophoretic force (repulsive force) were competing with each other, and an effective trapping force was much decreased, resulting in no trapping.

Next, we investigated NASSCA trapping. Figure 2 shows fluorescence spectrum during NIR laser irradiation (red) together with that without the irradiation (black). By using our optical setup, we successfully obtained an excimer emission peak around 490 nm while a monomer emission peak around 420 nm. Immediately after starting laser irradiation ($I = 100$ kW/cm^2), fluorescence gradually increased. As shown in Figure 2, the excimer peak obviously increased at 3 min from starting the laser irradiation. This result indicates that the local polymer concentration increased by NIR laser irradiation onto the B-Si. We estimated the concentration to be 70 w% with an intensity ratio between monomer and excimer emission. It should be noted that such condensation of the polymer was never observed in the absence of black silicon under the same irradiation condition. Namely, we succeeded in NASSCA optical trapping of polymer chains homogeneously dissolved in water. Furthermore, we successfully obtained new fluorescence peak around 520 nm from the trapped polymer chains. Although such fluorescence peak is frequently observed under special cases [7], our results suggest that NASSCA optical tweezers induce the interesting trapping behaviours for soft nanomaterials.

According to our theoretical calculation for spatial distribution of light electric field ($\boldsymbol{E}^2$), $\boldsymbol{E}^2$ should be enhanced with a factor of ca. 5 around nano-needles of a black silicon surface. The origin of the enhancement would be multi-scattering effect in the nano-structure. The present study strongly suggests that NASSCA can potentially manipulate nano-materials without suffering from thermal effects. This will open a new channel toward optical manipulation chemistry.

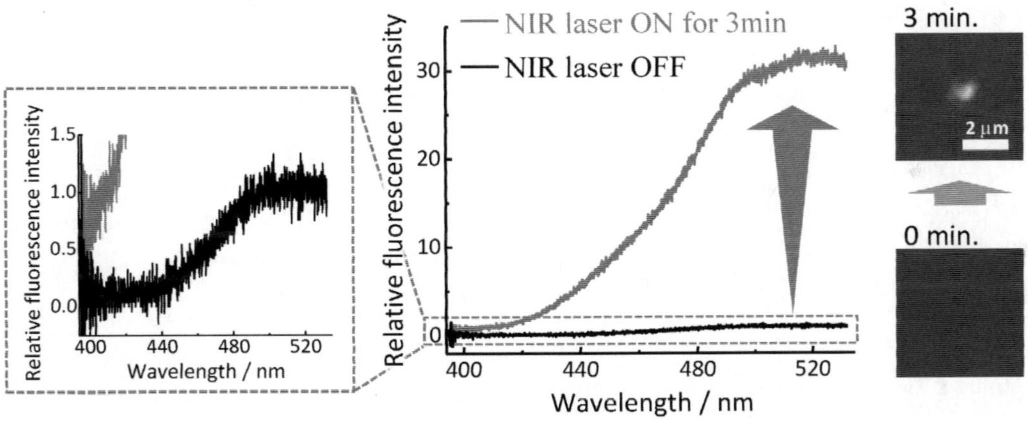

Figure 2. Fluorescence spectra obtained before NIR laser irradiation (black) and after NIR laser irradiation for 3minutes (red). These spectra show only excimer emission (around 490 nm) intensity dramatically increases in comparison with monomer fluorescence (375 ~ 450 nm) intensity. Images inset on the right side are fluorescence micrographs. The irradiation time of NIR laser is showed above the images.

Acknowledgement

This work was supported by JSPS KAKENHI Grant Numbers JP17K04974, JP16K17922 and JP16H06507 in Scientific Research on Innovative Areas "Nano-Material Manipulation and Structural Order Control with Optical Forces". This work was supported by JSPS KAKENHI Grant Numbers JP17K04974, JP16K17922 and JP16H06507 in Scientific Research on Innovative Areas "Nano-Material Manipulation and Structural Order Control with Optical Forces".

References

[1] Ashkin A, Dziedzic M J and Yamane T 1987 *Nature* **330** 769-71
[2] Ashkin A and Dziedzic M J 1987 *Science* **235** 1517-20
[3] Tsuboi Y 2016 *Nature Nanotech.* **11** 5-6
[4] Shoji T, Mototsuji A, Balčytis A, Linklater D, Juodkazis S and Tsuboi Y 2017 *Sci. Rep.* **7** 12298
[5] Takase M, Ajiki H, Mizumoto Y, Komeda K, Masanobu N, Nabika H, Yasuda S, Ishihara H and Murakoshi K 2013 *Nature Photon.* **7** 550
[6] Komatsu R, Balčytis A, Seniutinas G, Yamamura T, Nishijima Y and Juodkazis S 2015 *Sol. Energy Mater. Sol. Cells* **143** 72-7
[7] Ito F, Miyadera S, Matsuda H, Ishibashi Y, Ito S and Miyasaka H 2018 *Photochem. Photobiol. Sci.* **17** 910-916

IOP Conf. Series: Journal of Physics: Conf. Series **1220** (2019) 012042 doi:10.1088/1742-6596/1220/1/012042

Detection of optical force due to multiphoton absorption

S Nakamura, R Sunakawa, K Setoura, S Ito, H Miyasaka
Division of Frontier Materials Science and Center for Promotion of Advanced Interdisciplinary Research, Graduate School of Engineering Science, Osaka University, 1-3 Machikaneyama-cho, Toyonaka, 560-8531, Japan

sito@chem.es.osaka-u.ac.jp, miyasaka@chem.es.osaka-u.ac.jp

Abstract. Resonant optical manipulation using absorption force so far has been based on the linear (one-photon) absorption by target material, while optical forces due to multiphoton absorption have not been much investigated. As multiphoton absorption obeys different selection rule from that of one-photon absorption, and also shows non-linear dependence on light intensity, a larger variety of photo-mechanical responses of small particles can be expected by using multiphoton absorption force. In this study, we focused femtosecond laser pulses to a single polymer microparticle containing fluorescence dyes to exert multiphoton absorption force on the particle. We successfully observed the three-dimensional motion of the photo-irradiated microparticle due to the multiphoton absorption.

1. Introduction

Radiation force arising from the exchange of momentum between a small particle and photons can be divided into three terms: gradient, scattering, and absorption forces. The gradient force push the particle toward the spatial gradient of light field, while the scattering and absorption forces direct the particle along the propagation direction of light.[1][2] Among the three forces, the gradient force is usually utilized for the trapping and manipulation of small particles, which is well known as optical tweezer.[3][4][5] On the other hand, the scattering and absorption forces are useful for the transportation of small particles over a longer distance.[6]

In optical micro-manipulation, a laser beam the wavelength of which is not in resonance with target particles is usually employed as manipulation light to avoid photodamage of the particles due to photoabsorption, where the mechanical interaction between photons and particles is not usually so large and relatively high (MW/cm-1) light intensity is required. The momentum of photon is, however, resonantly transferred via the allowed optical transition; strong absorption force can be expected by tuning the wavelength of manipulation light to the peak absorption band of target particles and, this strong absorption force can be used for selective transportation of nanomaterials.[6]

Optical micro-manipulation using absorption force so far has been utilized linear one-photon absorption by target material. On the other hand, optical forces due to multiphoton absorption have not been well investigated. As the multiphoton absorption obeys different selection rule from that of the one-photon absorption, and shows non-linear dependence on light intensity, we can expect a larger variety of photo-mechanical responses of small particles by using the multiphoton absorption. In this study, we aimed to induce and detect the mechanical motion of polymer microparticles due to multiphoton absorption.

IOP Conf. Series: Journal of Physics: Conf. Series **1220** (2019) 012042 doi:10.1088/1742-6596/1220/1/012042

2. Experimental Section

2.1. Preparation of sample

Small particles of polystylene (PS) with fluorescent dyes (perylene and pyrene) were prepared as specimens. Perylene and pyrene were used as two-photon and three-photon absorbers, respectively. 500-nm sized (in diameter) PS particles (Polyscience, Inc.) were used as received. The particles were immersed in ethanol (Wako, 053-06531) solutions of each fluorescent dye. The stock solutions were kept in a dark area in an experimental room for 24 h, leading to the inclusion of each fluorescent dye into the PS particles with ethanol molecules by swelling. The particles containing the dye molecules were separated from the solution by centrifugation, then dispersed into ultrapure water. A drop of the colloidal suspension of the PS particles was injected into a 30-μL sample chamber consisting of two well-cleaned coverslips (Matsunami, 24 × 32 × 0.17 mm) and a 300 μm thick silicon-rubber sheet with a hole.

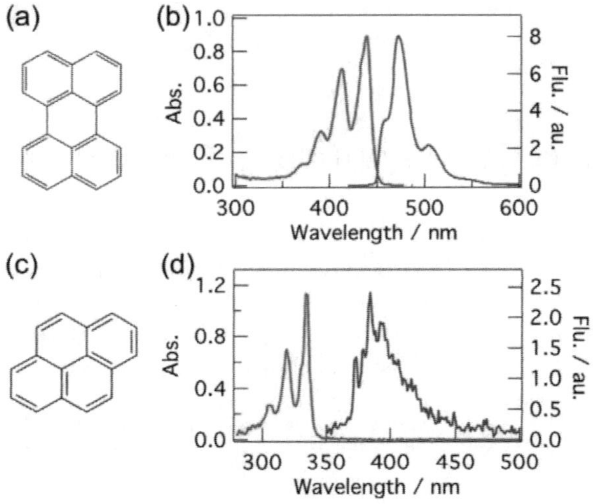

Fig.1. Structures of (a) perylene and (c) pyrene. (b) Absorption (red line) and fluorescence (blue line) spectra of perylene in benzene. (d) Absorption (red line) and fluorescence (blue line) spectra of of pyrene in hexane.

2.2. Optical setup

Each femtosecond (typically 70 fs time duration) light pulse from a mode-locked Ti:sapphire laser (Tsunami, Spectra Physics) operating at a wavelength of 800 nm and 80-MHz repetition was divided into two pulses by a half beam-splitter. A single PS particle is photo-irradiated with a pair of femtosecond light pulses with time-interval, Δt, under a tightly focused configuration by a high numerical aperture (NA) objective (N.A. 1.30/ x100, oil immersion, Olympus). These two successive light pulses were used for trapping the single polymer beads and inducing multiphoton excitation of the dyes in the beads. The relative time delay between the two light pulses, Δt, was controlled by using a stepping-motor-controlled optical delay stage. By controlling Δt, we could change multi-photon absorption strength under the same incident laser power because N-photon absorption strength is in proportion to the Nth power of light intensity.

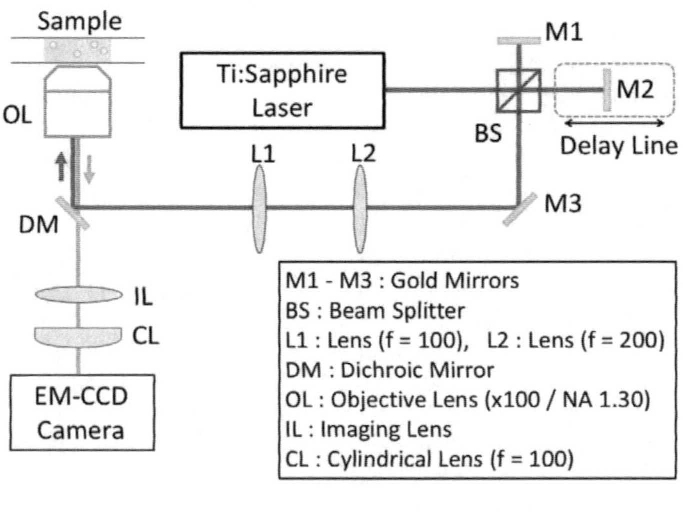

Fig. 2. Optical setup for inducing and detecting multiphoton absorption force.

2.3. Estimation of the Axial Position

For three-dimensional tracking of single fluorescent particles, we introduced a cylindrical lens into the detection optics to give the imaging system astigmatism.[7] Figure 3(a) shows fluorescence images of a single PS bead at 5 different z-positions obtained with the astigmatism imaging system. These fluorescence images clearly demonstrate that the shape (aspect ratio) of a PS bead are depending of its Z-position. We obtained fluorescence images of the PS bead while slightly changing the Z-position by using a piezo-electric stage. For each fluorescence image the X- and Y-widths are computed by 2D Gaussian fitting and the X- and Y-widths were plotted as a function of the Z-position (Figure 3(b)). The plots were curve-fitted by using the following equation (1) (red and blue solid lines in Figure 3(b)).[8] By using the plot, we determined the Z-position of a trapped PS particle.

$$\sigma(z) = \sigma_0 \sqrt{1 + \left(\frac{Z - Z_0}{d}\right)^2 + A\left(\frac{Z - Z_0}{d}\right)^3 + B\left(\frac{Z - Z_0}{d}\right)^4} \quad (1)$$

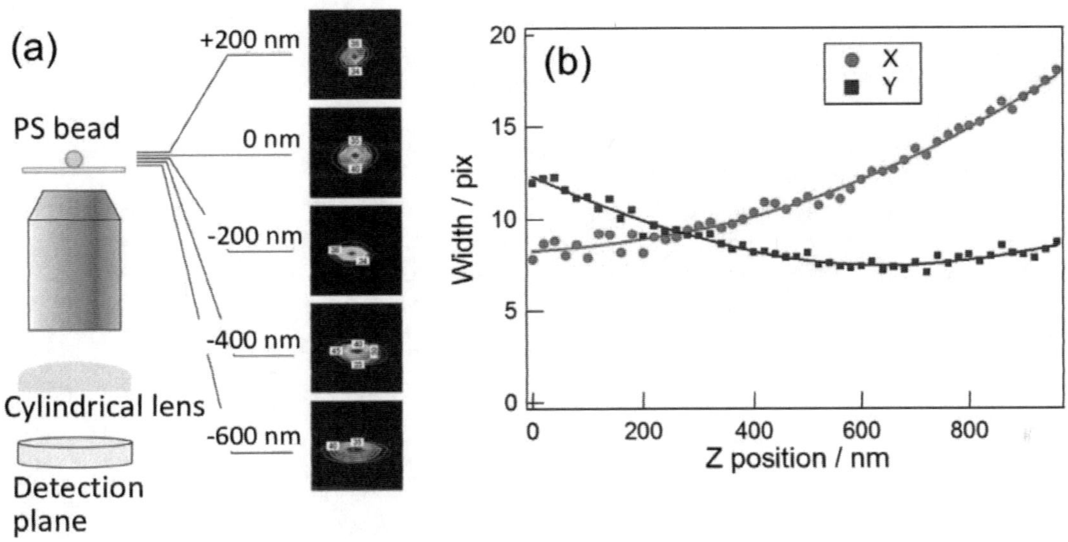

Fig.3. (a) Asymmetry in images of a 500-mm PS bead produced by a cylindrical lens. The bead contained perylene and it was photoexcited with the 800-nm fs light pulses through two-photon absorption. (b) Calibration curve for deducing the Z-position of each trapped PS particle; X- and Y-widths are plotted with respect to Z-position.

3. Result

Under the condition that Δt was larger than the pulse duration, the particle was optically trapped by the gradient force and slight multiphoton absorption took place by individual light pulses. The number of photons absorbed via multiphoton process increased with decreasing Δt up to the complete overlap of the pulse pair, leading to the increase in fluoresce intensity and change of fluorescence image of the particle due to the astigmatism. By analyzing the fluorescence image of the particles, we confirmed

IOP Conf. Series: Journal of Physics: Conf. Series **1220** (2019) 012042 doi:10.1088/1742-6596/1220/1/012042

that the particles were pushed in the direction of light propagation and the displacement increased with decreasing Δt up to $\Delta t = 0$ fs. Furthermore, by increasing Δt from $\Delta t = 0$ fs to the initial delay time, the trapping position of the particles returned to the original position. The particles with perylene (two-photon absorber) and pyrene (three-photon absorber) showed similar behaviors. The displacement caused by the multiphoton absorption was in the range from 50 - 200 nm at an input laser power of ca. 1 mW. It should be noted that the displacement varied particle by particle. This is ascribed to the heterogeneity in the number of dyes incorporated in a PS particle as well as the size of the PS particle. For quantitatively evaluating the multiphoton absorption force, detailed experimental parameters such as light intensity distribution at the focus, trapping stiffness of the gradient force, the number of dye molecules in a trapped PS bead, multiphoton absorption cross-section, etc. We are still devoting our effort to obtain such parameters for quantitative estimation of the multiphoton absorption force and the result will be published elsewhere in near future.

We thus demonstrated in the present study the mechanical motion of small particle due to multiphoton absorption force. As multiphoton absorption dominantly takes place only in the focal spot of laser beam under a tight focusing configuration, we can expect spatially selective micro-manipulation by the multiphoton absorption force. We hence foresee a larger variety of photo-mechanical responses of small particles by utilizing multiphoton (non-linear) photo-response.

References

[1] P N Lebedev 1901 *Annalen der Physik. Leipzig.* **6** 433.
[2] E F Nichol *et al* 1903 *Phys.Rev.* **17** 26.
[3] A Ashkin *et al* 1971 *Appl. Phys. Lett.* **19** 283.
[4] A Ashkin *et al* 1986 *Opt. Let.* **11** 288.
[5] A Ashkin *et al* 1987 *Nature* **330** 769.
[6] S Ito *et al* 2013 *Sci. Rep.* **3** 3047.
[7] H Pin Kao and A S Verkman 1994 *Biophysical Journal* **67** 1291.
[8] J W Shaevitz 2009 *International Journal of Optics* **6** 896208.

IOP Conf. Series: Journal of Physics: Conf. Series **1220** (2019) 012043 doi:10.1088/1742-6596/1220/1/012043

Anti-Stokes fluorescence from chlorophyll *a*

Hidetoshi Emura[1], Tomoyasu Noji[2], Masaharu Kondo[3], Yutaka Amao[1,2], and Mitsuru Sugisaki[1]

[1] Graduate School of Science, Osaka City University, Sumiyoshi, Osaka 558-8585, Japan
[2] OCARINA, Osaka City University, Sumiyoshi, Osaka 558-8585, Japan
[3] Materials Science and Engineering, Nagoya Institute of Technology, Showa-ku, Nagoya, Aichi 466-8555, Japan

E-mail: mitsuru@sci.osaka-cu.ac.jp

Abstract. Anti-Stokes fluorescence from chlorophyll *a* molecules dispersed in diethyl ether has been observed. From the excitation power dependence of the fluorescence spectrum, it is concluded that the anti-Stokes fluorescence appears via the linear optical process. To consider the relationship between the observed spectral shape and vibrational modes, the model spectral density has been determined based on the Raman spectrum and real-time coherent vibrational oscillation signals. The absorption spectrum observed in the experiment was well reproduced by the calculation, indicating the validity of the model spectral density. The involvement of the low-frequency vibrational modes in the anti-Stokes fluorescence process is discussed.

1. Introduction

Chlorophylls (Chl) and bacteriochlorophylls (BChl) play important roles in the initial steps of photosynthesis [1]. For example, BChl *a* is known as a key pigment of light-harvesting antennas of purple bacteria. Reflecting the periodic one-dimensional arrangement of BChl *a*, the excitation energy transfer between molecules following light harvest is well explained by introducing the concept of an exciton. Further, in photosystem II of higher plants, coherent propagation of excitons between Chl *a* molecules has been observed. It is therefore interesting to elucidate the relationship between the electronic structure and optical responses of (B)Chl, which covers a wide spectral range of sun light. Recently, clear anti-Stokes fluorescence from chloroplasts in *Parachlorella kessleri* cells and in leaves of *Zea mays* has been observed at room temperature [2,3], while the underlying mechanism, e.g., the influence and necessity of the protein environment surrounding pigments for the anti-Stokes fluorescence process, is still not fully understood. In the present study, we investigated the fluorescence spectra from Chl *a* molecules dispersed in diethyl ether over a wide spectral range. The relationship between the anti-Stokes fluorescence and molecular vibration modes is discussed.

2. Experimental

Chl *a* was purchased from Chlorophyll Research Institute Ltd. and used without further purification. As typically used [1], diethyl ether was employed as the solvent to prevent specimen degradation. The conventional absorption spectrum was recorded at room temperature on a commercial spectrophotometer (Shimazdu, UV-1850). The excitation light source used for the conventional fluorescence measurement was a He-Ne laser, while that for the anti-Stokes fluorescence measurement was a cw Ti:sapphire laser (Spectral Physics, 3900S) pumped by a 5-W laser diode (Spectral Physics,

IOP Conf. Series: Journal of Physics: Conf. Series **1220** (2019) 012043 doi:10.1088/1742-6596/1220/1/012043

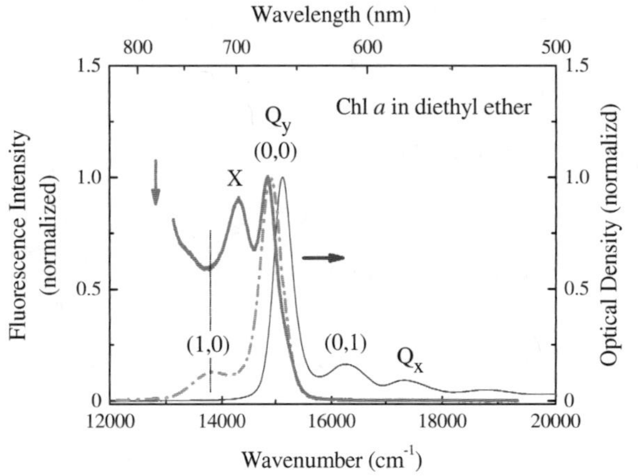

Figure 1. Absorption (thin blue curve), conventional fluorescence (orange dash-dotted curve), and anti-Stokes fluorescence (thick red curve) spectra of Chl *a* dissolved in diethyl ether. The sample concentration is 27 μM. The excitation energy of the conventional fluorescence is 15800 cm^{-1}, while that of the anti-Stokes fluorescence is 12800 cm^{-1}, which is indicated by the downward pointing arrow.

Millenia). The spectra were collected with a 30-cm single monochromator (Zolix, Z-300) and a cooled CCD camera (Mitty-CCD, H101411107-DS).

3. Results and discussion

Figure 1 shows the absorption and fluorescence spectra of Chl *a* dispersed in diethyl ether. The absorption peaks observed at 15100 and 17300 cm^{-1} are the so-called Q_y and Q_x bands, respectively. The small absorption peak at 16300 cm^{-1} originates from the quantization of vibrational levels of Q_y, i.e., the peak reflects the transition from the electronic ground state to the first vibrational excited level coupled with the electronic Q_y band. We note that the symbols (m, n) in Fig. 1 indicate the transition between the m-th vibrational level in the electronic ground state and the n-th vibrational level in the excited state. The dash-dotted curve in Fig. 1 shows the conventional fluorescence spectrum, where the excitation was with a He-Ne laser at 15800 cm^{-1}, which is slightly below the Q_x band. The main peak of the fluorescence spectrum at 14900 cm^{-1} is slightly redshifted compared to the 0-0 transition peak observed in the absorption spectrum due to the solvent reorganization. The fluorescence spectrum forms the mirror image of the absorption spectrum: the relative fluorescence intensity of the (0,0) and (1,0) peaks and the energy separation of these peaks are very close to those of the (0,0) and (0,1) peaks in the absorption spectrum. By contrast, the fluorescence spectrum upon the excitation below the (1,0) peak shows a significantly different shape, as shown by the thick solid curve in Fig. 1. Namely, another peak marked as X appears between the (0,0) and (1,0) peaks observed in the conventional fluorescence spectrum. In the following part of the present paper, we explore the possible mechanism of the significant change in the spectral shape.

The anti-Stokes fluorescence was measured at various excitation powers. The excitation power was changed over 2 orders of magnitude. The anti-Stokes fluorescence intensity also increases with the increase in excitation power. The relative intensities of the peaks at 14300 and 14900 cm^{-1} were almost constant within the excitation power range measured in the present study: these peaks linearly increase with the excitation power. This fact indicates that the anti-Stokes fluorescence originates from a one-photon process.

Figure 2 shows a series of fluorescence spectra measured at various excitation energies. It should be noted that each spectrum is magnified and that the baseline is vertically shifted for clarity. The top curve in Fig. 2a was measured upon the excitation at 13300 cm^{-1}, which is approximately 1600 cm^{-1} below the 0-0 transition energy of the Q_y band. The (0,0) peak at 14900 cm^{-1} is approximately twofold stronger than the peak X at 14300 cm^{-1}. With the decrease in excitation energy, the fluorescence intensity decreases. In addition, the spectral shape changes considerably: when the excitation is set to approximately 13000 cm^{-1}, the intensity of the fluorescence peak X is significantly enhanced. The (0,0) peak becomes relatively stronger again when excited at much lower energies. When the

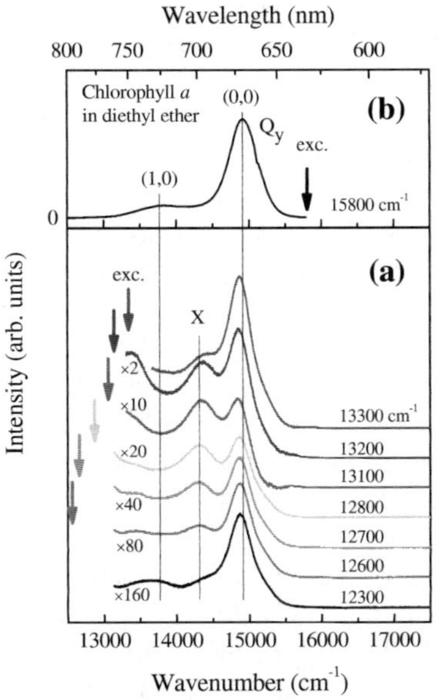

Figure 2. Fluorescence spectrum measured at various excitation energies, parts of which are indicated with the arrows. The conventional fluorescence spectrum is also shown in Fig. 2b for comparison.

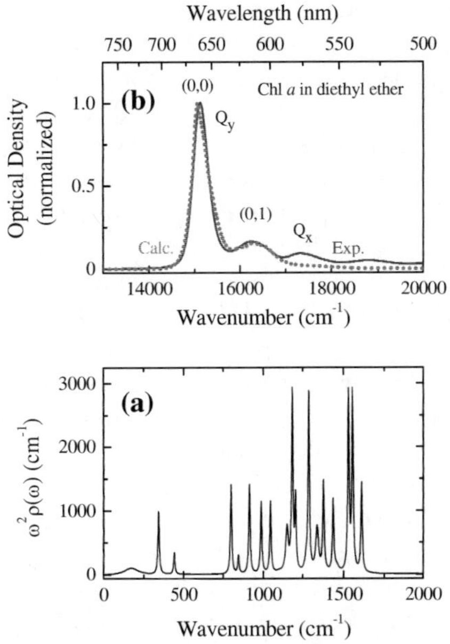

Figure 3. (a) Model spectral density used for the calculations. (b) Comparison of the absorption spectra obtained from the experiment (blue solid curve) and the calculation (red dotted curve).

excitation energy is detuned from 13000 cm^{-1}, the relative intensity monotonically decreases, and the peak X ultimately becomes very weak.

To investigate the contribution of vibrational oscillations to the anti-Stokes fluorescence process from Chl *a*, a model spectral density $\rho(\omega)$ was determined from the Raman spectra and real-time coherent vibrational oscillations reported in previous studies [4–6]. Reflecting the complicated structure of Chl *a*, it is well-known that many Raman peaks appear. In the present study, only the outstanding peaks are included in the model spectral density, as shown in Fig. 3a. In addition, the influence from the solvent is included by introducing a broad low-frequency peak into the spectral density. In the present study, the line shape of each phonon mode is simply approximated with a Lorentzian function [7].

The steady-state absorption spectrum $\sigma_{\mathrm{A}}(\omega)$ is then calculated using the line broadening function $g(t)$ [8,9], i.e.,

$$\sigma_{\mathrm{A}}(\omega) \propto \mathrm{Re} \int_0^{\infty} dt \exp\left[i(\omega - \omega_{\mathrm{eg}})t - g(t)\right]. \quad (1)$$

The cut-off frequency is set to be 1800 cm^{-1} when the numerical integrations are performed. The absorption spectrum of Chl *a* calculated in this way is shown by the dotted curve in Fig. 3b. The experimentally observed absorption spectrum (solid curve in Fig. 3b) is well reproduced, indicating the validity of the spectral density constructed here. The separation between the 0-0 and 0-1 vibrational transitions is approximately 1200 cm^{-1}, which mainly reflects the distribution of the vibrational modes of Chl *a*. Namely, as seen in Fig. 3a, most of the vibrational modes are in the frequency range of 800 cm^{-1} to 1600 cm^{-1}, which determines the energy separation between the (0,0) and (0,1) peaks of the absorption spectrum or between the (0,0) and (1,0) peaks of the ordinary fluorescence spectrum upon the excitation above the Q_y band. It should be mentioned that the vibrational peaks (0,0) and (0,1) are clearly separated compared to those observed in BChl *a* [9], which indicates that the interaction between the Chl *a* molecules and the surrounding solvent is weaker than that between BChl *a* and the solvent.

In the case of the excitation on the anti-Stokes side of the 0-0 transition energy, the molecule seems to dominantly couple with rather low frequency modes of approximately 400 cm^{-1}. The detailed mechanism of this phenomenon is still unclear, but it has been shown that the coupling strength of the vibrational modes depends significantly on the excitation energy [5,6]. The unusual spectral shape observed in the anti-Stokes fluorescence could suggest that the fluorescence appears before the thermal equilibrium in the Q_y band is attained. It should also be noted that no significant change in the spectral shape around the peak X was observed at sample concentrations between 15 and 90 μM. Therefore, we consider that the aggregation of Chl *a* molecules is a less plausible origin of the anti-Stokes fluorescence. To prove these hypotheses, it is necessary to measure time-resolved fluorescence over a wide excitation energy range as well as the solvent and temperature dependences, but these are left as the subjects for future studies.

To summarize, we have observed the anti-Stokes fluorescence from Chl *a* molecules in an organic solvent. This fact indicates that the anti-Stokes fluorescence can emerge without any assistance from a protein environment. The model spectral density that leads to the spectral shapes of conventional absorption and fluorescence spectra has been determined. It is plausible that among the many vibrational modes, the low-frequency modes of approximately 400 cm^{-1} have a significant influence on the spectral shape of the anti-Stokes fluorescence.

Acknowledgments

The authors thank Mamoru Nango and Takehisa Dewa for valuable discussions. This work was supported by a Grant-in-Aid for Scientific Research from the Japanese Ministry of Education, Culture, Sports, Science, and Technology.

References

[1] Blankenship R E 2002 *Molecular Mechanisms of Photosynthesis* (Oxford: Blackwell Science Inc)

[2] Hasegawa M, Shiina T, Terazima M and Kumazaki S 2010 Selective excitation of photosystems in chloroplasts inside plant leaves observed by near-infrared laser-based fluorescence spectral microscopy *Plant Cell Physiol.* **51** 225–38

[3] Hasegawa M, Yoshida T, Yabuta M, Terazima M and Kumazaki S 2011 Anti-stokes fluorescence spectra of chloroplasts in *Parachlorella kessleri* and maize at room temperature as characterized by near-infrared continuous-wave laser fluorescence microscopy and absorption microscopy *J. Phys. Chem. B* **115** 4184–94

[4] Koyama Y, Umemoto Y, Akamatsu A, Uehara K and Tanaka M 1986 Raman spectra of chlorophyll forms *J. Mol. Struct.* **146** 273–87

[5] Du J, Nakata K, Jiang Y, Tokunaga E and Kobayashi T 2011 Spectral modulation observed in Chl-*a* by ultrafast laser spectroscopy *Opt. Express* **19** 22480

[6] Kosumi D, Nishiguchi T, Sugisaki M and Hashimoto H 2015 Ultrafast coherent spectroscopic investigation on photosynthetic pigment chlorophyll *a* utilizing 20 fs pulses *J. Photochem. Photobiol. A Chem.* **313** 72–8

[7] Jimenez R, Salazar G, Yin J, Joo T and Romesberg F E 2004 Protein dynamics and the immunological evolution of molecular recognition *Proc. Natl. Acad. Sci. USA* **101** 3803–8

[8] Mukamel S 1995 *Principles of nonlinear optical spectroscopy* (New York, Oxford: Oxford University Press)

[9] Sugisaki M, Fujii R, Cogdell R J and Hashimoto H 2008 Linear and nonlinear optical responses in bacteriochlorophyll *a Photosynth. Res.* **95** 309–16

IOP Conf. Series: Journal of Physics: Conf. Series **1220** (2019) 012044 doi:10.1088/1742-6596/1220/1/012044

Pump-probe study of β'-(BEDT-TTF)(TCNQ) crystal near antiferromagnetic transition

Hirofumi Mino[1], Shiho Tokuda[2] and Masatoshi Sakai[3]

[1] College of Liberal Arts and Sciences, Chiba University, 1-33 Yayoi-cho, Inage-ku, 263-8522, Chiba, Japan
[2] Graduate School of Education, Chiba University, 1-33 Yayoi-cho, Inage-ku, 263-8522, Chiba, Japan
[3] Graduate School of Engineering, Chiba University, 1-33 Yayoi-cho, Inage-ku, 263-8522, Chiba, Japan

E-mail: mino@faculty.chiba-u.jp

Abstract. The photo-excited state properties of an organic Mott-insulator β'-(BEDT-TTF)(TCNQ) crystal were investigated by pump-probe measurements. Considerable change in the pump-probe reflection signal with a wide spectral range was observed near the antiferromagnetic (AF) transition temperature of the TCNQ layers. In the time response, approach to the AF temperature induced the appearance of a slow decay signal whose lifetime was estimated to be about 3 msec maximum. These results indicate that a sustained mechanism exists for photo-excited states near the AF transition, which is probably related to a critical slowing down phenomenon.

1. Introduction

The study of photo-excited states in organic Mott-insulators has attracted a lot of interest in recent years from the viewpoint of a photo-induced insulator-metal transition. The differences in the insulator-metal transition by photo induction and thermal induction have been intensively studied for many years [1]. Femtosecond spectroscopy has revealed that the photo-induced insulator-metal transition occurs extremely quickly, typically in less than several picoseconds. It is expected that this feature will be used for high-speed devices such as optical switches and memories. Besides the insulator-metal transition, the Mott-insulator shows various phase transitions caused by changes in temperature and pressure: antiferromagnetic (AF), superconductive, charge order, relaxer phases, and so on. However, near those phase transitions, few reports have considered the change of optical transition or photo-excited states by visible region spectroscopy. Furthermore, pump-probe studies investigating slow responses for the photo-excited states in time regions over a microsecond are relatively few compared to the fast response, in the femtosecond or picosecond time region.

In this study, we report pump-probe experimental results indicating a large signal in the visible spectra region, with a long lifetime reaching several milliseconds in an organic Mott-insulator β'-(BEDT-TTF)(TCNQ) crystal near the AF transition. This crystal consists of bis(ethylenedithio)-tetrathiafulvalene (BEDT-TTF) as the donor and 7,7,8,8-tetracyanoquinodimethane (TCNQ) as the acceptor. They form a charge-transfer complex. The crystal is a layered structure, stacking two-dimensional BEDT-TTF sheets and one-dimensional TCNQ chains sheets. The β'-(BEDT-TTF)(TCNQ)

IOP Conf. Series: Journal of Physics: Conf. Series **1220** (2019) 012044 doi:10.1088/1742-6596/1220/1/012044

crystal is Mott-insulator at room temperature, and the metal-insulator transition occurs at 330 K [2]. It is known that the AF transition occurs on each BEDT-TTF layer and TCNQ layer at 20 K and 3.5 K, respectively [3, 4]. Many studies have reported its electronic state about the phase transition. However, there are few studies of electronic state change induced by photo-excitation. Therefore, to examine the phenomena that reflect the state change by photo-excitation, we used the pump-probe reflectance spectrum measurement and the pump-probe time-resolved measurement. Our results showed a large pump-probe signal approaching the AF transition temperature (=T_N) of TCNQ.

2. Experiment

Using the pump-probe spectrum measurement, we first investigated the reflectivity change in the visible light region in the β'-(BEDT-TTF)(TCNQ) crystal. Semiconductor CW lasers with wavelengths of 405 nm and 532 nm were used as the pump beam. Their intensity on the sample surface was 500 mW/cm^2 and 111 mW/cm^2, respectively. A white LED was used as a probe light source. The sample was placed in a cryostat and the temperature was changed from 20 K to 1.6 K. The pump-beam spot was coincident with the probe-light spot on the sample surface. The pump beam was modulated by an optical chopper at a frequency of 130 Hz. The differential reflectivity spectra ($\Delta R/R$) were detected using a spectrometer with a photomultiplier tube and lock-in amplifier.

We then used the pump-probe time-resolved measurement to investigate the relaxation process of photo-excited states. A 405-nm semiconductor CW laser and He-Ne laser (633 nm) were used as the pump and probe beam, respectively. They were changed into pulsed lasers using an acousto-optic (AO) modulator. The width of each pulse, pump and probe was 70 μsec, and the interval of each pulse was 10 msec. The time delay between pump pulses and probe pulses was controlled by a pulse generator. These beams were focused on the same positon on the sample surface, and their intensity was 560 mW/cm^2 and 1.4 mW/cm^2, respectively. The intensity change of the probe pulse was detected using a photomultiplier tube and a lock-in amplifier. The temperature was changed from 15 K to 2.8 K.

3. Results and Discussion

Figure 1 shows the differential reflectivity spectra obtained by photo-excitation with the 405-nm (3.06 eV) laser. Positive values of $\Delta R/R$ indicate an increase of reflectivity under photo-excitation. At 20 K, the reflectivity change appears, and it increases as the temperature becomes lower. It is around 0.2% in the spectrum region from 1.9–2.9 eV at 9.0 K, and significantly increases at 4.2 K (around 0.5%). This result suggests that the optical transition change in the broadband region is induced by photo-excitation as the temperature approaches the T_N of the TCNQ layers. The $\Delta R/R$ signal intensity systematically decreases with increasing temperature, and is difficult to observe above 20 K. In addition, there is no signal intensity peak at 20 K, which is the T_N of the BEDT-TTF layers. These results suggest that the one-dimensional electric structure of the TCNQ layers may be strongly related to the $\Delta R/R$ signals.

To examine the temporal behavior of the reflectivity change, He-Ne laser pulses chopped by the AO modulator were used as a probe light source. Figure 2 shows time-resolved pump-probe signals with 405-nm laser pulse excitation at various temperatures from 15 K to 2.8 K. It is evident that the photo-

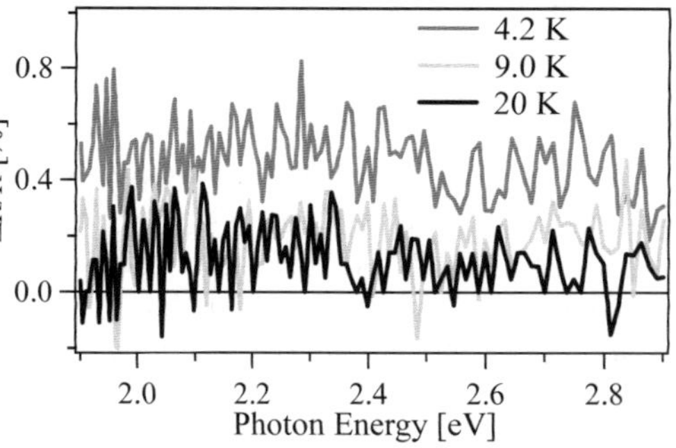

Figure 1. The reflectivity change with a white LED obtained by 405-nm laser excitation.

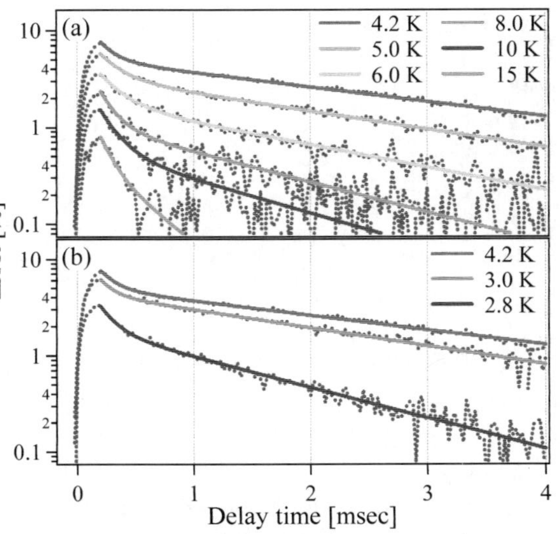

Figure 2. Dotted lines are the time resolved pump-probe signals at various temperatures (a) from 15 K to 4.2 K, and (b) from 4.2 K to 2.8 K. Solid lines are the fitting curves with double exponential decay.

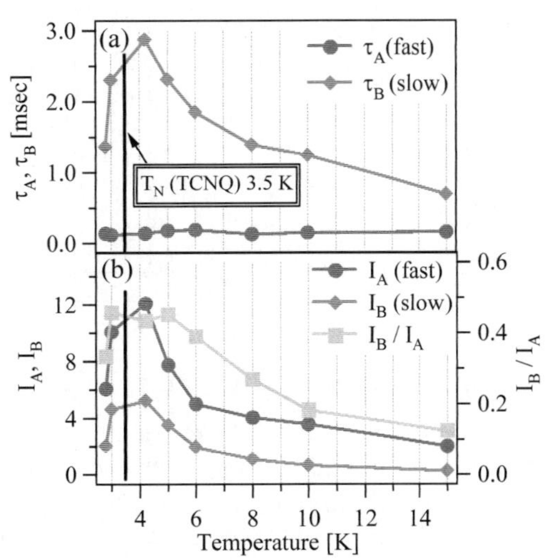

Figure 3. (a) Fast (τ_A) and slow (τ_B) reaction times, and (b) the intensity I_A, I_B and related intensity I_B/I_A plotted against temperature. The vertical solid line indicates the T_N of the TCNQ layers.

excited state relaxation processes are strongly dependent on temperature. All signals show decay profiles, and signal intensity increases as temperature decreases from 15 K to 4.2 K. This result is in agreement with Fig. 1. In contrast, the signal intensity decreases as temperature decreases from 4.2 K to 2.8 K. It is evident that the largest signal is obtained near the T_N. The $\Delta R/R$ signal may thus correlate strongly with the AF transition in the TCNQ layers. These data are well fitted by a double exponential curve $I_A \exp(-t/\tau_A) + I_B \exp(-t/\tau_B)$. Here the I_A (I_B) and τ_A (τ_B) together indicate the intensity and the relaxation time of the fast (slow) relaxation components. This means that photo-excited states are ruled by two different relaxation mechanisms. In Fig. 3(a), the τ_A and τ_B are plotted against the temperature. The τ_A without depending on temperature is about 0.15 msec. On the other hand, the τ_B becomes longer as the temperature approaches the T_N of the TCNQ layers, and the longest relaxation time of 2.93 msec was obtained at 4.2 K. The relaxation behavior of τ_B shows that there is the mechanism lengthening the photo-excited state lifetime. It appears to be related to electron spin ordering for the AF transition. In Fig. 3(b), both I_A and I_B increase near the T_N. This means that the fast relaxation component is also affected by the AF transition. However, I_B/I_A increases near the T_N, indicating that the slow relaxation component is more effective than the fast one. Here, we consider the relaxation time order of both the τ_A and τ_B. Even the fast one (τ_A) shows sub-msec response. This time order is substantially longer than the general recombination lifetime of photo-excited electrons and holes that are optically allowed. Therefore the photo-excited states resulting in the $\Delta R/R$ signals are considered to be optically inactive, even though the origin of the two different relaxations is not currently clear.

Fig. 4 shows the reflectivity change spectra obtained by photo-excitation with a 532-nm (2.33 eV) laser. The average value of the reflectivity change was about 0.3%, from 1.9 eV to 2.9 eV at around the T_N (4.2 K and 3.1 K). Under higher or lower temperatures than the T_N, the signal intensity was reduced. In addition, the pump-probe signals appeared in both higher and lower energy regions than the laser excitation energy, and the spectral shape of the signals was very similar to those in Fig. 1. These results indicate that similar photo-excited states were induced by both 405-nm and 532-nm laser excitation.

Here, we think about the energy of intramolecular transition in each BEDT-TTF and TCNQ molecule. A TCNQ crystal strongly absorbs around 2.8-eV light [5], and a BEDT-TTF crystal starts absorbing close to 2.1-eV light [6]. Assuming that the intramolecular transition in our sample is similar to those

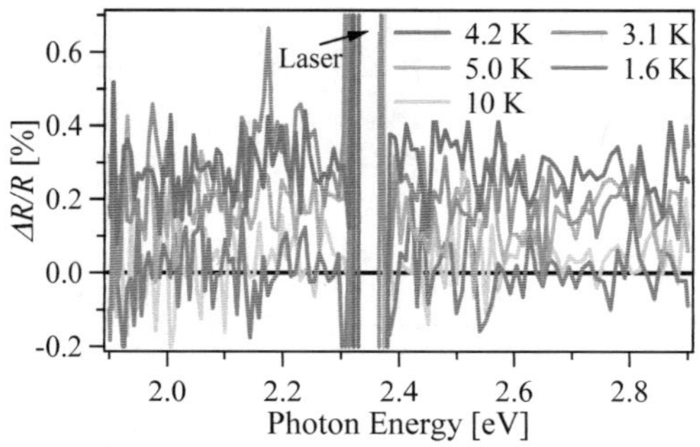

Figure 4. The reflectivity change with a white LED obtained by 532-nm laser excitation.

of the TCNQ and BEDT-TTF crystals, it would seem that there is a difference in the pump-probe signals between 405-nm and 532-nm laser excitation. However, the result in Fig. 1 is similar to that in Fig. 4. This means that the pump-probe signal may not strongly depend on the intramolecular transition of the TCNQ molecule. Otherwise, there is a possibility that the above assumption is not appropriate.

It is also necessary to think about the intermolecular transition between the BEDT-TTF and TCNQ molecules. Taking into account that the lowest intermolecular transition of β'-(BEDT-TTF)(TCNQ) occurs at about 1.18 eV [7], the pump-beam energy fully satisfied the energy range of the intermolecular transition, and might have been able to induce the intermolecular transition.

Here, we consider that both the pump-probe signal intensity and relaxation time have maximum values near the T_N. In magnetic materials, it is well known that the electron-spin on each site has a long-range correlation near the critical point (second order phase transition), and that the relaxation time to equilibrium state becomes very long at the critical temperature [8]. This phenomenon is called the critical slowing down. According to Fig. 3, the relaxation time of the long relaxation component (τ_B) drastically increases as the temperature approaches T_N. This result may be qualitatively explained by the critical slowing down. The photo-excitation may give rise to the long-range fluctuation in the AF spin ordering. It is also expected that the increase in the lifetime of photo-excited states would induce photo-carrier accumulation or molecular distortion, resulting in the broadband region spectrum change. The reason why the above-mentioned critical phenomena are not observed at the T_N of the BEDT-TTF layers (20 K) is not clear. The pump-probe investigation of other similar Mott-insulators near the AF transition would serve to clear up this and other questions regarding the origins of the critical phenomena.

4. Summary

Pump-probe measurements were performed in a β'-(BEDT-TTF) (TCNQ) crystal close to the AF transition temperature. Pump-probe signals reaching a broad spectrum range and showing large increases when approaching the T_N of the TCNQ layers were obtained with both the 405-nm and 532-nm CW laser excitation. The dynamical behavior of the photo-excited state was investigated using time-resolved measurement, and the existence of slow relaxation component which may indicate critical slowing down at around the T_N of the TCNQ layers was clarified.

References

[1] Nasu K ed 2004 *Photoinduced Phase Transitions* (Singapore: World Scientific)
[2] Mori T and Inokuchi H 1986 *Solid State Commun.* **59** 355
[3] Iwasa Y, Mizuhashi K, Koda T, Tokura Y and Saito G 1994 *Phys. Rev. B* **49** 3580
[4] Kawamoto A, Miyagawa K, Shimizu A and Kanoda K 1997 *Synth. Met.* **85** 1601
[5] Girlanda R *et al* 1986 *IL Nuovo Cimento D* **7** 469
[6] Kozlov M E, Tanaka Y, Tokumoto M and Tani T 1994 *Chem. Phys. Lett.* **223** 318
[7] Torrance J B, Vazquez J E, Mayerle J J and Lee V Y 1981 *Phys. Rev. Lett.* **46** 253
[8] Hohenberg P C and Halperin B I 1977 *Rev. Mod. Phys.* **49** 435

IOP Conf. Series: Journal of Physics: Conf. Series **1220** (2019) 012045 doi:10.1088/1742-6596/1220/1/012045

Transient grating spectroscopy of β-carotene pumped with spectrally chirped pulses

Shintaro Ooi[1], **Sigehito Mitoma**[1], **Mamoru Nango**[2], **Yutaka Amao**[1, 2], **and Mitsuru Sugisaki**[1]

[1] Graduate School of Science, Osaka City University, Sumiyoshi, Osaka 558-8585, Japan
[2] OCARINA, Osaka City University, Sumiyoshi, Osaka 558-8585, Japan

E-mail: mitsuru@sci.osaka-cu.ac.jp

Abstract. Nonlinear optical responses of β-carotene were investigated upon excitation using chirped pulses. We especially focus on spectrally resolved transient grating (TG) signals to discuss the influence of the spectral chirp over a wide spectral range. A significant change in the TG signal was observed when negatively chirped pulses were used. The experimental results were qualitatively reproduced by employing the Brownian oscillator model. The relationship between the wave packet motion in the excited state and spectral chirp is discussed. It is concluded that the third order nonlinear optical response reflects the competition between the spectral chirp and the energy separation of the potential curves.

1. Introduction

In the early events of photosynthesis, carotenoids absorb blue/green light, which matches the spectral maximum of solar radiation. The electronic excitation of carotenoids upon photoexcitation is then transferred to (bacterio)chlorophylls and forms excitonic states. In the early 2000s, it was demonstrated that the efficiency of the excitation energy transfer from carotenoids to bacteriochlorophylls in bacterial light-harvesting pigment-protein complexes can be manipulated using spectrally and temporally shaped pulses [1]. This fact indicates that exciton formation is controllable by designing the phase and amplitude of the spectral components of the excitation light. However, the underlying mechanism between the complicated shape of the optical pulses and the resulting efficiency of photosynthesis is still not clear. In the present study, we investigate how a simple spectral chirp influences the optical responses of carotenoid molecules. The experimental results are analyzed using the Brownian oscillator model. The temporal population change with the spectral chirp is discussed.

2. Experimental details

The setup for the nonlinear optical measurements has been reported before [2,3]. In the present study, a slightly modified configuration was employed for better operation. The excitation light source was a home-built non-collinear optical parametric amplifier (NOPA) pumped using a femtosecond Ti:sapphire regenerative amplifier (Coherent, Legend). The output pulses from the NOPA were passed through a prism compressor to compress or control the group velocity dispersion of the spectral components. The temporal width and phase of the excitation pulses were determined by means of the

IOP Conf. Series: Journal of Physics: Conf. Series **1220** (2019) 012045 doi:10.1088/1742-6596/1220/1/012045

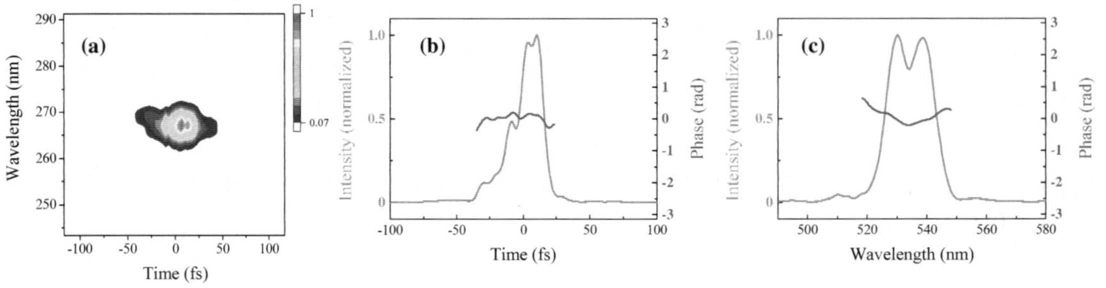

Figure 1. (a) Temporal evolution of the sum-frequency spectra of pump pulses measured in the non-collinear configuration, i.e. the SH-FROG trace. (b) and (c) Temporal and spectral profiles through the maximum intensity of pump pulses. The phases retrieved from the FROG traces are also shown.

so-called second harmonic (SH) frequency-resolved optical gating (FROG) technique, as shown in Fig. 1.

β-Carotene was purchased from Wako Pure Chemical Industries Ltd. and was purified by recrystallization from a benzene solution. For the optical measurements, β-carotene was dissolved in tetrahydrofuran (THF). The concentration of the sample was adjusted to an optical density of ~1 at its absorption maximum. All optical measurements were performed at room temperature.

3. Results and discussion

Figure 2 shows the temporal change of the spectrally resolved transient grating (TG) signals from β-carotene measured under (a) positive, (b) zero, and (c) negative chirp conditions. Here, the first and second pump pulses were irradiated on the sample simultaneously, and the TG spectra were measured at various delay times T of the third pulse with respect to the second pulse. The strong signals at the origin of time are due to the coherent artifact, which are truncated in the figure and are not discussed in the present study. The TG signals following the coherent artifact are weak but increases with the evolution of the population time T. The TG signals reach the maximum intensity at approximately 400 fs and then slowly begin to decay. The temporal profile shown in Fig. 2a under the positively chirped pulse excitation is very similar to that under the zero-chirp excitation in Fig. 2b. However, the TG

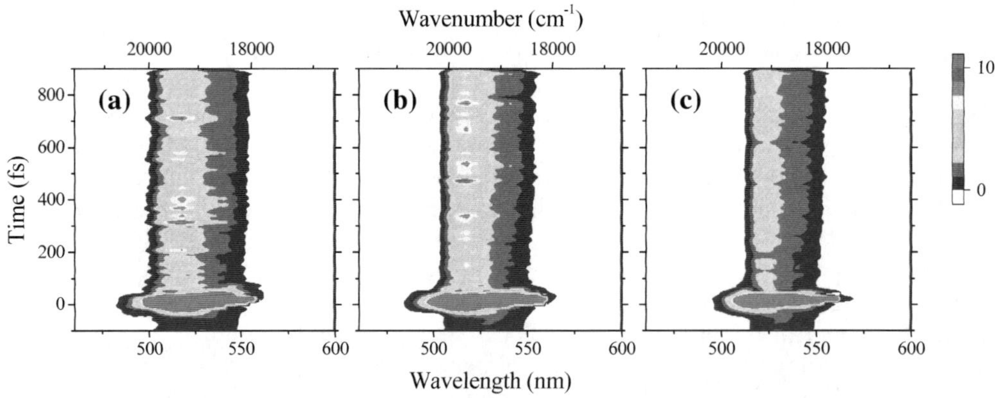

Figure 2. Spectrally resolved TG signals from β-carotene measured with (a) positively, (b) zero, and (c) negatively chirped pulses. The ordinate indicates the temporal separation between the second and third pulses, i.e., the population time T.

IOP Conf. Series: Journal of Physics: Conf. Series **1220** (2019) 012045 doi:10.1088/1742-6596/1220/1/012045

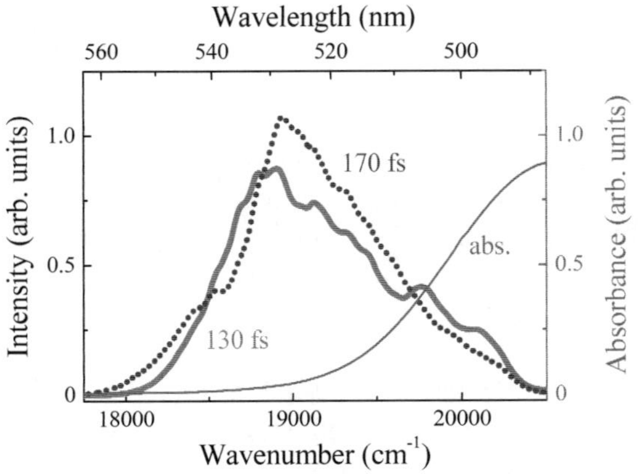

Figure 3. TG spectra of β-carotene at T=130 fs (thick solid curve) and 170 fs (dotted curve) under the zero-chirp excitation. The spectral shape and peak energy changes with T. The stational absorption spectrum (thin solid curve) is also shown for comparison.

signal is approximately 30% weaker with the negatively chirped pulses, as shown in Fig. 2c.

As shown in Fig. 3, the TG spectrum changes its shape during the very early population times. Specifically, not only the increase in intensity but also the peak energy shift was observed. When the excitation energy is below the absorption edge, the TG signal monotonically decreases as time evolves [4]. However, reflecting the relaxation of the population from the S_2 to the S_1 state, the TG signal increases in intensity, especially on the higher energy side, up to approximately 400 fs. This is why the TG spectral shows a blueshift.

Nonlinear polarizability based on the Brownian oscillator model [5] was calculated as reported in our previous studies [2,4], but in the present study the spectral chirp of the excitation pulse was also taken into consideration. As the electronic state of β-carotene, the ground state S_0, the one-photon forbidden state S_1, the lowest optically allowed state S_2, intermediate state S_x, which is located between S_1 and S_2, and the more excited states accessible from S_1, S_x, and S_2 were considered when drawing the Feynman diagrams to calculate the response function of each optical process. The response function was calculated from the spectral density and the line-broadening function that indicates the time evolution of the transition frequency correlation. The material parameters used for the present calculation are listed in our previous paper [2].

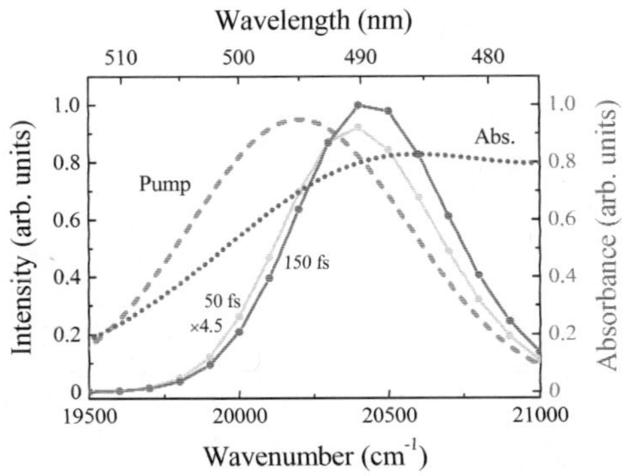

Figure 4. The calculated TG signals at T=50 fs and 150 fs. The spectral shape of the pump pulse and the absorption spectrum are also shown, which determine the spectral shape of the TG signal. Note that the TG signal at T=50 fs is magnified by 4.5 times for clarity.

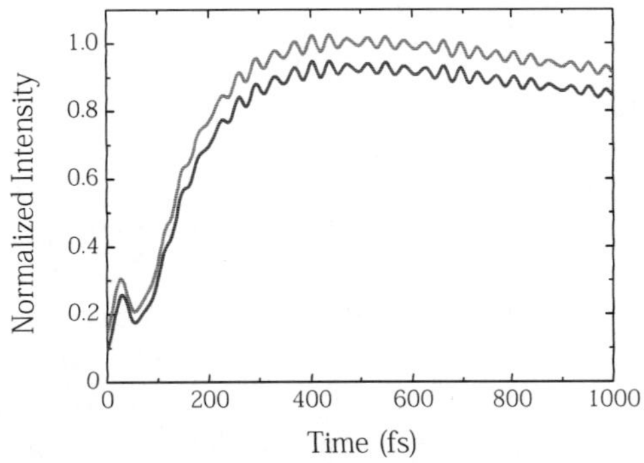

Figure 5. The calculated temporal profiles of the TG signals excited with positively (red curve) and negatively (blue curve) chirped pulses.

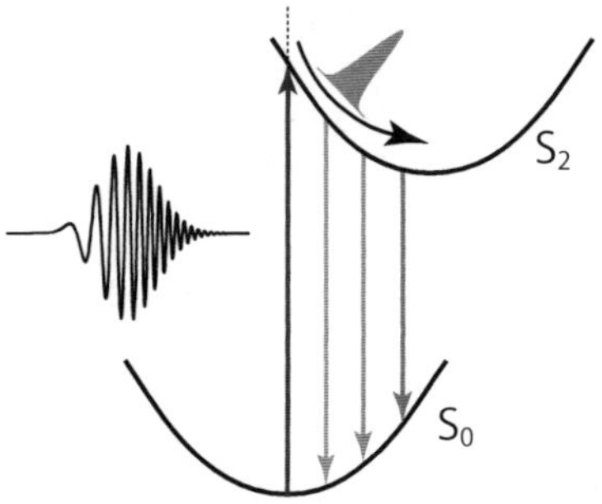

Figure 6. A schematic drawing of the potential curves of β-carotene and a pump pulse. The wave packet initially excited in S_2 is efficiently damped when a negatively chirped pulse is used, which results in a weak TG signal.

The spectral shift observed in the experiment was qualitatively reproduced as shown in Fig. 4. However, in comparison with the experiment, the rise of the calculated TG spectrum is much faster, and significant spectral shift was seen on the slightly higher-energy side. The reason for the quantitative inconsistency between the experiment and the calculations is not clear at present. A possible interpretation is inappropriate evaluation of the decay rate from S_2 to S_1. It might be necessary to consider the involvement of another intermediate state to reach better agreement.

Figure 5 shows the calculated temporal profiles of the TG signals under the positively and negatively chirped pulse excitations. Again, the experiment and the calculation show qualitative agreement: the TG signal under negative chirp excitation is weak. This is because the wave-packet in the S_2 state is efficiently dumped into the S_0 state when a negatively chirped pulse is used, as schematically shown in Fig. 6: the constitution of the response function that reflects the population in the excited state becomes small, resulting in the weak signal.

4. Summary

We have observed significant changes in the spectral shape and intensity of the TG signals of β-carotene upon the excitation using zero and chirped pulses. The experimental results have been qualitatively reproduced by employing the Brownian oscillator model. We conclude that the reduction in intensity of the nonlinear optical response under the negatively chirped pulse excitation is most probably due to the efficient dumping of the population from S_2 to S_0. In future work, it will be necessary to determine the real shape of each potential curve to attain better quantitative agreement between the experiment and the calculations.

Acknowledgments

The authors would like to thank Selvakumar V. Nair for helpful discussions. This work was supported in part by JSPS KAKENHI Grant Number 15H03679.

References

[1] Herek J L, Wohlleben W, Cogdell R J, Zeidler D and Motzkus M 2002 Quantum control of energy flow in light harvesting *Nature* **417** 533–5

[2] Sugisaki M, Yanagi K, Cogdell R J and Hashimoto H 2007 Unified explanation for linear and nonlinear optical responses in β-carotene: a sub-20-fs degenerate four-wave mixing spectroscopic study *Phys. Rev. B* **75** 155110/1-11

[3] Sugisaki M, Kosumi D, Saito K, Cogdell R J and Hashimoto H 2012 Generation of coherently coupled vibronic oscillations in carotenoids *Phys. Rev. B* **85** 245408/1-10

[4] Sugisaki M, Fujiwara M, Nair S V, Ruda H E, Cogdell R J and Hashimoto H 2009 Excitation-energy dependence of transient grating spectroscopy in β-carotene *Phys. Rev. B* **80** 035118/1-10

[5] Mukamel S 1995 *Principles of nonlinear optical spectroscopy* (New York, Oxford: Oxford University Press)

IOP Conf. Series: Journal of Physics: Conf. Series **1220** (2019) 012046 doi:10.1088/1742-6596/1220/1/012046

Raman scattering mediated by Majorana fermions in Kitaev nanoribbons

Kosuke Suzuki and Shoji Yamamoto
Department of Physics, Hokkaido University, Sapporo 060-0810, Japan
E-mail: kosuke@phys.sci.hokudai.ac.jp

Abstract. We study magnetic Raman, i.e. inelastic light, scattering of Kitaev's quantum spin liquids realizable in honeycomb nanoribbons. Elementary excitations in the Kitaev spin liquids consist of itinerant Majorana (matter) fermions and localized gauge fluxes (visons). Matter fermions are characterized by the projective point (gauged rotation) group inherent in the lattice. Inelastic light scattering is mediated by a pair of matter fermions without any vison excitation. We can selectively "observe" matter fermions of the same symmetry and those of different symmetries depending on the polarization of incoming and/or outgoing lights. A pair of matter fermions with particular translational quantum numbers $\pm k$ cannot be Raman active.

1. Introduction

Nanoribbon electronic materials, whether their electrons are itinerant or localized, have rich physics. Band gaps of graphene nanoribbons, for instance, depend on their bulk and edge shapes [1]. Nanoribbons of our current interest consist of localized Kramers doublets on a honeycomb lattice with Kitaev type interactions [2]. The Kitaev spin model has an exactly solvable "liquid" ground state, where competing interactions prevent any conventional magnetic long-range order from growing even at absolute zero [3]. In the Kitaev quantum spin liquid, spin degrees of freedom fractionalize into more fundamental entities—matter fermions and localized $\mathbb{Z}_2$ gauge fluxes. Magnetic Raman scattering is such a useful probe as to detect matter fermion excitations in pair separately from localized gauge flux excitations. We make a projective-symmetry-group analysis of momentum-resolved intensities of Raman scattering mediated by matter fermions to reveal how the Raman spectrum is structured as a function of the geometry of nanoribbons.

2. Model

The Kitaev spin Hamiltonian on a honeycomb nanoribbon reads

$$\mathcal{H} = -\sum_{\lambda=x,y,z}\sum_{<i,j>_\lambda} J_\lambda \sigma_i^\lambda \sigma_j^\lambda;\ [\sigma_i^\lambda, \sigma_j^\mu] = 2\mathrm{i}\delta_{ij}\varepsilon_{\lambda\mu\nu}\sigma_i^\nu, \tag{1}$$

where $<i,j>_\lambda$ runs over nearest neighbors with λ depending on the $<i,j>$ pair [cf. Fig. 1]. Two integers (p,q) determine the edge shape and the width of the nanoribbon [1]. We write spin variables with two Majorana fermions,

$$\sigma_i^\lambda = \mathrm{i}\eta_i^\lambda c_i;\ \{c_i, c_j\} = 2\delta_{ij},\ \{\eta_i^\lambda, \eta_j^{\lambda'}\} = 2\delta_{ij}\delta_{\lambda\lambda'},\ \{\eta_i^\lambda, c_j\} = 0,\ c_i^\dagger = c_i,\ (\eta_i^\lambda)^\dagger = \eta_i^\lambda, \tag{2}$$

with $\{A,B\} = AB + BA$. Since the fermionic Hilbert space becomes twice the original spin Hilbert space, we demand that the Majorana fermions should satisfy the condition $\eta_i^x\eta_i^y\eta_i^z c_i = 1$. The fermionic Hamiltonian reads

$$\mathcal{H} = \mathrm{i}\sum_{\lambda=x,y,z}\sum_{<i,j>_\lambda} J_\lambda \hat{u}_{<i,j>_\lambda} c_i c_j;\ \hat{u}_{<i,j>_\lambda} \equiv \mathrm{i}\eta_i^\lambda \eta_j^\lambda. \tag{3}$$

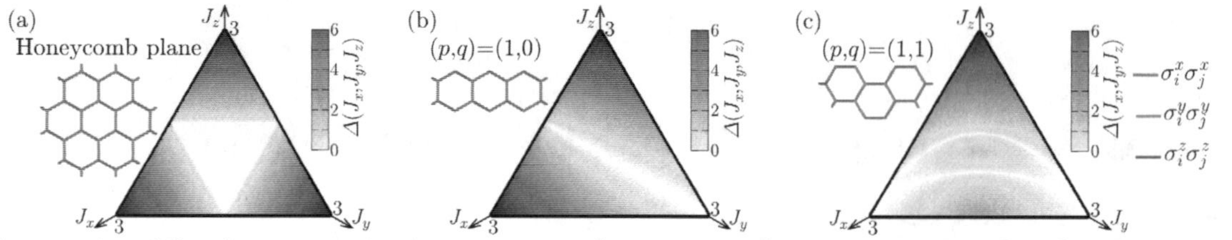

Figure 1. The lowest excitation energy of a matter fermion against the flux-free gauge background as a function of J_λ's on the plane of $J_x = J_y = J_z = 3$ for the Kitaev models on the honeycomb plane (a), (1,0)-nanoribbon (b), and (1,1)-nanoribbon (c).

Bond operators $\hat{u}_{<i,j>_\lambda}$ commute with the Hamiltonian so they can be treated as $\mathbb{Z}_2$ classical variables, $u_{<i,j>_\lambda} = \pm 1$. Given a set of classical variables $u_{<i,j>_\lambda}$, we have a matter fermion quadratic Hamiltonian diagonalizable straightforwardly. We set a ground state of gauge variables, where gauge flux operators defined on each closed loop C

$$\hat{W}_C = \prod_{<i,j>_\lambda \in C} \sigma_i^\lambda \sigma_j^\lambda = \prod_{<i,j>_\lambda \in C} (-\mathrm{i}\hat{u}_{<i,j>_\lambda}), \tag{4}$$

all take an eigenvalue of unity [2]. We diagonalize the gauge-ground Kitaev Hamiltonian into

$$\mathcal{H} = \sum_k \sum_{n=1}^{p+2q+1} \epsilon_{k,n}\left(\alpha_{k,n}^\dagger \alpha_{k,n} - \frac{1}{2}\right);\ c_i = \sum_k \sum_{n=1}^{p+2q+1} \left(\psi_{k,n;i}\alpha_{k,n} + \psi_{k,n;i}^* \alpha_{k,n}^\dagger\right). \tag{5}$$

The quasiparticle $\alpha_{k,n}^\dagger$ has the excitation energy $\epsilon_{k,n}$ labeled with a band index n and momentum k belonging to the first Brillouin zone. The ground state is the quasiparticle vacuum, $\alpha_{k,n}|0\rangle = 0$. Similarly to the honeycomb-lattice Kitaev spin liquid [4], Kitaev nanoribbons each have a QSL ground state of their own. Whether or not matter fermion excitations are gapped from the ground state depends on the lattice geometry and relative exchange coupling constants (Fig.1).

3. Magnetic Raman scattering

We consider a magnetic Raman operator of the Loudon-Fleury type [5]

$$R = -\sum_{\lambda=x,y,z} \sum_{<i,j>_\lambda} J_\lambda (\boldsymbol{e}_{\rm in} \cdot \boldsymbol{d}_{i,j})(\boldsymbol{e}_{\rm sc} \cdot \boldsymbol{d}_{i,j}) \sigma_i^\lambda \sigma_j^\lambda = \mathrm{i} \sum_{\lambda=x,y,z} \sum_{<i,j>_\lambda} J_\lambda (\boldsymbol{e}_{\rm in} \cdot \boldsymbol{d}_{i,j})(\boldsymbol{e}_{\rm sc} \cdot \boldsymbol{d}_{i,j}) u_{<i,j>_\lambda} c_i c_j, \tag{6}$$

where $\boldsymbol{e}_{\rm in}$ ($\boldsymbol{e}_{\rm sc}$) is the polarization vector of incident (scattered) light and $\boldsymbol{d}_{i,j}$ is the vector connecting the ith and jth sites. The magnetic Raman operator commutes with bond operators $u_{<i,j>_\lambda}$. It does not change any configuration of $u_{<i,j>_\lambda}$ but excites matter fermions only in pair. Neglecting the Rayleigh terms, we write the magnetic Raman operator as

$$R = \sum_k \sum_{n,n'=1}^{p+2q+1} \left(A_{k,n,n'}\alpha_{k,n}^\dagger \alpha_{k,n'} + A_{k,n,n'}^* \alpha_{k,n'}^\dagger \alpha_{k,n} + B_{k,n,n'}\alpha_{k,n}^\dagger \alpha_{-k,n'}^\dagger + B_{k,n,n'}^* \alpha_{-k,n'}\alpha_{k,n}\right), \tag{7}$$

where $A_{k,n,n'}$ and $B_{k,n,n'}$ are obtained by applying the unitary transformation Eq. (5) to Eq. (6). Note that the total momentum of two matter fermions mediating Raman scattering is zero under the long-wavelength approximation. The magnetic Raman intensity reads

$$I(\omega) = \frac{1}{2\pi\hbar L}\int_{-\infty}^{\infty} dt e^{\mathrm{i}\omega t}\langle 0|e^{\frac{\mathrm{i}\mathcal{H}t}{\hbar}} R e^{-\frac{\mathrm{i}\mathcal{H}t}{\hbar}} R|0\rangle = \frac{2}{L}\sum_k \sum_{n,n'=1}^{p+2q+1} \delta(\hbar\omega - \epsilon_{k,n} - \epsilon_{-k,n'})|B_{k,n,n'}|^2, \tag{8}$$

where L is the total number of sites.

We set J_λ's all equal to unity and show in Fig. 2 the magnetic Raman spectra together with the matter fermion density of states. Sharp peaks of Raman spectra correspond to the van Hove singularities. Because the (1,0) Kitaev nanoribbon has gapless excitations of the matter fermion, its Raman spectrum of the outgoing light polarization sc2 has finite intensity from the

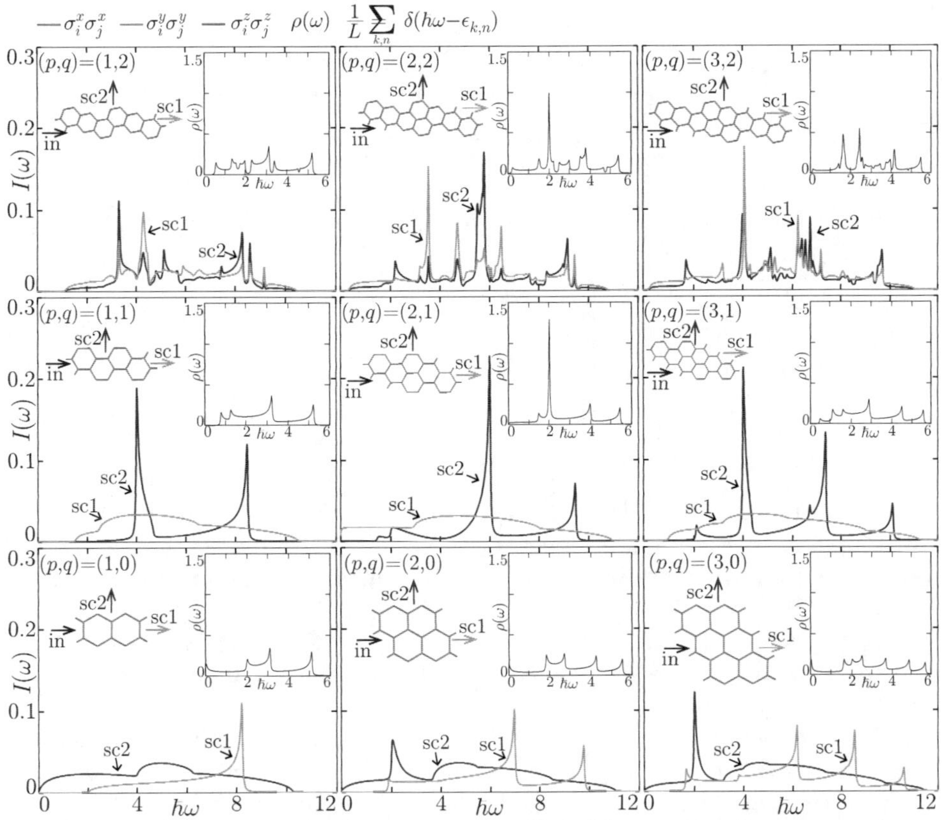

Figure 2. Magnetic Raman scattering spectra mediated by matter fermions, whose density of states are shown in the insets, of the gauge-ground Kitaev nanoribbons. Given a polarization vector of incoming light (in), we calculate two cases of outgoing light polarization (sc1, sc2).

energy $\hbar\omega = 0$. In the gapped case, the Raman spectra have no intensity in the energy region less than twice the lowest excitation energy of matter fermions.

Unlike the case with the Kitaev honeycomb model [5], the Raman spectra depend on the polarization. Taking the $(1, 0)$ Kitaev nanoribbon as an example, we discuss the polarization dependence from the point of view of its symmetry. We consider the gauge transformations $c_i \to -c_i$, which read $u_{<i,j>_\lambda} \to -u_{<i,j>_\lambda}$ with sites j being the nearest neighbors of site i [Fig. 3(a)]. The gauge-ground Kitaev Hamiltonian is invariant to certain gauged rotations. Such symmetry operations form what is called a projective symmetry group (PSG) [6, 7]. Figure 3(b) demonstrates symmetry operations within the PSG

$$\widetilde{\mathbf{D}}_2 = \{E, \Lambda_\mathrm{A}\Lambda_\mathrm{B}, \Lambda_\mathrm{A}C_{2x}, \Lambda_\mathrm{B}C_{2x}, C_{2y}, \Lambda_\mathrm{A}\Lambda_\mathrm{B}C_{2y}, \Lambda_\mathrm{A}C_{2z}, \Lambda_\mathrm{B}C_{2z}\}. \tag{9}$$

Matter fermion eigenstates are generally characterized by the irreducible representations Γ_1, Γ_2, Γ_3, and Γ_4 of the double group of $\mathbf{C}_2$

$$\widetilde{\mathbf{C}}_2 = \{E, \Lambda_\mathrm{A}\Lambda_\mathrm{B}, \Lambda_\mathrm{A}C_{2x}, \Lambda_\mathrm{B}C_{2x}\}. \tag{10}$$

Only the zone-center ($k = 0$) and zone-boundary ($k = \pm\pi$) eigenstates have the full symmetry $\widetilde{\mathbf{D}}_2$. The two matter fermion bands belong to the Γ_3 and Γ_4 representations [Fig. 4(a)]. The two-particle excited states are characterized by their product representations

$$\Gamma_3 \times \Gamma_4 = \Gamma_1,\ \Gamma_3 \times \Gamma_3 = \Gamma_4 \times \Gamma_4 = \Gamma_2. \tag{11}$$

We show in Fig. 4(b) the momentum-resolved magnetic Raman spectra

$$I(k,\omega) = 2 \sum_{n,n'=1}^{p+2q+1} \delta(\hbar\omega - \epsilon_{k,n} - \epsilon_{-k,n'})|B_{k,n,n'}|^2 \tag{12}$$

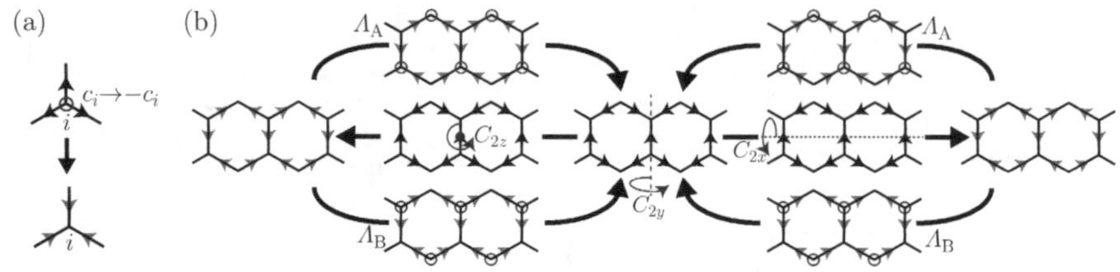

Figure 3. (a) Gauge transformation $c_i \to -c_i$. (b) Symmetry operations $\in \widetilde{\mathbf{D}}_2$. $u_{<i,j>_\lambda}$ reads ± 1 for the $i \rightleftharpoons j$ arrows, respectively.

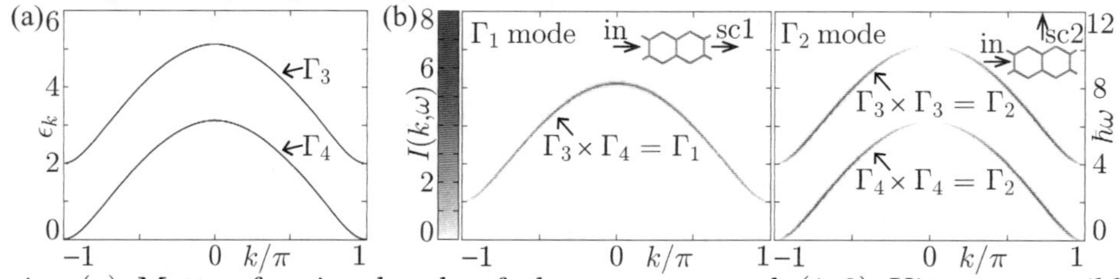

Figure 4. (a) Matter fermion bands of the gauge-ground $(1, 0)$ Kitaev nanoribbon. (b) Momentum-resolved magnetic Raman spectra of the gauge-ground $(1, 0)$ Kitaev nanoribbon and the product representations corresponding to the matter fermion excitations in pair.

with their belonging product representation. The scattered light polarizations sc1 and sc2 correspond to the Γ_1 and Γ_2 representations, respectively. The Γ_1 mode detects pair fermions of different representations, whereas the Γ_2 mode reveals those of the same representation. The Γ_2 mode trivially disappears at $k = 0$ and π because of the Pauli exclusion principle. That is why the Raman spectrum of polarization sc2 is not strongly featured without any sharp peak. Pair fermion excitations at the zone center and boundary are described by the higher symmetry group $\widetilde{\mathbf{D}}_2$. Interband two particle excitations at $k = \pi$ are characterized by the antisymmetric response to $\Lambda_A C_{2z}$ and $\Lambda_B C_{2z}$. Since the operation C_{2z} behaves as the space inversion in the present two-dimensional plane, this "ungerade" excitation is forbidden in Raman scattering. Therefore, the Raman spectrum of polarization sc1 has only one peak arising from interband two particle excitations at $k = 0$.

4. Concluding remarks

A polarization-independent magnetic Raman spectrum is characteristic of the Kitaev spin liquid on the two-dimensional honeycomb lattice [5] describable with the two-dimensional point group $\mathbf{C}_{6v}$, possessing *only one* Raman-active representation, E_2, in *two dimensions* [8, 9]. PSG of the gauge-ground $(1, 0)$ Kitaev nanoribbon merely has *one-dimensional* representations and *two* of which are Raman active, resulting in a strong polarization dependence of the Raman spectrum. The present Raman scattering is mediated by matter fermions in pair and a PSG analysis on them reveals the polarization dependence quantitatively.

References

[1] Ezawa M 2006 *Phys. Rev.* B **73** 045432
[2] Kitaev A 2006 *Ann. Phys* **321** 2
[3] Savary L and Balents L 2017 *Rep. Prog. Phys* **80** 016502
[4] Baskaran G, Mandal S and Shankar R 2007 *Phys. Rev.* Lett **98** 247201
[5] Knolle J, Chern G W, Kovrizhin D L, Moessner R and Perkins N B 2014 *Phys. Rev.* Lett **113** 187201
[6] Wen X G 2002 *Phys. Rev.* B **65** 165113
[7] Mellado P, Petrova O and Tchernyshyov O 2015 *Phys. Rev.* B **91** 041103
[8] Cépas O, Haerter J O and Lhuillier C 2008 *Phys. Rev.* B **77** 172406
[9] Perreault B, Knolle J, Perkins N B and Burnell F J 2015 *Phys. Rev.* B **92** 094439

Anomalous Hanle curves induced by in-plane nuclear field in single self-assembled InAlAs and InAs nanostructures

S. Yamamoto[1], R. Matsusaki[1], R. Kaji[1], and S. Adachi[1]
[1] Department of Applied Physics, Hokkaido University, Sapporo 060-8628, Japan
E-mail: yama-st@eis.hokudai.ac.jp

Abstract. We studied the formation mechanism of in-plane nuclear field by anomalous Hanle effect measurements in single self-assembled $In_{0.75}Al_{0.25}As/Al_{0.3}Ga_{0.7}As$ quantum dots and InAs/GaAs quantum rings. The observed anomalous Hanle curves indicated quite large width and hysteretic behavior to the externally applied magnetic field. These anomalies appears in both materials whose electron g factors have opposite sign each other, and cannot be reproduced by a traditional spin dynamics model. In this study, we show that a proposed spin dynamics model can explain the formation of a large in-plane nuclear field regardless of the sign of electron g factor. The model is based on the dynamic nuclear spin polarization mechanism including the effects of nuclear quadrupole interaction.

1. Introduction

In the semiconductor nanostructures, the photon angular momentum can be transferred onto the lattice nuclei via the localized electron spin by hyperfine interaction (HFI), and thus a large nuclear spin polarization (NSP) can be generated [1, 2]. The NSP affects an electron spin as a nuclear field (Overhauser field) $\boldsymbol{B}_{\mathrm{n}}$, and conversely, the electron spin acts on the lattice nuclei as a Knight field $\boldsymbol{B}_{\mathrm{e}}$. This reciprocal interplay complicates the spin physics in semiconductor nanostructures. Therefore, the engineering of NSP is required for the realization of potential applications such as spin storage and optical manipulation of a confined electron spin.

Because HFI induces simultaneous flip-flop of the electron and nuclear spins generally [1], it has been believed that the direction of NSP, hence the direction of B_{n}, is only parallel to the injected electron spin. However, recent experiments revealed that the large in-plane nuclear field $B_{\mathrm{n},x}$ which is perpendicular to the photo-injected electron spin can be generated under the external in-plane magnetic field B_x [3, 4]. Further, an independent experiment with single strain-free quantum dot (QD) suggested that the formation of $B_{\mathrm{n},x}$ is closely associated with quadrupole interaction (QI) [5], which is the coupling between an electric quadrupole moment of nucleus with spin $I>1/2$ and the electric field gradient caused by residual strain and/or random alloying in the self-assembled (SA) nanostructures.

Although the model of $B_{\mathrm{n},x}$ formation was proposed recently and the qualitative agreement has been achieved [6], further experimental and theoretical investigations should be carried out for complete understanding of the QI effects. In this paper, we compare the $B_{\mathrm{n},x}$ generated in two different materials where the sign of electron g factor $\mathrm{g_e}$, one of the most crucial parameters for the direction of B_{n}, is opposite each other.

IOP Conf. Series: Journal of Physics: Conf. Series **1220** (2019) 012047 doi:10.1088/1742-6596/1220/1/012047

2. Sample and Experimental Setup

Single SA $In_{0.75}Al_{0.25}As/Al_{0.3}Ga_{0.7}As$ QDs and SA InAs/GaAs quantum rings (QRs) grown by molecular beam epitaxy on (100)-GaAs substrate were used in this study. From the independent experiments, it was confirmed that the sign of g_e of the conduction band in InAlAs QDs and InAs QRs are positive and negative respectively [7]. After fabrication of small mesa structures, time-integrated micro-photoluminescence (μ-PL) measurements under B_x up to 1 T were carried out at 6 K. Typical PL spectra of both single nanostructures are shown in Fig. 1. In this paper, we focus on the positively charged exciton X^+ which showed no fine-structure splitting. This is because the degree of circular polarization (DCP) ρ_c of X^+ PL is determined essentially by z component of the electron spin polarization $\langle S_z \rangle$ through the relation $\rho_c = 2\langle S_z \rangle$. Here, the DCP is defined as $\rho_c = (I^- - I^+)/(I^- + I^+)$ where $I^{+(-)}$ is the $\sigma^{+(-)}$ polarized PL intensity. A continuous wave Ti:Sapphire laser was used for excitation to the tail of the wetting layer. The corresponding wavelengths were ~730 nm for InAlAs QDs and ~865 nm for InAs QRs. The excitation light was adjusted to the circularly polarized one by combination of a linear polarizer, a half waveplate, and a quarter waveplate. The simultaneous acquisition of the σ^+ and σ^- PL components was achieved with a beam displacer. Each displaced PL component is focused on a different detection area of the CCD, and then, DCP can be acquired by a single exposure process in highly accurate manner. Details of the apparatus are found in Ref. [8].

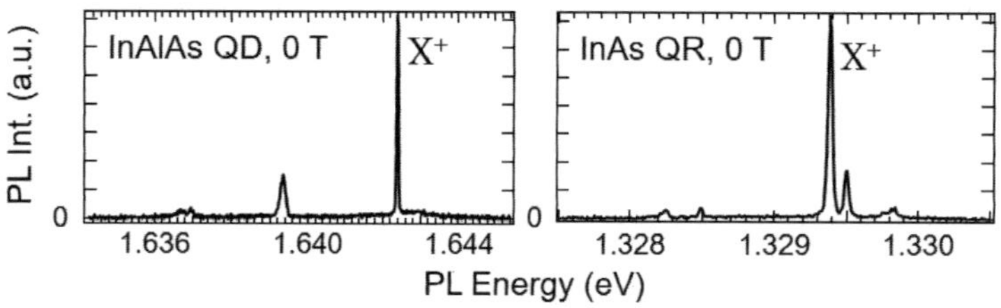

Figure 1. PL spectra of single InAlAs QD and single InAs QR at 6 K and 0 T.

3. Hanle effect measurements

As is well known, the precession of the electron spin $\langle \boldsymbol{S} \rangle$ due to a magnetic field in Voigt geometry is described by Bloch equation. The steady-state $\langle S_z \rangle$ decays with increasing B_x, and the depolarization curve (Hanle curve) shows a Lorentzian function of B_x if $|\boldsymbol{B}_\mathrm{n}| = 0$. Further, the shape of Hanle curve is modified reflecting the direction and the magnitude of B_n induced by HFI [1]. In the traditional model, $\boldsymbol{B}_\mathrm{n}$ induced by *nuclear spin cooling* due to $\boldsymbol{B}_\mathrm{e}$ is written as

$$\boldsymbol{B}_\mathrm{n} = b_\mathrm{n}[(\boldsymbol{B}_\mathrm{T}^{(\mathrm{n})} \cdot \langle \boldsymbol{S} \rangle)\boldsymbol{B}_\mathrm{T}^{(\mathrm{n})}]/(|\boldsymbol{B}_\mathrm{T}^{(\mathrm{n})}|^2 + \xi B_\mathrm{L}^2), \quad (1)$$

where $\boldsymbol{B}_\mathrm{T}^{(\mathrm{n})}$ is a sum of externally applied field $\boldsymbol{B}_\mathrm{ext}$ and $\boldsymbol{B}_\mathrm{e}$, B_L is a dipolar field among neighboring nuclei, and ξ=2-3 is a linkage coefficient. The factor b_n is a material-dependent parameter which is proportional to $1/g_e$. Figure 2(a) shows $\langle S_z \rangle$ calculated with the traditional model. The upper (lower) panel corresponds to the case of negative (positive) g_e. In the figure, the dashed lines indicate Lorentzian shaped Hanle curves. The *normal* Hanle curve is deformed only in the narrow B_x region and the deformation is different depending on the sign of g_e. Former is related to the magnitude of $\boldsymbol{B}_\mathrm{e}$ (~ a few mT), and latter can be explained by considering that the direction of $B_{\mathrm{n},x}$ depends on the sign of g_e through b_n in Eq. (1); namely amplification (compensation) of B_x occurs when g_e is negative (positive).

Figure 2(b) and (c) show the observed *anomalous* Hanle curves in InAlAs QD and InAs QR, respectively. The dashed curves are normal Lorentzian ones expected from the spin lifetime T_s=0.5 ns and g_e=+0.35 (-0.42) for InAlAs QD (InAs QR). It is obvious that the observed

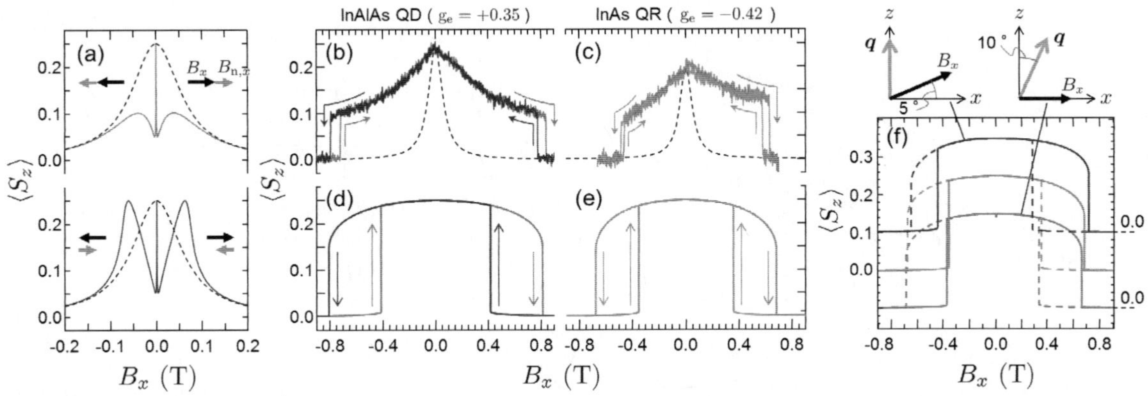

Figure 2. (a) The calculated Hanle curves by traditional theory (upper: $g_e<0$, lower: $g_e>0$). The dashed line shows Lorentzian shaped Hanle curve without B_n and the pairs of arrows show the direction of B_x and $B_{n,x}$ schematically. (b), (c) The observed anomalous Hanle curves in InAlAs QD and InAs QR at 6 K. The dashed line shows Lorentzian shaped Hanle curves. (d), (e) The calculated anomalous Hanle curves with $g_e=+0.35$ in (c) for InAlAs and $g_e=-0.42$ in (d) for InAs QR. (f) The calculated anomalous Hanle curves with the tilted B_x (top), with the normal condition (middle), and with the tilted QI axis (bottom). Top insets are the configuration of the principal axis of QI ($\boldsymbol{q}$) and B_x.

curves deviate significantly from the Lorentzian and have quite large width despite the opposite sign of g_e. Further, the critical fields B_c at which $\langle S_z \rangle$ changes abruptly are seen in both curves and the value of B_c depends on the sweep direction of B_x (i.e. hysteresis). Highly-preserved $\langle S_z \rangle$ suggests the compensation of B_x by $B_{n,x}$, and the hysteresis originates from the bistability of NSP, hence B_n. Similar compensation of external field by B_n has been well-studied in z-direction [2]. Compared with the traditional theory, the most curious point is that the $B_{n,x}$ was always anti-parallel to B_x regardless of the sign of g_e.

Figure 2(d) and (e) show the computed $\langle S_z \rangle$ according to the model of $B_{n,x}$ formation which was proposed in our previous study [6]. Both results reproduce the above-mentioned anomalies qualitatively, and therefore, the model proves to explain well the formation of $B_{n,x}$ in spite of the opposite sign of g_e. Our model is based on the dynamic nuclear spin polarization where the dynamics of NSP $\langle I_k \rangle$ along $k = x, y, z$ direction can be written as

$$\frac{d\langle I_k \rangle}{dt} = \frac{1}{T_{\mathrm{NF},k}}[Q(S_k - S_{\mathrm{eq},k}) - \langle I_k \rangle] - \frac{1}{T_{\mathrm{ND},k}}\langle I_k \rangle, \tag{2}$$

where $1/T_{\mathrm{NF},k}$, $1/T_{\mathrm{ND},k}$, S_k, and $S_{\mathrm{eq},k}$ are k components of formation and relaxation rate of NSP, electron spin polarization, and thermal equilibrium value of S_k, respectively. Q is a momentum conversion coefficient from electron to nuclei. Note that, $1/T_{\mathrm{NF},k}$ includes $\langle I_k \rangle$ self-consistently and thus NSP shows nonlinear response such as bistability [2].

In addition, the effects of QI have to be considered. The introduced effects into our model are (i) suppression of NSP precession and (ii) sign inversion between out-of- and in-plane nuclear g factor g_n. The first QI effect (i) originates from unequally-spaced splitting of nuclear energy levels [quadrupolar splitting (left inset of Fig. 3)]. As seen in Fig. 3, the spin pairs with $|I_z| > 1/2$ indicate zero or quite small splittings. Namely, the *effective* g_n decreases, and thus the precession of NSP is suppressed even under a large B_x. The second QI effect (ii) is supposed to arise from nuclear spin state-mixing induced by the combination of QI and Zeeman splitting due to B_x. Since the direction of $B_{n,k}$ depends on $g_{n,k}$ as well as g_e, the sign inversion between $g_{n,x(y)}$ and $g_{n,z}$ could realize the $B_{n,x}$ formation regardless of the sign of g_e. Note that g_e is nearly isotropic because of the symmetrical nature of conduction band.

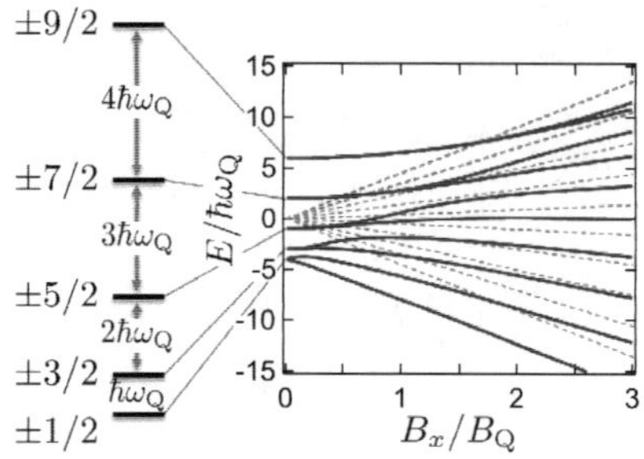

Figure 3. The energy splitting of a single In nucleus (I=9/2) under QI and Zeeman interaction in Voigt geometry. $\boldsymbol{q}$ is assumed to be along z axis. The dashed lines indicate the Zeeman splitting without QI. The left inset indicates quadrupolar splitting under zero magnetic field. Here, the horizontal (vertical) axis is normalized by the quadrupolar field B_Q (quadrupolar splitting energy $\hbar\omega_\mathrm{Q}$).

Finally, we should comment about the asymmetry of the observed curve in Fig. 2(c) while the corresponding calculation [Fig. 2(e)] is symmetrical. The possible origin of the asymmetry is inclination of the principal axis $\boldsymbol{q}$ of QI from the crystal growth axis (z axis) and/or deviation of external magnetic field from the sample plane (x-y plane). In general, $\boldsymbol{q}$ is determined by strain axis and lies along almost parallel to z axis in SA nanostructures. Also, misalignment of the apparatus could make the external field deviated from x-y plane. Figure 2(f) shows the computed $\langle S_z \rangle$ with a tilted $\boldsymbol{q}$ by 10 degrees (red) ,with a tilted B_x by 5 degrees (blue) and with the normal condition (middle) where B_x and $\boldsymbol{q}$ point strictly x and z axes respectively as with Fig. 2(e). The respective configurations are schematically depicted in the insets. Both cases induce the asymmetry in Hanle curve, but there are some differences. The inclination of $\boldsymbol{q}$ induces a horizontal shift of whole curves, while the tilting of B_x induces horizontal shifts of the inner and outer B_c in the opposite direction each other, and thus the hysteresis width becomes different depending on the sign of B_x. Therefore, it is likely that the observed asymmetry in Fig. 2(c) comes from the inclination of B_x due to the misalignment of the sample mounting.

4. Summary

We observed the anomalous Hanle curves in InAlAs QDs and InAs QRs, which indicated the formation of considerable $B_{\mathrm{n},x}$ regardless of the sign of $\mathrm{g_e}$. The observed anomalies in the curves could not be explained by the traditional theory, and we indicated that our model of $B_{\mathrm{n},x}$ formation could reproduce experimental results qualitatively. In the model, the strain-induced QI plays a key role to generate $B_{\mathrm{n},x}$. Moreover, the tilting effect of B_x and QI axis which induced the asymmetry to the curves was discussed.

Acknowledgments

This work is supported in part by JSPS KAKENHI (Grant No. JP26800162) and the Asahi Glass Foundation (Japan).

References

[1] Fleisher V G and Merkulov I A 1984 *Optical Orientation (Modern Problems in Condensed Matter Sciences)* vol 8 ed F Meier and B P Zakharchenya (North-Holland: Elsevier Science Publishers B.V.) chaps. 2 and 5
[2] Recent optical investigation of nuclear spin physics in QDs are reviewed comprehensively: Urbaszek B, Marie X, Amand T, Krebs O, Voisin P, Maletinsky P, Högele A and Imamoglu A 2013 *Rev. Mod. Phys.* **85** 79
[3] Krebs O, Maletinsky P, Amand T, Urbaszek B, Lemaître A, Voisin P, Marie X and Imamoglu A 2010 *Phys. Rev. Lett.* **104** 056603
[4] Nilsson J, Bouet L, Bennett A J, Amand T, Stevenson R M, Farrer I, Ritchie D A, Kunz S, Marie X, Shields A J and Urbaszek B 2013 *Phys. Rev.* B **88** 085306
[5] Sallen G, Kunz S, Amand T, Bouet L, Kuroda T, Mano T, Paget D, Krebs O, Marie X, Sakoda K and Urbaszek B 2014 *Nat. Commun.* **5** 3268
[6] Yamamoto S, Matsusaki R, Kaji R and Adachi S 2018 *Phys. Rev.* B **97** 075309
[7] Matsusaki R, Kaji R, Yamamoto S and Adachi S 2018 *Appl. Phys. Lett.* (submitted)
[8] Matsusaki R, Kaji R, Yamamoto S, Sasakura H and Adachi S 2018 *Appl. Phys. Express* (in press)

IOP Conf. Series: Journal of Physics: Conf. Series **1220** (2019) 012048 doi:10.1088/1742-6596/1220/1/012048

Photocurrent characteristics of nanostructured thin films consisting of surface-modified silicon nanoparticles

Ryu-ichi Yoshikado, Takayuki Hasegawa, Yoshihito Tanaka, Shuhei Tsubota, and Seiichi Sato

Graduate School of Material Science, University of Hyogo, Koto, Kamigori, Ako-gun, Hyogo 678-1297, Japan

E-mail: t_hase@sci.u-hyogo.ac.jp

Abstract. We have studied optical and electrical properties in thin films consisting of the network of silicon nanoparticles modified with mercaptosuccinic acid (MSA). A photocurrent measurement system was set up to observe the photocurrent signal as functions of excitation photon energy and power. In addition, we observed the absorption spectra of the thin film and solution of the nanoparticles to clarify the contribution of the nanoparticles structure to the optical transition property. It was found that the photocurrent has an excitation-energy dependence while the profile is different from that of a bulk silicon photodiode. This behaviour suggests the specific photogenerated-carrier transport originating from the structure of the thin film of MSA-modified silicon nanoparticles.

1. Introduction

Semiconductor nanoparticles have attracted considerable attention in fundamental semiconductor physics and functional materials. A thin film consisting of a network of nanoparticles enables us to control transport properties of optically- and electrically-injected carriers[1,2], which is promising to develop novel electric and optoelectronic devices. Moreover, the carrier transport process in a subpicosecond range is sensitive to the structure of the thin film because the length of the carrier motion is comparable to the nanoparticle size and interparticle spacings[3]. This characteristic is beneficial in the development of terahertz optoelectronic devices[4].

A silicon nanoparticles (SiNPs) thin film has a great advantage in technological and economical aspects. There are numerous reports on the optical and electrical properties in SiNPs thin films while most of the studies were conducted in the structure consisting of SiNPs embedded in a silicon oxide[3]. The investigation of various structures of the SiNPs thin film is essential to the comprehensive understanding. Recently, Sato *et al.* fabricated thin films consisting of SiNPs surface-modified with a thin functional group using a mechanochemical process. This fabrication method enables us to produce various nanoparticles environments by selecting the modifier, which is expected to bring out a new functionality of the SiNPs thin film. In fact, the thin film consisting of the SiNPs modified with a short chain carboxylic acid, mercaptosuccinic acid (MSA), exhibits a distinct transport property of electrically-injected carriers[5].

In this work, we studied transport property of photogenerated carriers in the thin film consisting of the MSA-modified SiNPs at various photoexcitation conditions. We set up a photocurrent measurement system to observe the photocurrent signal as functions of excitation photon energy and

IOP Conf. Series: Journal of Physics: Conf. Series **1220** (2019) 012048 doi:10.1088/1742-6596/1220/1/012048

power, and discussed the photocurrent response from the aspect of the structure of the MSA-modified SiNPs thin film.

2. Experimental

MSA-modified SiNPs were produced by wet grinding of the silicon powder in MSA-containing super dehydrated methanol using a planetary ball mill[5]. The average diameter of the SiNPs was estimated to ~20 nm by a scanning electron microscope; The quantum size effect is not expected in the diameter[6]. The thin film was fabricated by a drop casting process of the SiNPs dispersed in distilled water to a glass plate. For the electrical characterization, the droplet was deposited on a SiO_2 film with Au electrodes separated by a distance of 5 μm (The gap area was 5×600 μm^2). In measurements of current-voltage characteristics, the light source was a laser diode with a photon energy of 1.9 eV. In measurements of the excitation energy dependence of the photocurrent signal, the light source was produced by the combination of a Xe lamp and a single monochromator with a spectral resolution of ~7 nm. The excitation light was focused around the electrode gap region by using an objective lens. Moreover, the reflected and scattered excitation lights by the sample surface were monitored using a CMOS camera to optimize the light focusing. The intensity of the excitation light was below ~1 μW to suppress a saturation behaviour of the photocurrent as described later. In addition, the excitation light was chopped by 5 Hz to detect the photocurrent signal by using a lock-in amplifier. The schematic view of the optical setup is shown in Fig. 3(a). All the measurements were conducted at room temperature.

3. Results and Discussion

The solid curves in Fig. 1 show the absorption spectra in the solution and thin film of the MSA-modified SiNPs. The offset profile of that in the thin-film sample is attributed to the scattering of the excitation light by a textured sample surface. The absorption spectra consist of a gradually increasing comportment that is similar to the spectrum in a bulk silicon[7] and a peak structure at ~2.5 eV. Such a peak structure is not observed in the SiNPs modified with ethoxy or propoxy groups fabricated by the same process (not shown here). Moreover, as shown by the dots in Fig. 1, the absorption in a MSA solution is negligible in the present energy range. Thus, the peak structure is specific to the MSA-modified SiNPs.

Figure 2 shows the current-voltage characteristics under the light illumination and dark conditions in the thin-film sample using the 1.9 eV-laser diode with the excitation power of ~3 mW.

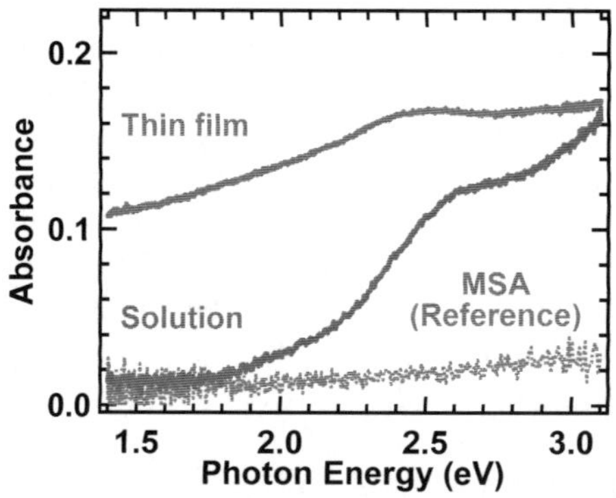

Figure 1. Absorption spectra in the solution and thin film of the MSA-modified SiNPs, where the spectrum in the thin-film sample is magnified by 0.7. The dots indicates the absorption spectrum in a MSA solution dissolved in methanol as a reference.

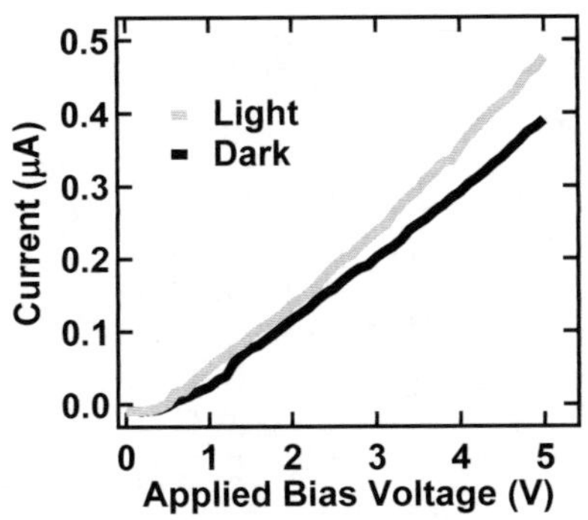

Figure 2. Current-voltage characteristics under the light illumination and dark conditions in the MSA-modified SiNPs thin film. The energy and power of the excitation light were 1.9 eV and 3 mW, respectively.

The photocurrent signal that corresponds to the difference between the two curves is clearly seen at the applied bias voltage above ~3 V (6 kV/cm). The behaviour indicates that the transport process of photogenerated carriers is enhanced above ~3 V, because the number of photogenerated carriers is generally less dependent on the internal electric field.

To comprehensively evaluate the transport process of photogenerated carriers in the MSA-modified SiNPs thin film, we observed the photocurrent signal as functions of excitation photon energy and power using the optical setup shown in Fig. 3(a). Figure 3(b) shows the spectrum of the photosensitivity that is defined as the photocurrent value divided by the excitation power. The estimation of the photosensitivity is as follows. The inset in Fig. 3(b) shows the excitation-power dependences of the photocurrent value at the excitation photon energies of 1.8 and 2.5 eV. It is clear that the photocurrent values show saturation behaviours with an increase in excitation power. This behaviour is possibly attributed to reflect the volume of the nanoparticles in the electrode gap region. The dashed curves show the phenomenological fitted results using the equation of $I_{pc}(x)=A[1-\exp(-Bx)]$, where A is a maximum photocurrent value and B is a rate of the saturation. In the estimation of the photosensitivity, we assumed the slope of the fitting curve at the origin to the photosensitivity as shown by the lines. The estimated photosensitivity spectrum is increased with a decrease in photon energy, which will be discussed in the next paragraph. In addition, the spectrum has a periodical component with the period of ~0.28 eV. We confirmed that this periodical component disappears when the excitation light is defocused. Moreover, the observed peak wavelengths can be calculated by the equation of $\lambda_n=2L/n$, where λ_n is the peak wavelength, L is the gap distance that of 5 μm, and n is an odd number. These facts suggest that the electrode gap structure is the origin of the periodical component; namely, the electrode gap structure affects the fluence of the excitation light to the SiNPs thin film.

As described in Fig. 3(b), the photosensitivity is increased with a decrease in photon energy. The rate of increase in photosensitivity from 3.0 eV to 1.5 eV is ~7. The rate is much higher than that of a bulk silicon photodiode, which is ~2 (Hamamatsu Photonics, S1336 series)[8]. The prominent near-infrared component in the present spectrum is attributed to the nature of the transport process in the network of the SiNPs nanoparticles; A similar spectral shape has reported in a SiNPs thin film[1]. Another possibility of the interpretation for the spectral shape is the contribution of sulfur atoms. A sulfur-doped silicon photodiode shows an enhancement of the photosensitivity in an infrared range, which is originated from optical transitions of sulfur-related states[9]. We note that sulfur is one of the element of MSA and the sulfur atoms are considered to couple to outermost silicon atoms in the MSA-modified SiNPs[5]. The contribution of the MSA modification is suggested also to the absorption spectra with the peak structures described in Fig. 1.

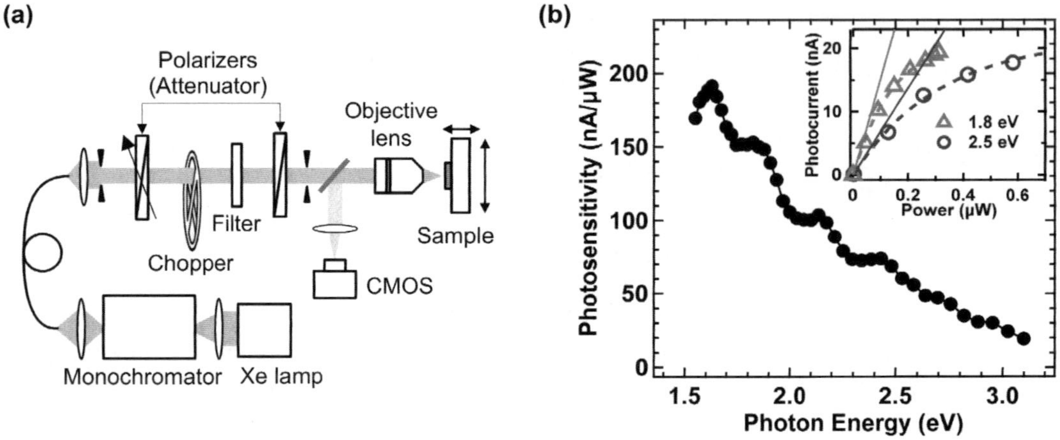

Figure 3. (a)Schematic view of the optical setup for the photocurrent measurement. (b)Photosensitivity spectrum in the MSA-modified SiNPs thin film under the applied bias voltage of 5 V (10 kV/cm). The inset shows the excitation-power dependences of the photocurrent value at the excitation photon energies of 1.8 and 2.5 eV. The dashed curves indicate the phenomenological fittings and the slopes of the solid lines correspond to the photosensitivities in a low excitation regime.

4. Summary
We have studied the optical and electrical properties of the thin film and solution of the MSA-modified SiNPs from the aspects of the absorption and photosensitivity spectra. The absorption spectra show the peak structure at ~2.5 eV, which is specific to the MSA modification. From the measurements of the photocurrent signals as functions of excitation power and photon energy and the consideration of the saturation behaviour, we precisely estimated the photosensitivity spectrum in the range from visible to near infrared. We found that the photosensitivity in the MSA-modified SiNPs thin film is further increased as approaching the near infrared range compared to that of a bulk silicon photodiode. The mechanism of the present transport process is attributed to the network structure of the nanoparticles or the additional optical transitions originating from sulfur-related states.

References
[1] Rezgui D B, Gourbilleau F, Maestre D, Palais O, Sibai A, Lemiti M and Brémond G 2012 *J. Appl. Phys.* **112** 024324
[2] Gutsch S, Laube J, Hartel M A, Hiller D, Zakharov N, Werner P and Zacharias M 2013 *J. Appl. Phys.* **113** 133703
[3] Němec H, Zajac V, Kužel P, Malý P,Gutsch S, Hiller D and Zacharias M 2015 *Phys. Rev. B* **91** 195443
[4] Daghestani S N, Persheyev S, Cataluna A M, Ross G and Rose J M 2011 *Semi. Sci. Technol.* **26** 075015
[5] Sato S, Dobashi T and Matsuda S 2015 *Chem. Eng. J.* **268** 356
[6] Matsumoto T, Suzuki J, Ohnuma M, Kanemitsu Y and Masumoto Y 2001, *Phys. Rev. B* **63** 195322
[7] Green A M 2008 *Solar Ene. Mat. & Solar Cells* **92** 1305
[8] Hamamatsu.com, https://www.hamamatsu.com/us/en/product/type/S1336-8BQ/index.html
[9] Li C, Zhao J, Yu X, Chen Q, Feng J, Han P and Sun H 2017 *IEEE Sens. J.* **17** 2367

Q-factor dependence of angle-resolved transmission spectra in CuCl microcavities

T Yokoyama, T Nishida, G Oohata, and K Mizoguchi*

Department of Physical Science, Osaka Prefecture University, Sakai, 599-8531, Japan

E-mail: k.mizoguchi@p.s.osakafu-u.ac.jp

We have investigated the angle-resolved transmission spectra of the CuCl λ- and $\lambda/2$-microcavities composed of cavity layers sandwiched by distributed Bragg reflectors (DBRs) with various numbers of periods of the DBR layer which affect the degree of confinement of the photons in the microcavity, so-called quality factors (Q factors). The angle-resolved transmission spectra of the CuCl microcavities drastically change with the Q factor and the thickness of the cavity layer. The incident angle dependence of the peak energies obtained from the angle-resolved transmission spectra indicates that the cavity polariton modes are observed in the CuCl microcavities, except for the λ-microcavity with 2 periods of the DBR layer. The change of the angle-resolved transmission spectra in the microcavities will result from the difference in the coupling state between the excitons and photons in the microcavities.

1. Introduction

A microcavity which is composed of a cavity layer sandwiched by mirrors with high reflectivity has a property to confine photons in a cavity layer [1]. In a semiconductor microcavity which includes a semiconductor layer in a cavity layer, cavity polaritons are generated in the microcavity when excitons and photons in a cavity layer will strongly interact with each other [2-4]. When the degree of confinement of photons in the microcavities which is referred to as the quality factor (Q factor) is changed, it is expected that the coupling state between photons and excitons in the semiconductor microcavities will be modified, and then the transmission and reflection spectra of the microcavities will dramatically change with the Q factor.

In this study, we have investigated the angle-resolved transmission (ART) spectra of the CuCl microcavities that consists of the cavity layers with the thickness of λ ($\lambda/2$) and the distributed Bragg reflectors (DBRs) with various numbers of periods of the DBR layer. Here, we have confirmed that the Q factor of the CuCl microcavities depends on the number of periods of the DBR layer. It is found that the ART spectra of the microcavities change with the Q factor and the thickness of the cavity layer. We discuss the change of the ART spectra observed in the microcavities from the viewpoint of the coupling state between the excitons and cavity photons.

2. Experimental Methods

The schematic structure of the fabricated CuCl microcavity is shown in figure 1. The CuCl λ- and $\lambda/2$-microcavities were fabricated on (0001)-Al_2O_3 substrates by using a vacuum deposition method. The cavity layers in the fabricated microcavities are sandwiched between two DBRs. Here, λ is given by λ_{ex}/n_b, in which λ_{ex} is the resonant wavelength of the Z_3 exciton of CuCl and n_b is the background

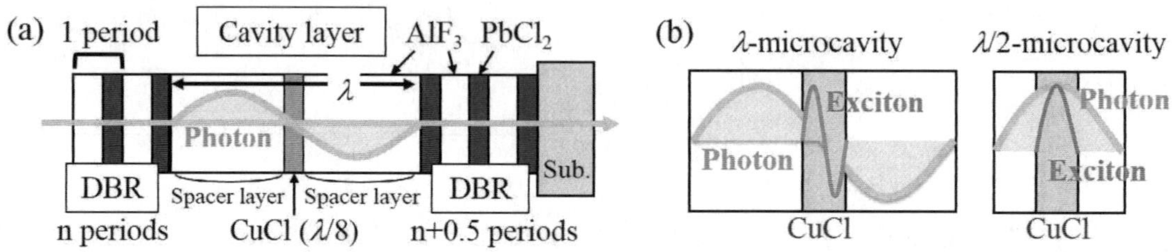

Figure 1. (a) Schematic structure of the fabricated CuCl microcavity. (b) Schematic diagrams of the spatial distributions of the electric field of the cavity photon (orange curve) and the wave function of the exciton (green curve) in the cavity layers of the CuCl λ- and $\lambda/2$-microcavities.

refractive index of CuCl [2,3]. In the CuCl λ ($\lambda/2$)-microcavities, the cavity layer with the thickness of λ ($\lambda/2$) consisted of a CuCl active layer with thickness of $\lambda/8$ sandwiched by AlF_3 spacer layers. The constituent layers of the DBRs were composed of $PbCl_2$ and AlF_3 with a layer thickness of $\lambda/4$, where the thickness of one period of the DBR was $\lambda/2$. The Q factor of the microcavity, which was represented by the ratio of the peak energy to the spectral width of the cavity photon mode, was controlled by changing the number of periods of the DBR [2-4]. The Q factors of the fabricated microcavities obtained by measuring the cavity photon modes at room temperature are described as annotations in figure 2. The ART spectra were observed at 10 K using a 32-cm single monochromator combined with a charge-coupled device camera and a Xe lamp as the probe light source. The spectral resolution and the angle resolution were approximately 0.4 nm and 2 degree, respectively.

3. Experimental results and Discussion

The ART spectra observed in the λ-microcavities with 2, 3 and 5 periods of the DBR layer, and those in the $\lambda/2$-microcavities with 2, 3 and 5 periods are shown in figure 2. In the λ-microcavities with 3 and 5 periods, three peaks are clearly observed. The three peaks are attributed to be the cavity polariton modes, because the peak energies of these peaks depend on the incident angle θ, which corresponds to the dispersion relation of the cavity polaritons [2-4]. Moreover, the spectral profiles of these peaks change between the symmetric and asymmetric structures with the increase of the incident angle. On the other hand, in the λ-microcavity with 2 periods, two dips near the exciton energies appear on the broad peak of which the peak energy changes with the incident angle. The changes of the spectral profiles between the λ-microcavities with different periods of the DBR layer suggest that the coupling state between the photons and excitons in the cavity layer depends on the Q factor.

In all the $\lambda/2$-microcavities shown in figures 2(d), (e) and (f), three peaks are obviously observed. It is noted that although the Q factor of the $\lambda/2$-microcavity with 2 periods is smaller than that of the λ-microcavity with 2 periods, three peaks dependent on the incident angle appear in the $\lambda/2$-microcavity with 2 periods. This result implies that the photons and excitons in the $\lambda/2$-microcavities will be strongly coupled to each other as compared with the λ-microcavities.

Next, we have investigated the incident angle dependence of the peak (dip) energies observed in the microcavities. Since the spectral profiles of the peaks and dips observed in the ART spectra change between the symmetric and asymmetric structures with the incident angle, the ART spectra are fitted by the following equation which is represented by the superposition of Fano functions [5],

$$T(E) = \sum_{n=1}^{3} F_n(E) + Background \;, \quad F_n(E) = A_n \times \left\{ \frac{\left(\frac{E-E_{0n}}{\gamma_n/2}+q_n\right)^2}{\left(\frac{E-E_{0n}}{\gamma_n/2}\right)^2+1} - 1 \right\}. \qquad (1)$$

Here, A_n, E_{0n}, and γ_n are the amplitude, the peak energy, and the full width at half maximum (FWHM), respectively. q_n is asymmetric parameter, so-called Fano parameter. The red solid curves in figure 2 indicate the fitted curves, which are in good agreement with experimental results. The peak energies obtained from the fitting procedure are plotted as a function of incident angle in figure 3. The incident angle dependence of the peak energies in the microcavities except the λ-microcavity with 2 periods shows the anti-crossing behavior around the Z_3 and $Z_{1,2}$ exciton energies of CuCl. In the λ-microcavity with 2 periods, the dip energies are almost independent on the incident angle, while the peak energy of the broad peak changes with the incident angle. This result indicates that the anti-crossing behavior does not appear in this microcavity; consequently, the dips around the exciton energies will be observed by the weak coupling between the photons and excitons, such as the polaritons observed in CuCl crystals.

Finally, we discuss the coupling constant between the photons and excitons in the microcavities. In order to estimate the coupling constants, the incident angle dependence is fitted by using the following effective Hamiltonian H [2-4],

$$H = \begin{pmatrix} E_{\text{cav}}(\theta) & \Omega_3/2 & \Omega_{1,2}/2 \\ \Omega_3/2 & E_{Z3} & 0 \\ \Omega_{1,2}/2 & 0 & E_{Z1,2} \end{pmatrix}, \quad E_{\text{cav}}(\theta) = E_{\text{ph}}\left(1 - \frac{\sin^2\theta}{n_{\text{eff}}^2}\right)^{-\frac{1}{2}}. \qquad (2)$$

Here, E_{cav} is the peak energy of the cavity photon mode which depends on the incident angle, and E_{ph} is the cavity photon energy at the normal incidence. n_{eff} is the effective refractive index in the cavity layer.

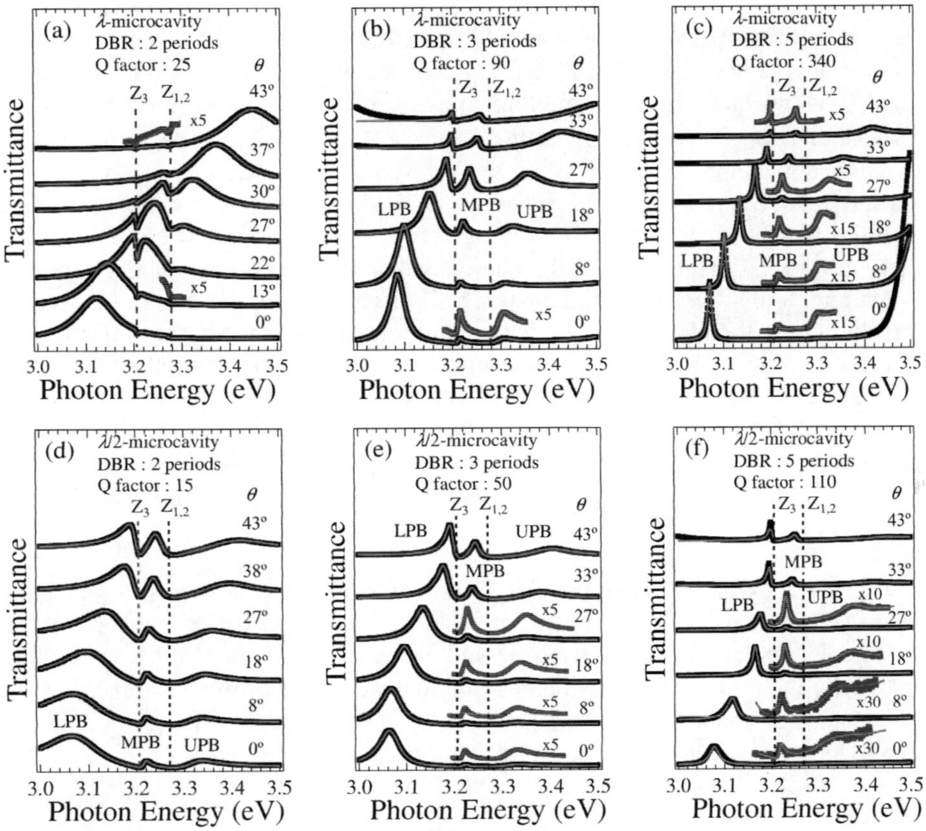

Figure 2. Transmission spectra of the CuCl λ-microcavities with 2 periods of the DBR layer (a), 3 periods (b), and 5 periods (c), and those of the CuCl $\lambda/2$-microcavities with 2 periods (d), 3 periods (e) and 5 periods (f) at various incident angles of the probe light. The black dotted curves indicate the experimental results and the blue dotted curves show the transmission spectra multiplied by appropriate constants. The red solid curves show the fitted curves. The vertical dashed lines indicate the Z_3 and $Z_{1,2}$ exciton energies of CuCl. θ is the incident angle of the probe light. LPB, MPB and UPB indicate the lower, middle, and upper polariton branches, respectively.

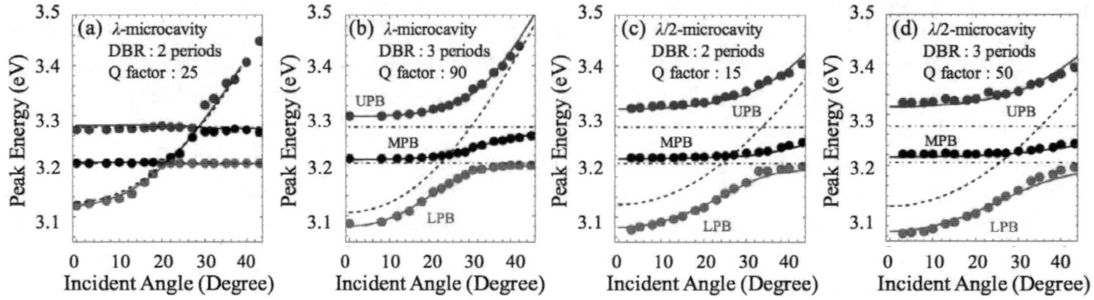

Figure 3. Incident angle dependence of the peak energies in the CuCl λ-microcavities with 2 periods (a) and 3 periods (b) and that in the CuCl $\lambda/2$-microcavities with 2 periods (c) and 3 periods (d). The solid curves indicate the fitted curves by using the effective Hamiltonian. The horizontal dashed-dotted lines indicate the Z_3 and $Z_{1,2}$ exciton energies of CuCl. The dashed curves show the dispersion relation of the cavity photon.

Ω_3 ($\Omega_{1,2}$) is the coupling constants between the cavity photon and the Z_3 ($Z_{1,2}$) exciton. E_{Z3} and $E_{Z1,2}$ are the Z_3 and $Z_{1,2}$ exciton energies, respectively. The solid curves in figure 3 indicate the fitted curves which well reproduce the incident angle dependence of the peak energies. From the fitting procedure, the coupling constants of Ω_3 and $\Omega_{1,2}$ in the $\lambda/2$-microcavities are estimated to be approximately 50 and 95 meV, respectively, while those in the λ-microcavities except the λ-microcavity with 2 periods are approximately 45 and 85 meV, respectively. In the λ-microcavity with 2 periods, the coupling constants of Ω_3 and $\Omega_{1,2}$ are less than approximately 40 meV. Although these values obtained in this microcavity show no characteristic of the strong coupling state, we consider that the observation of the dips around the exciton energies will be due to the weak coupling between the photons and excitons. Moreover, it is noted that the coupling constants of the $\lambda/2$-microcavities are slightly larger than those of the λ-microcavities. Assuming that the coupling constant will originate from the degree of overlap between the spatial distribution of the electric fields of the cavity photons in the cavity layers and the wave functions of the excitons with various principal quantum numbers ($n\geq1$) in the active layers as shown in figure 1(b), we have calculated the overlap integral between the electric field of the cavity photon and the wave functions of the excitons. The calculated results indicate that the values of the overlap integral in the $\lambda/2$-microcavities are larger than those in λ-microcavities, which is consistent with the characteristics of the coupling constants obtained in the $\lambda/2$- and λ-microcavities.

4. Conclusion

We have investigated the ART spectra of the CuCl λ- and $\lambda/2$-microcavities with various periods of the DBR layer. The ART spectra of the microcavities change with the Q factor and the thickness of the cavity layer. The cavity polariton modes are observed in the ART spectra of the microcavities, except for the λ-microcavity with 2 periods. The incident angle dependence of the peak energies observed in the microcavities indicates that the CuCl microcavities except the λ-microcavity with 2 periods express the strong coupling state between the photons and excitons.

Acknowledgments
This work was supported by JSPS KAKENHI Grant Number JP16H04003.

References

[1] Weisbuch C, Nishioka M, Ishikawa A, and Arakawa Y 1992 *Phys. Rev. Lett.* **69** 3314
[2] Oohata G, Nishioka T, Kim D, Ishihara H, and Nakayama M 2008 *Phys. Rev. B* **78** 233304
[3] Yoshino S, Oohata G, Shim Y, Ishihara H, and Mizogichi K 2013 *Phys. Rev. B* **88** 205311
[4] Nakayama M, and Ueda M 2017 *Phys. Rev. B* **95** 125315
[5] Fano U 1961 *Phys. Rev.* **124** 1866

IOP Conf. Series: Journal of Physics: Conf. Series **1220** (2019) 012050 doi:10.1088/1742-6596/1220/1/012050

Spectral properties of chiral electromagnetic near fields created by chiral plasmonic nanostructures

S. Hashiyada[1]*, K. Endo[2], T. Narushima[1,3], Y. Togawa[2,4] and H. Okamoto[1,3]*

[1] Center for Mesoscopic Sciences, Institute for Molecular Science, Okazaki, Aichi 444-8585, Japan
[2] Department of Physics and Electronics, Osaka Prefecture University, Sakai, Osaka 599-8531, Japan
[3] School of Physical Sciences, Department of Structural Molecular Sciences, The Graduate University for Advanced Studies (Sokendai), Okazaki, Aichi 444-8585, Japan
[4] Chirality Research Center (CResCent), Hiroshima University, Higashi-Hiroshima, Hiroshima 739-8526, Japan

*E-mail: shun_hashiyada@anet.soken.ac.jp, aho@ims.ac.jp

Abstract. The surface-enhanced chiral-optical spectroscopy is based on the interaction of chiral molecules with chiral electromagnetic near field localized on a chiral plasmonic nanostructure. It is of fundamental importance to reveal the spectral characteristics of chiral near fields for maximizing the chiral interaction. Here we investigate relations between near field and far field polarization characteristics of the chiral plasmonic nanostructures, using electromagnetic simulations. We found that spectral features of chiral near fields created by the nanostructures intercorrelate with those of far field optical rotation. This finding may provide us a method to characterize and design the chiral near field.

1. Introduction

The advent of artificially engineered nanomaterials has allowed light to be manipulated in ways previously unseen in nature. One of the most intriguing phenomena arising from this is the creation of electromagnetic (EM) near fields around chiral plasmonic nanostructures which display greater chiral asymmetry than circularly polarized light in free space (referred to as superchiral fields) [1,2]. Spectroscopic applications of superchiral near fields, named "superchiral spectroscopy", affords new opportunities for the ultrasensitive characterization of chiral molecules [1,3]. In superchiral spectroscopy, asymmetric spectral shifts induced by chiral molecules are measured; they arise from asymmetry of effective refractive indices for left- and right-handed superchiral near fields. This spectroscopy is based on the same principle as that for surface plasmon resonance sensor [4], that is, dependence of spectral positions of plasmon resonances on dielectric environments at the surfaces of nanostructures. The asymmetry of chirality-induced spectral peak shift should be maximized when the chiral asymmetry of EM near field is maximized. However, the spectral characteristics of superchiral near fields remain unclear.

In this study, we investigate the spectral properties of chiral asymmetry of EM near field created by chiral plasmonic nanostructures, using rigorous EM field simulation based on the finite-difference time-domain (FDTD) method. We evaluate the chiral near field with the ellipticity angle (that corresponds to circular dichroism, CD), which is experimentally obtained by near field polarimetry measurements [5].

Our simulation results show that spectral features of the ellipticity angle of near field created by the nanostructures intercorrelate with those of far field optical rotation (OR). We also show that this relation obtained with the numerical simulation on the chiral plasmonic nanostructure can be explained by a qualitative model where the excited plasmon on the nanostructure is represented as an oscillating chiral electric-magnetic dipole [6]. These results suggest that spectral features of chiral near fields (more specifically, ellipticity) can be evaluated approximately from those of far field OR. Our finding may provide a method to characterize and design the chiral near field.

2. Finite-difference time-domain (FDTD) simulation

We simulated the EM fields for two-dimensional square lattice gold nano-gammadion arrays on a glass substrate (Fig. 1a) [1,7]. We used commercial FDTD simulation software (Poynting for Optics, Fujitsu Ltd., Japan) for this purpose. Drude-Lorentz model parameter was used for the dielectric function ($\varepsilon_{\text{gold}}$) of gold. For the glass, $\varepsilon_{\text{glass}} = 2.25$ (wavelength independent). The computational domain was defined with periodic boundary condition in *x*,*y* to mimic the arrays, and with perfectly matched layer condition in *z*. A plane wave of linearly polarized EM field was irradiated from the substrate side of the nanostructure (+*z* direction), and the polarization characteristics (the ellipticity angle η and the azimuth angle θ) of the electric field in an *xy*-plane above the gold surface ("P_{FF}" and "P_{NF}" in Fig. 1a) were evaluated. We can calculate $\eta = \tan^{-1}(S_3/S_0)/2$ and $\theta = \sin^{-1}(S_2/S_1)/2$ using the Stokes parameters, $S_0 = |\tilde{E}_x|^2 + |\tilde{E}_y|^2$, $S_1 = |\tilde{E}_x|^2 - |\tilde{E}_y|^2$, $S_2 = 2\text{Re}(\tilde{E}_x^*\tilde{E}_y)$, and $S_3 = 2\text{Im}(\tilde{E}_x^*\tilde{E}_y)$. Quantities with tildes are complex.

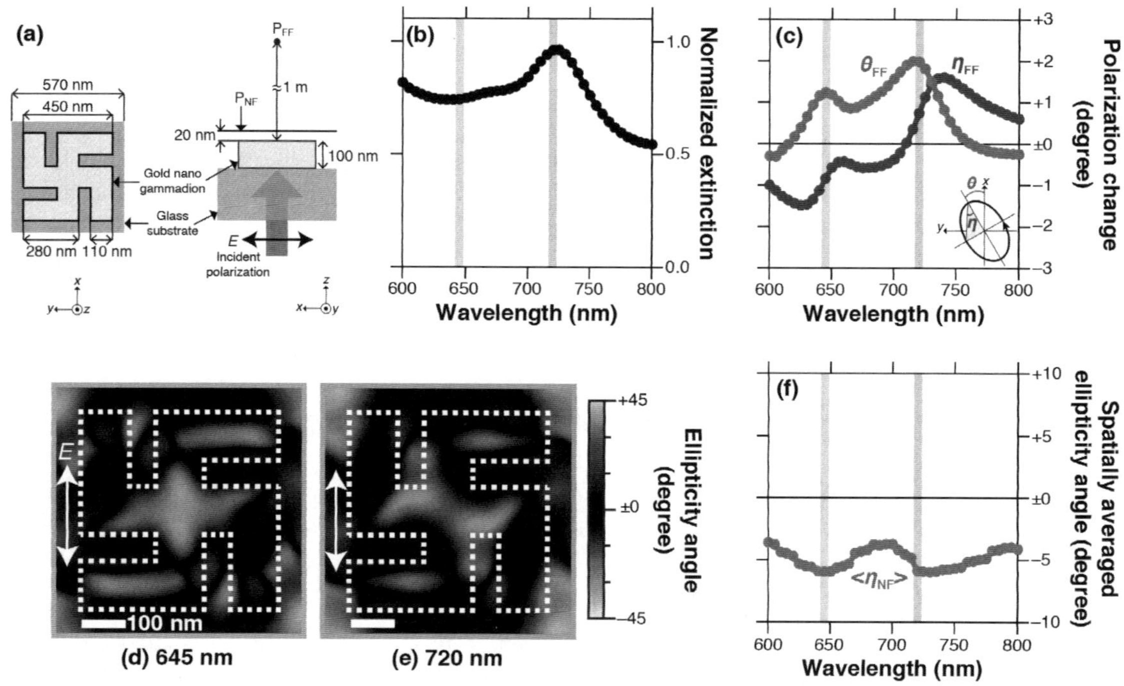

Figure 1. (**a**) The model of the FDTD simulation. The near field and far field polarization characteristics of two-dimensional square lattice gold nano gammadion arrays on a glass substrate were evaluated at the plane "P_{NF}" and at the position "P_{FF}", respectively. (**b,c**) Far field extinction (**b**), circular dichroism η_{FF} (blue) and optical rotation θ_{FF} (red) (**c**) spectra of the nanostructures. (**d,e**) Maps of near field ellipticity angle η_{NF} of the electric field generated near the nanostructure using linearly polarized light (polarization direction indicated by arrows) for the wavelength of 645 nm (**d**) and 720 nm (**e**). (**f**) spectrum of the areal averaged value of near field ellipticity angle $\langle\eta_{NF}\rangle$ calculated from the η_{NF} maps (Figs. 1d,e for instance).

We first investigated the far field polarization characteristics of the gammadions at the position "P_{FF}". As shown in Fig. 1b, gammadions showed plasmon resonances with peaks at ~670 nm and at ~720 nm. At around these wavelengths, gammadions exhibited OR (given by θ_{FF}) and CD (given by η_{FF}) in the far-field regime (shown in Fig. 1c). The spectral features of the far field CD and OR were not changed, even if the direction of the light incidence was reversed (i.e., $-z$ direction). This result indicates that the observed polarization effect from gammadions on a glass substrate is reciprocal [7], and similar to that observed for three-dimensional chiral materials such as chiral molecules and chiral crystals.

We, then, probed the near field polarization characteristics of the gammadions in the evaluation area A at the plane "P_{NF}". Figures 1d,e show maps of the ellipticity angle (η_{NF}) of the electric near field generated by the gammadions with linearly polarized light at around the plasmon resonance wavelengths. The η_{NF} signal from the central region of the gammadion always showed a prominent signal, and its sign (handedness of chirality) was governed by the handedness of the structural chirality of the nanostructure. In practical applications of chiral near fields, where the fields interact with molecules that are uniformly distributed on the nanostructured substrate, the areal averaged value rather than the local value of η_{NF} should be important. We here calculated the areal averaged value of η_{NF}, $\langle\eta_{NF}\rangle \equiv \int \eta_{NF}(\mathbf{r})dA/A$, from the η_{NF} maps, where A is the area of the unit cell whose size is 570×570 nm^2. As shown in Figure 1f, the $\langle\eta_{NF}\rangle$ spectrum showed two negative peaks at ~645 and ~720 nm. Interestingly, peak wavelengths of the $\langle\eta_{NF}\rangle$ spectrum were consistent with those of the far field OR spectrum, while the signs of the $\langle\eta_{NF}\rangle$ peaks were opposite to those of the far field OR peaks. We thus found that the spectral features of chiral near fields (ellipticity) created by chiral plasmonic nanostructures intercorrelate with those of far field OR.

3. Chiral dipole model

To discuss the intercorrelation between near and far field polarization characteristics, we simulated $\langle\eta_{NF}\rangle$ based on a simple model where the excited plasmon on the chiral nanostructure is represented as an oscillating chiral electric-magnetic dipole [6]. A chiral dipole induced by a monochromatic EM field can be described by an electric $\mathbf{p}$ and magnetic $\mathbf{m}$ dipole moments:

$$\tilde{\mathbf{p}} = \varepsilon_0\tilde{\alpha}_e\tilde{\mathbf{E}} + i(\tilde{\alpha}_c/c)\tilde{\mathbf{H}} \qquad \tilde{\mathbf{m}} = \mu_0\tilde{\alpha}_m\tilde{\mathbf{H}} - i(\tilde{\alpha}_c/c)\tilde{\mathbf{E}}, \tag{1}$$

where $\tilde{\alpha}_e$, $\tilde{\alpha}_m$, and $\tilde{\alpha}_c$ are the isotropic electric, magnetic, and electric-magnetic polarizability, respectively. $\tilde{\mathbf{E}}$ and $\tilde{\mathbf{H}}$ are the local electric and magnetic fields at the position of the dipole. It should be noted that the real and the imaginary parts of the complex electric-magnetic polarizability $\tilde{\alpha}_c$ determine spectral features of the far field OR (θ_{FF}) and CD (η_{FF}), respectively.

Figure 2 shows the model we considered. Suppose that the chiral dipole is induced by the linearly polarized plane wave electric field propagating in $+z$ direction, $\tilde{\mathbf{E}}_{in} = {}^t(E_0,0,0)e^{i(kz-\omega t)}$, where ω and k represent the angular frequency and the wavenumber, respectively. The electric near fields ($kr << 1$) induced by electric and magnetic dipoles can be described approximately by [8]

$$\tilde{\mathbf{E}}_p(\mathbf{r}) = [3\mathbf{r}(\mathbf{r}\cdot\tilde{\mathbf{p}}) - r^2\tilde{\mathbf{p}}]/(4\pi\varepsilon_0 r^5) \qquad \tilde{\mathbf{E}}_m(\mathbf{r}) = -ikc(\mathbf{r}\times\tilde{\mathbf{m}})/(4\pi r^3). \tag{2}$$

The total electric fields in xy-plane near the dipole are then expressed as

$$\tilde{\mathbf{E}}_{total}(\mathbf{r}) = \tilde{\mathbf{E}}_{in} + \tilde{\mathbf{E}}_p(\mathbf{r}) + \tilde{\mathbf{E}}_m(\mathbf{r}) \approx E_0e^{-i\omega t}\begin{pmatrix} 1 + \tilde{\alpha}_e f_x(\mathbf{r}) + i\tilde{\alpha}_c g(\mathbf{r}) + i\tilde{\alpha}_m h(\mathbf{r}) \\ i\tilde{\alpha}_c f_y(\mathbf{r}) + \tilde{\alpha}_e g(\mathbf{r}) - \tilde{\alpha}_c h(\mathbf{r}) \end{pmatrix}, \tag{3}$$

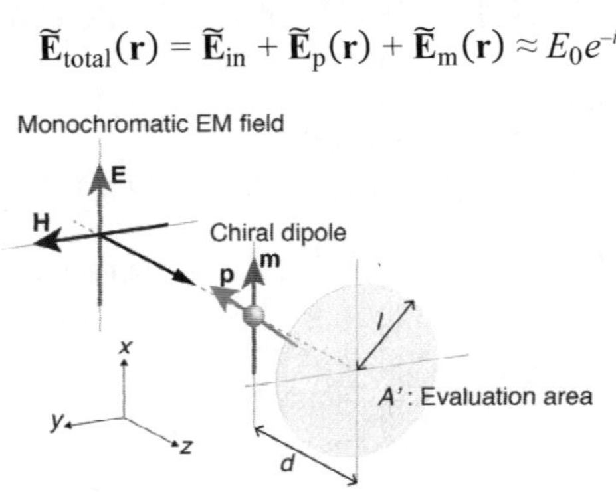

Figure 2. The model for the polarization analysis of the electric field around an oscillating chiral electric-magnetic dipole. The evaluation area A' (= πl^2) is located at a plane d away from the dipole.

where $f_x(\mathbf{r}) = (3x^2 - r^2)/(4\pi r^5)$, $f_y(\mathbf{r}) = (3y^2 - r^2)/(4\pi r^5)$, $g(\mathbf{r}) = 3xy/(4\pi r^5)$, and $h(\mathbf{r}) = kz/(4\pi r^5)$. Using equation (3), we obtained the ellipticity angle ($\eta_{\mathrm{NF}}(\mathbf{r})$) of the electric field near the chiral dipole

$$\eta_{\mathrm{NF}}(\mathbf{r}) \propto \mathrm{Im}\left(i\tilde{\alpha}_{\mathrm{c}} f_{\mathrm{y}}(\mathbf{r}) + \tilde{\alpha}_{\mathrm{e}} g(\mathbf{r}) - \tilde{\alpha}_{\mathrm{c}} h(\mathbf{r})\right) = \mathrm{Re}(\tilde{\alpha}_{\mathrm{c}}) f_{\mathrm{y}}(\mathbf{r}) + \mathrm{Im}(\tilde{\alpha}_{\mathrm{e}}) g(\mathbf{r}) - \mathrm{Im}(\tilde{\alpha}_{\mathrm{c}}) h(\mathbf{r}). \quad (4)$$

For deriving equation (4), we used three approximations: (i) $\eta_{\mathrm{NF}}(\mathbf{r}) \approx (S_3(\mathbf{r})/S_0(\mathbf{r}))/2$ because η_{NF} should be small near the dipole ($|\eta_{\mathrm{NF}}| << 0.5$ rad $\approx 28.6^{\circ}$), (ii) the properties of local $\eta_{\mathrm{NF}}(\mathbf{r})$ can be determined by $S_3(\mathbf{r})$ because $S_0(\mathbf{r})$ (the intensity of the total electric field) does not vary rapidly in the evaluation area *A'* shown in Fig. 2, and (iii) the quadratic terms of polarizabilities can be neglected because these terms vanish rapidly with the distance *d*. Here, *A'* is the area of the circle with a radius *l*, located at a plane *d* away from the dipole. Finally, we obtained the areal averaged value of η_{NF} in the near field region ($kr = k(l^2 + d^2)^{1/2} \sim 0.1$)

$$\begin{aligned} \langle \eta_{\mathrm{NF}} \rangle &\propto \mathrm{Re}(\tilde{\alpha}_{\mathrm{c}}) \int f_{\mathrm{y}}(\mathbf{r})\, dA' + \mathrm{Im}(\tilde{\alpha}_{\mathrm{e}}) \int g(\mathbf{r})\, dA' - \mathrm{Im}(\tilde{\alpha}_{\mathrm{c}}) \int h(\mathbf{r})\, dA' \\ &= -[\sin\theta \sin 2\theta\, \mathrm{Re}(\tilde{\alpha}_{\mathrm{c}}) + 4kd(1 - \cos\theta)\mathrm{Im}(\tilde{\alpha}_{\mathrm{c}})]/(8d) \\ &\approx -\sin\theta \sin 2\theta\, \mathrm{Re}(\tilde{\alpha}_{\mathrm{c}})/(8d) \\ &\propto -\mathrm{Re}(\tilde{\alpha}_{\mathrm{c}}), \end{aligned} \quad (5)$$

where $\theta = \tan^{-1}(l/d)$. Equation (5) indicates that $\langle \eta_{\mathrm{NF}} \rangle$ is correlated approximately with the far field OR (given by $\mathrm{Re}(\tilde{\alpha}_{\mathrm{c}})$) with opposite sign. This result suggests the that spectral features of near field ellipticity can be evaluated from those of far field OR. This result is consistent with that obtained with the FDTD simulation.

4. Conclusion

In the present study, we have qualitatively discussed relations between near and far field polarization characteristics of chiral plasmonic nanostructures. Our theoretical considerations revealed that spectral features of chiral near fields (ellipticity) intercorrelate with those of far field OR. This result may provide a method to characterize chiral near fields based on far field observables. This also suggests a way to design chiral near fields, as the far field OR spectrum of chiral nanostructures can be designed by the geometric design of the nanostructure [9].

Acknowledgements

This work was supported by Grants-in-Aid for Scientific Research (KAKENHI; Nos. JP17H07330, JP16H06506, JP15H02161) from the Japan Society for the Promotion of Science (JSPS) and by JSPS Core-to-Core Program (A. Advanced Research Networks), and the Photon Frontier Network Program of the Ministry of Education, Culture, Sports, Science, and Technology (MEXT), Japan.

References

[1] Hendry E, Carpy T, Johnston J, Popland M, Mikhaylovskiy R V, Lapthorn A J, Kelly S M, Barron L D, Gadegaard N and Kadodwala M 2010 *Nat. Nanotechnol.* **5** 783
[2] Schäferling M, Dregely D, Hentschel M and Giessen H 2012 *Phy. Rev.* X **2** 031010
[3] Tullius R, Karimullah A S, Rodier M, Fitzpatrick B, Gadegaard N, Barron L D, Rotello V M, Cooke G, Lapthorn A and Kadodwala M 2015 *J. Am. Chem. Soc.* **137** 8380
[4] Homola J, Yee S S and Gauglitz G 1999 *Sens. Actuators, B: Chem.* **54** 3
[5] Hashiyada S, Narushima T and Okamoto H 2018 *ACS Photon.* **5** 1486
[6] Barron L D 2004 *Molecular Light Scattering and Optical Activity* (Cambridge: Cambridge University Press)
[7] Kuwata-Gonokami M, Saito N, Ino Y, Kauranen M, Jefimovs K, Vallius T, Turunen J and Svirko Y 2005 *Phy. Rev. Lett.* **95** 227401
[8] Jackson J D 1999 *Classical Electrodynamics* 3rd edn (New York: Wiley)
[9] Karimullah A S, Jack C, Tullius R, Rotello V M, Cooke G, Gadegaard N, Barron L D and Kadodwala M 2015 *Adv. Mater.* **27** 5610

IOP Conf. Series: Journal of Physics: Conf. Series **1220** (2019) 012051 doi:10.1088/1742-6596/1220/1/012051

Synthesis and characterization of quantum dot chains: Self-assembly of mercaptopropionic-acid-capped quantum dots conjugated with short single-stranded DNA

M Oda[1, 2], K Sasano[2], A Nishi[2], T Zako[3] and T Tani[4]

[1] Department of Basic Sciences, Faculty of Engineering, [2] Department of Applied Science for Integrated System Engineering, Graduate School of Engineering, Kyushu Institute of Technology, 1-1 Sensui-cho, Tobata-ku, Kitakyushu, Fukuoka, 804-8550 Japan
[3] Department of Chemistry and Biology, Graduate School of Science and Engineering, Ehime University, 2-5 Bunkyo-cho, Matsuyama, Ehime 790-8577 Japan
[4] Department of Applied Physics, Tokyo University of Agriculture and Technology, 2-24-16 Naka-cho, Kogane-i, Tokyo 184-8588, Japan
Email: odamasa@mns.kyutech.ac.jp

Abstract. In this paper, we describe the synthesis and characterization of quantum dot (QD) chains. We used small QD-DNA conjugates, i.e., short-ligand-capped QDs conjugated with short (15-mers) single-stranded DNA, as the building blocks. The small conjugates were separated based on the valence of the DNA, i.e., the amount of DNA per particle, using agarose gel electrophoresis. The conjugates formed self-assemblies composed of closely spaced (inter-particle spacing of 0.6–2 nm) QDs. We report the results of our investigation of the electrophoresis and hybridization of the conjugation.

1. Introduction

DNA-functionalized quantum dots (QDs) are QDs conjugated with single-stranded (ss) DNA oligomers. They have been used as building blocks for self-assemblies in a variety of fields, such as materials science and engineering. In particular, self-assemblies composed of closely spaced QDs have received considerable attention due to their novel optical and electronic properties, which arise from energy transfer/coupling between neighboring QDs [1,2].

To increase the efficiency of energy transfer/coupling in the assemblies, it is necessary to decrease the inter-particle spacing between adjacent QDs. Short-ligand-capped QDs conjugated with short ssDNA particles are believed to be suitable for this purpose, although there are some practical difficulties. For example, it is easy to quench their photoluminescence (PL) during synthesis procedures [3] and difficult to control the valence of the DNA [4] due to the small charge and molecular size of short DNA.

We recently investigated the synthesis of one-dimensional (1D), closely spaced QD chains based on short mercaptopropionic-acid (MPA)-capped QDs conjugated with short (15-mer) ssDNA. In this paper, we present the results of this investigation.

2. Experiments

2.1. Conjugation of ssDNA onto QD surface

IOP Conf. Series: Journal of Physics: Conf. Series **1220** (2019) 012051 doi:10.1088/1742-6596/1220/1/012051

DNA-functionalized QDs were fabricated according to the following procedure: first, CdSe/ZnS core-shell QDs capped with trioctylphosphineoxide (TOPO) were synthesized via pyrolytic decomposition of organometallic compounds (or QDs provided by NS Materials Inc.) [5]. Next, the TOPO ligand was exchanged with MPA in chloroform using Pong's method [6]. The QDs were rendered water-soluble, so that a droplet containing QDs capped with MPA floated on chloroform. The droplet was washed several times with fresh chloroform, and then diluted in borate buffer solution (pH 8.3). QDs capped with MPA are known to be stable in the pH range between 7.0 and 9.5 [6]. Thus, all procedures described in §2.1–2.3 were performed at a pH of 8.3.

Separately, thiolated short ssDNA oligomers (5'-HS-$(CH_2)_6$-TACGCCACCAGCTCC-3'; Eurofins Genomics K.K.) were dispersed in borate solution. Then, 3 mL of mixed borate solution containing the QDs capped with MPA (10^{-5}–10^{-7} mol/L) and the ssDNA (10^{-5}–10^{-7} mol/L) was prepared and left for 12 hours at room temperature to conjugate the ssDNA oligomers onto QD surface.

2.2. Agarose gel electrophoresis

Agarose gel electrophoresis was used to separate the DNA-functionalized QDs based on the DNA valence (N_{DNA}). Just prior to the electrophoresis, the DNA-functionalized QD solution (100 μL–1 mL) was concentrated with a centrifugal filter device to a volume of 50 μL. Then, electrophoresis was performed using agarose gel (NuSieve GTG agarose; Lonza) in 0.5× tris-borate buffer solution with an electrophoresis system (Mupid-2plus; Takara Bio Inc.). Some parts of the gel containing the separated QDs were cut and mashed after electrophoresis, then loaded in separate microtubes with 500 μL borate buffer solution. We extracted the QDs from the gel by rotating the microtubes using a 12 cm rotator at 20 rpm for 12 h.

2.3. Hybridization of DNA-functionalized QDs

Here, 1 mL of the DNA-functionalized QD solutions were mixed with 1 mL of solutions containing similarly prepared QDs conjugated with complementary DNA oligomers (5'-HS-$(CH_2)_6$-ATGCGGTGGTCGAGG-3'). We fabricated the QD self-assemblies based on DNA hybridization by gradually cooling the mixed solutions to room temperature following by heating them to 75 °C.

2.4. Characterization of QD self-assemblies

We characterized the structures of the QD self-assemblies using transmission electron microscopy (TEM; H-9000; Hitachi, Ltd.). The samples for the TEM analysis were obtained as follows: first, TEM grids (NPC15; Okenshoji Co., Ltd.) were hydrophilized by plasma treatment. Droplets of self-assembled QD solutions were dropped onto plastic films, and then covered by TEM grids for 20 minutes to adsorb the assemblies onto the TEM grid. Finally, the grids were dried by evacuation for 2 days.

3. Results and discussions

DNA-functionalized QDs with $N_{DNA} = 2$ can be used to build 1D chains, as shown in Figure 1(a). First, we describe our results regarding the optimization of the mixing molar ratio of ssDNA to QDs used for the conjugation process. Figure 1(b) shows a typical TEM image of QD self-assemblies fabricated by using DNA-functionalized QDs prepared with a molar ratio of 8:1. This was the best ratio for the fabrication of chain-like assemblies in the case without using electrophoresis, although the assemblies had branched structures, as shown in Figure 1(b). The branched structures suggest that QDs with $N_{DNA} = 3$ or more were also fabricated, even when using the best molar ratio for obtaining those with $N_{DNA} = 2$. Monomers and dimers of QDs were obtained in the ratio of 1:1–2:1 (not shown here).

The apparent ratio dependence of the assembled structure confirms that both the conjugation of thiol-modified DNA and crosslinking of the DNA-functionalized QDs works properly. The average inter-particle spacing was estimated to be 0.6 nm, although it was several times shorter than the DNA molecule. We believe that this shortness is due to the deformation of the DNA by drying in a vacuum and/or adsorbing into the grid prior to the TEM observations.

We also found a novel structure in assemblies prepared at the molar ratio of 8:1, although its production

IOP Conf. Series: Journal of Physics: Conf. Series **1220** (2019) 012051 doi:10.1088/1742-6596/1220/1/012051

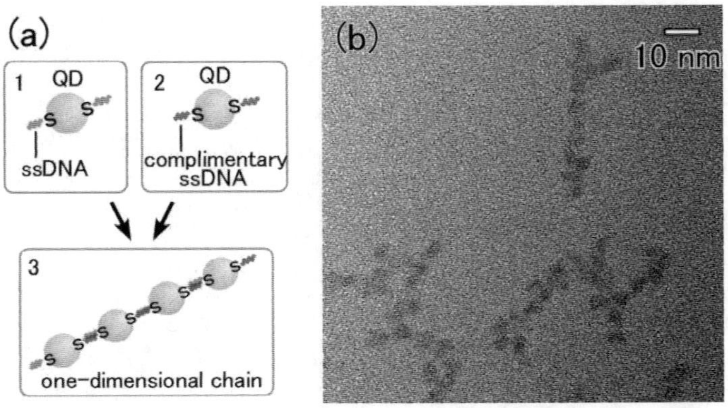

Figures 1. (a) Schematic diagram of the formation process of one-dimensional closely spaced quantum dots (QDs). (b) Typical transmission electron microscopy (TEM) image of QD self-assemblies fabricated from QD-DNA conjugates. The molar mixing ratio, r, of the sample was 8. Average diameter of the QDs was 4.0 ± 0.4 nm.

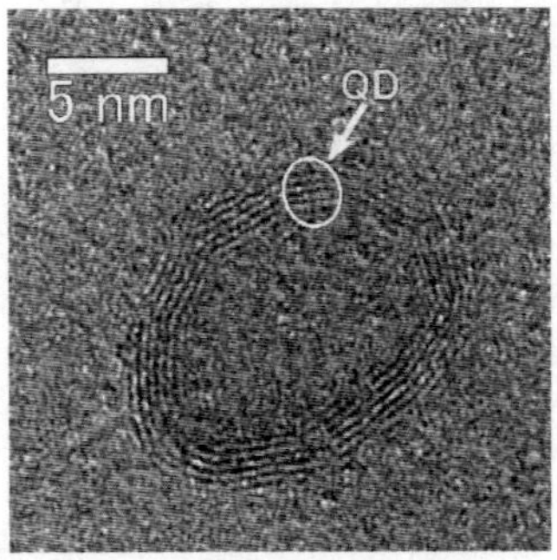

Figure 2. TEM image of circle-like QD structures. Average diameter of the QDs was 2.8 ± 0.4 nm. The lattice spacing was 0.37 nm, which matches with the (100) plane of bulk cadmium selenide (CdSe).

efficiency and reproducibility were quite low. It seems that the QDs formed circle-like structures with short inter-particle distance, and that their lattice fringes aligned along the circle, i.e., the crystal orientations of the QDs aligned with the circle, as shown in Figure 2. Further investigations of the structure may provide important information not only on how best to fabricate well-aligned QD assemblies but also on optical properties of QD assemblies exhibiting highly efficient energy transfer/coupling.

Next, we describe the electrophoresis separation of the DNA-functionalized QDs before hybridization. Figure 3 shows a typical PL image of agarose gel after electrophoresis. The QDs capped with MPA (without DNA conjugation) migrated towards an anode, as shown in lane 1, which means that the MPA ligands attached to the surfaces of the QDs have negative charges. A broad PL band appeared, and the most luminous position, i.e. the place most densely populated by QDs, was around A on line 1. We also measured the PL spectra of the QDs at different migration distances in the broad PL band (not shown here) and observed only a small change in the peak PL energy, although the PL energy depended on the size of the QD. Therefore, the origin of the broadness of the PL band is assumed to be charge dispersion due to the numerical dispersion of MPA ligands on the surfaces of the QDs.

It has been noted that PL quenching of the QDs sometimes occurs during and after electrophoresis when a buffer is used. We found that buffers containing ethylenediaminetetraacetic acid (EDTA), such as tris/borate/EDTA (TBE) buffer, caused such quenching. The DNA-functionalized QDs prepared at mixing ratios of 4:1 and 8:1 migrated in lanes 2 and 3, respectively. The QDs were separated into α, β and γ bands, as shown in Figure 3, although the tails of the bands overlapped slightly. The probability of separation into slow migration bands (β and γ) increased with the mixing ratio. Thus, we assume that the migration speed decreases with increasing N_{DNA}, as shown in Figure 3, although this migration order was opposite to that of conjugates between streptavidin-coated QDs and relatively long DNA [4]. The bandwidth was not as broad as that of the QDs in lane 1, which suggests that the charge on the short DNA oligomer was larger than the total charge of the MPA ligands bound to the surfaces of the QDs.

Then, we describe the structural features of self-assemblies of the QDs after hybridization. Figures 4 (a) and (b) show typical TEM images of the self-assemblies of the QDs extracted from band α (prepared at the molar ratio of 8:1) and similarly prepared QDs conjugated with complementary DNA oligomers. Figure 4 (c) shows a histogram of the particle number per assembly obtained from the TEM images. As expected, the most frequent particle number was 2 (the black bar in Figure 4 (c)), whicfh confirms that the main component in band α was QDs with $N_{\mathrm{DNA}} = 1$. We also observed 1D QD chains without blanched structures, as shown in Figure 4 (b) and by the red bars in Figure 4 (c). The average inter-

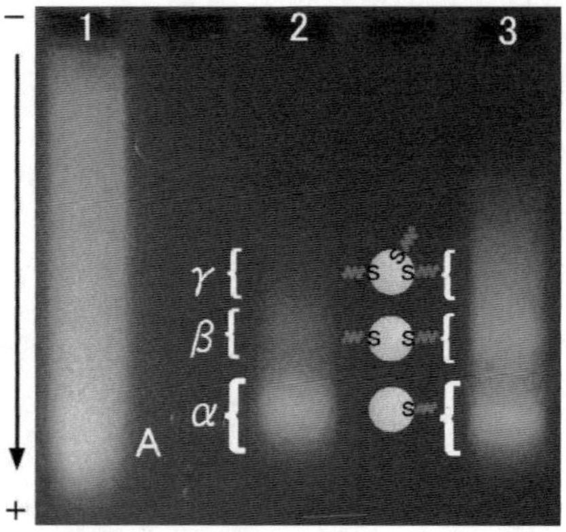

Figures 3. Photoluminescence (PL) image of agarose gel under 365-nm light irradiation. The QDs without DNA, with DNA (r = 4), and with DNA (r = 8) migrated in lanes 1, 2 and 3.

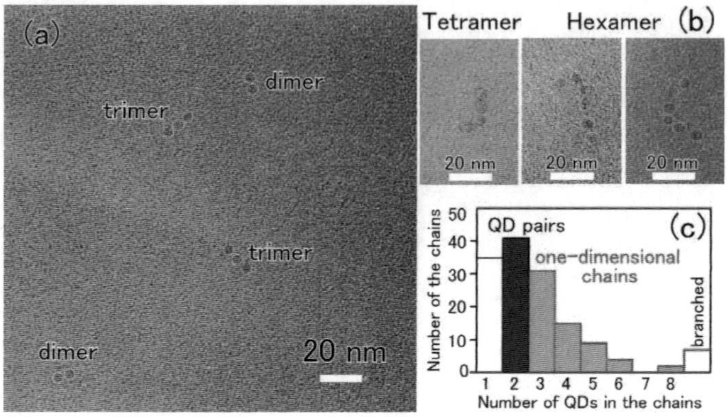

Figures 4. (a) and (b) show typical TEM images of self-assemblies of the QD-DNA conjugates extracted from band α (r = 8). The lines represent the outlines of the particle. (c) Histogram of the particle number per assembly. Average diameter of the QDs was 5.8 ± 0.8 nm.

particle spacing was estimated to be approximately 1–2 nm, although accurate estimation was difficult due to the low spatial resolution of the TEM images. The low resolution was probably due to the existing of agarose molecules. The mixing of the QDs with $N_{DNA} = 2$ into the band α is believed to contribute to the formation of the 1D closely spaced QD chains without blanched structures.

We observed QD chains with blanched structures in self-assemblies fabricated using QDs extracted from band β (not shown here). The shape of the QD chains was similar to those in Figure 1 (b). This suggests that band β did not only contain QDs with $N_{DNA} = 2$, but also those with $N_{DNA} = 3$ (and 1). Thus, it is necessary to improve the separation resolution to fabricate 1D assemblies using QDs from band β. We recently found that we can narrow each band by removing excess MPA ligands from the MPA-capped QDs prior to conjugating them with the ssDNA oligomers The improvement will be described elsewhere in the near future.

4. Summary

We conjugated short-ligand-capped QDs, i.e., MPA-capped QDs with thiol-modified short (15-mer) ssDNA. The small QD-DNA conjugates were used as building blocks for self-assemblies of 1D QD chains. We fabricated closely spaced assemblies with inter-particle spacings of 0.6–2 nm. The small QD-DNA conjugates were roughly separated by N_{DNA} using agarose gel electrophoresis. Some of the conjugates obtained after the electrophoresis formed 1D closely spaced chains without branched structures. We did not achieve precise control of the shape and particle number. We plan to improve this method for fabricating 1D closely spaced chains in the near future.

Acknowledgments

This work was partly supported by grants-in aid from the Ministry of Education, Science, Sports and Culture, Nos. 25286015 / 16K04878.

References

[1] Tikhomirov G, Hoogland S, Lee P, Fischer A, Sargent E, and Kelley S, 2011, *Nat. Nanotechnol.*, **6**, 485
[2] Koole R, Liljeroth P, Donega C, Vanmaekelbergh D, and Meijerink A, 2006, *JACS*, **128**, 10436
[3] Banerjee A, Pons T, Lequeux N, and Dubertret B, 2016, *Interface Focus*. **6**, 20160064
[4] Carstairs H, Lymperopoulos K, Kapanidis A, Bath J, and Turberfield A, 2009, *ChemBioChem*, **10**, 1781
[5] Hashizume K, Matsubayashi M, Vacha M, and Tani T, 2002, *J. Lumin.*, **98**, 49
[6] Pong B, Trout B, and Lee J, 2008, *Langmuir*, **24**, 5270

Bi^{3+} Heterovalent Doping in $APbBr_3$ Lead Halide Perovskite Single Crystals: Urbach Tail and Photon Recycling Effect

Mizuki Hoyano[1], Ryo Akashi[1], Kenichi Oto[1], Yoshihiko Kanemitsu[2] and Yasuhiro Yamada[1]*

[1]Chiba Univ., Yayoicho 1-33, Inage, Chiba, 263-8522, Japan
[2]Kyoto Univ., Gokasho, Uji, Kyoto, 611-0011, Japan

*corresponding author e-mail: yasuyamada@chiba-u.jp

Abstract. We focused on Bi^{3+} as a promising candidate for donor impurity in lead halide perovskites and fabricated electron-doped $CH_3NH_3PbBr_3$ single crystals. Room-temperature optical measurement revealed the increase in Urbach tail and suppression of photon recycling effect by Bi doping, which results in photoluminescence (PL) blue shift. Furthermore, we conducted optical reflectance and PL measurements at cryogenic temperature, where the thermal broadening is supressed. At low temperature, exciton dominates the optical properties and the photon recycling effect is negligible. From the PL intensity and spectral width, we quantitatively evaluated the impact of Bi-doping on crystal inhomogeneity.

1. Introduction

Lead Halide perovskites $APbX_3$ ($A = CH_3NH_3^+$, $CH(NH_2)_2^+$; $X = I^-$, Br^-, Cl^-) have attracted attention in recent years as a promising class of solar-cell materials that displays high solar light-to-electricity conversion efficiency with low-cost and easy manufacturing method [1]. These materials exhibit excellent crystalline quality with low defect density [2], resulting in high photoluminescence (PL) quantum efficiency [3,4] and long carrier diffusion length [5]. Furthermore, photon recycling effect has been reported in lead halide perovskites [4,6-9]. The photon recycling effect is the repeated light emission and re-absorption inside crystal, which occurs only in high-quality semiconductor materials possessing long carrier lifetime, sharp band-edge absorption (small Urbach tail) [10,11] and small Stokes shift [12] such as GaAs and CdTe. Owing to the above excellent features, lead halide perovskites are expected to be applied to a variety of electrical and optical devices.

Carrier doping by impurity substitution is a very important technology for the further development of the fundamental research and device application of this unique semiconductor material. However, the carrier doping technique has not been established in lead halide perovskites. It is necessary to fabricate carrier-doped halide perovskites and evaluate their electrical and optical properties. We focused on Bi^{3+} as a promising candidate for donor impurity, whose ionic radius is close to that of Pb^{2+}. We recently succeeded to fabricate Bi-doped $CH_3NH_3PbBr_3$ single crystals and studied their material properties [13]. In this study, we evaluated the optical properties of Bi-doped $CH_3NH_3PbBr_3$ single crystals by comparing room- and cryogenic-temperature measurements.

2. Preparation of Bi-doped $CH_3NH_3PbBr_3$ single crystals

IOP Conf. Series: Journal of Physics: Conf. Series **1220** (2019) 012052 doi:10.1088/1742-6596/1220/1/012052

We fabricated $CH_3NH_3PbBr_3$ single crystals with different Bi concentration as shown in Figure 1(a). As the Bi concentration increases, the color of the single crystals changes from orange to dark red. The antisolvent vapor-assisted crystallization method [14] was used to fabricate high-quality single crystals. CH_3NH_3Br, $PbBr_2$ and $BiBr_3$ were dissolved in *N,N*-dimethylformamide (DMF). As schematically illustrated in figure 1(b), the solution was poured in a small vial which was placed in a larger vial with toluene. As the toluene is evaporated by heating on a hot plate at 40 ℃ and mixed with the solution in the inner vial, the solubility decreases and rectangular parallelepiped single crystals were precipitated because inclusion of toluene vapor into DMF reduces solubility of $CH_3NH_3PbBr_3$.
On the basis of element analysis, we confirmed that the Bi/Pb ratio in crystal was proportional to the Bi/Pb ratio in feed solution. The crystal grown with the 1 % Bi/Pb ratio in feed solution contains 0.018% Bi/Pb ratio [13]. Electrical conductivity measurement on the 0.09 % (Bi/Pb ratio in the crystal) doped crystal revealed that doped carriers are electrons and electron mobility is 48 cm^2/Vs [13]. However, we found that the doped electron density is six-order-magnitude smaller than Bi^{3+} density in crystal, suggesting that Bi^{3+} ions do not supply electrons effectively [13].

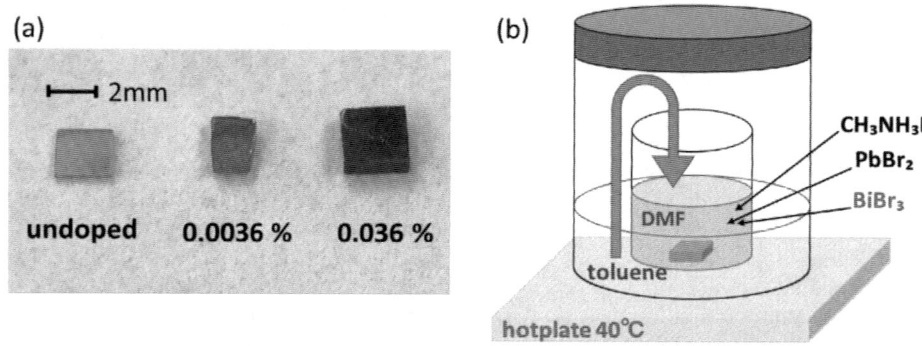

Figure 1. (a) Photograph of undoped and Bi-doped $CH_3NH_3PbBr_3$ single crystals. (b) Schematic illustration of anti-solvent vapor assisted growth technique.

3. Optical properties of Bi-doped $CH_3NH_3PbBr_3$ single crystals

Figure 2 shows the optical reflectance, absorption and PL spectra of undoped and Bi-doped crystals (0.036%: Bi/Pb ratio in crystal) at room temperature. Since thick crystals were used in the optical absorption measurement, only the absorption tail below the band gap energy was observed. The absorption spectrum draws a larger Urbach tail on the low energy side with increasing Bi concentration [13]. This indicates the lowered crystallinity by Bi^{3+} doping. In the undoped crystal, the optical reflectance peak energy (2.32 eV), corresponding to band gap energy, is different from the PL peak energy (2.27 eV) as shown in figure 2(a). This discrepancy is well explained by photon recycling effect. In undoped crystals, the photocarriers generated near the surface diffuse into the interior region of crystal. Since the high-energy light above bandgap energy emitted from the interior region is strongly re-absorbed due to the small Stokes shift [12], the PL peak energy is redshifted compared with the original PL. On the other hand, the PL peak energy of Bi-doped crystal agrees well with the reflectance peak energy. We confirmed that the carrier diffusion length is shortened due to reduction in PL lifetime by Bi doping [13]. Thus, the emission from the surface of the crystal is dominant in Bi-doped crystals. This leads to the suppression of photon recycling effect, and thus the PL peak of Bi-doped crystals coincides with the band gap energy.

Since room-temperature PL properties are dominated by the thermal broadening and photon recycling effect, it is rather difficult to evaluate the impact of Bi doping on the optical properties. Therefore, we conducted optical measurements at 4.4K as shown in Figure 3. The optical reflectance spectra are antisymmetric, which is a common feature of exciton resonance. From this observation, we estimated the free exciton resonance energy as 2.25 eV. We confirmed that the exciton resonance energy is unchanged by Bi doping, indicating that Bi doping gives no impact on the band gap energy at low temperature. Unlike the case of room temperature PL, the main PL peak almost agrees with the exciton resonance energy derived from the reflectance.

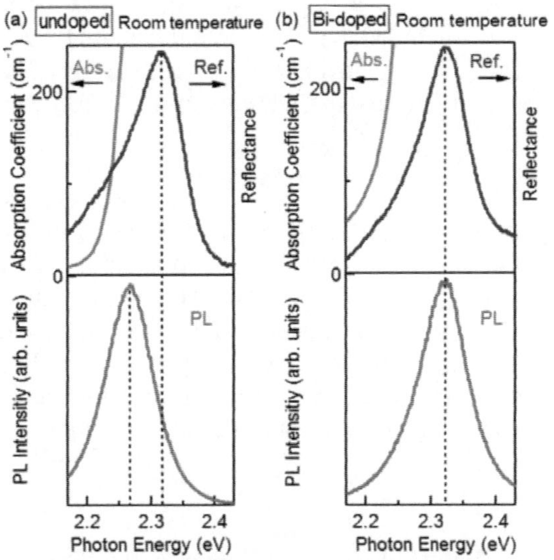

Figure 2. Reflectance spectra (blue), optical absorption (green) and PL spectra (red) of (a) undoped and (b) Bi-doped $CH_3NH_3PbBr_3$ single crystals at room temperature.

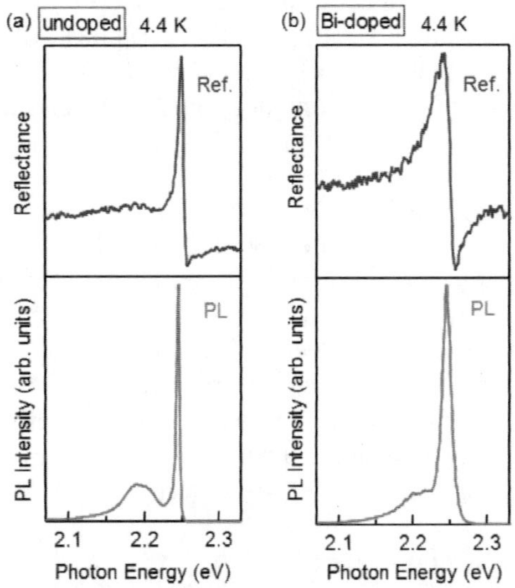

Figure 3. Reflectance spectra (blue) and PL spectra (red) of (a) undoped and (b) Bi-doped $CH_3NH_3PbBr_3$ single crystals at 4.4K.

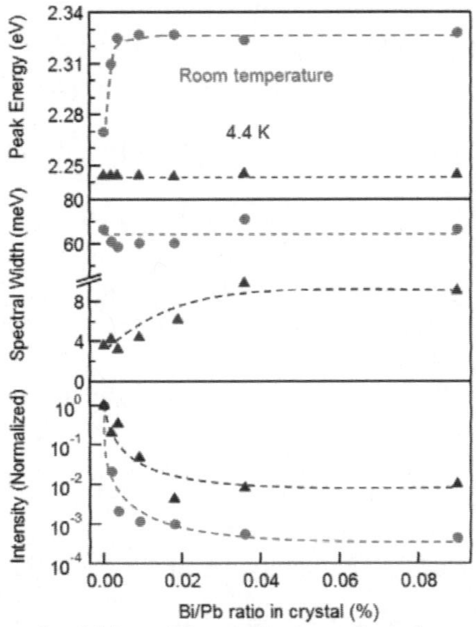

Figure 4. PL peak energy, spectral width and intensity as a function of the Bi/Pb ratio in crystal at 4.4 K (blue) and room temperature (red). Dashed curves are eye-guides.

Figure 4 summarizes the Bi-concentration dependence of PL peak energy, spectral width, and intensity at room temperature and 4.4 K. PL intensities are normalized by that of undoped sample. At 4.4 K, there is no change in PL peak energy by Bi doping unlike room-temperature PL. This is because the exciton absorption becomes dominant and the photon recycling effect is negligible at low temperature. The PL

IOP Conf. Series: Journal of Physics: Conf. Series **1220** (2019) 012052 doi:10.1088/1742-6596/1220/1/012052

spectral width at 4.4 K becomes broader as the Bi concentration increases due to increased inhomogeneous broadening, while room-temperature PL width is rather dominated by thermal broadening. The PL intensity is strongly reduced by Bi doping both at room temperature and 4.4 K. This suggests that a large number of deep defect levels are created by Bi doping and the non-radiative recombination is enhanced. Based on these observations, we concluded that the Bi doping causes defect formation rather than direct carrier supply. However, the formation mechanism and energy levels of the deep defect states remain unclear. Further studies are needed to clarify the features of the deep defect states.

4. Conclusion

In this study, the optical properties of Bi-doped $CH_3NH_3PbBr_3$ were evaluated at room temperature and cryogenic temperature. We qualitatively evaluate the Bi-doping effects from the broadening of PL spectral width by Bi doping. From the decrease in PL intensity, we can conclude that Bi doping induces a large number of deep defect levels rather than simple electron doping. We consider that Bi doping forms defects in the lattice structure of $CH_3NH_3PbBr_3$ rather than supplying carriers by substituting Pb^{2+} with Bi^{3+}, and greatly influences the PL lifetime and PL quantum efficiency. Further studies are needed to develop efficient carrier doping technology such as improvement of crystal growth technique and exploring the other dopant impurity.

Acknowledgments
Part of this work was supported by JST-CREST (Grant No. JPMJCR16N3), the Collaborative Research Program of Institute for Chemical Research, Kyoto University (grant # 2017-15).

References

[1] Kojima A, Teshima K, Shirai Y and Miyasaka T 2009 *J. Am. Chem. Soc.* **131** 6050-6051
[2] Lian Z, Yan Q, Gao T, Ding J, Lv Q, Ning C, Li Q and Sun J -L 2016 *J. Am. Chem. Soc.* **138** 9409-9412
[3] Deschler F, Price M, Pathak S, Klintberg L E, Jarausch D -D, Higler R, Hüttner S, Leijtens T, Stranks S D, Snaith H J, Atatüre M, Phillips R T and Friend R H 2014 *J. Phys. Chem. Lett.* **5** 1421-1426
[4] Yamada T, Yamada Y, Nakaike Y, Wakamiya A and Kanemitsu Y 2017 *Phys. Rev. Appl.* **7** 014001
[5] Dong Q, Fang Y, Shao Y, Mulligan P, Qiu J, Cao L and Huang J 2015 *Science* **347** 967-970
[6] Yamada Y,Yamada T, Phuong L Q, Maruyama N, Nishimura H, Wakamiya A, Murata Y and Kanemitsu Y 2015 *J. Am. Chem. Soc.* **137** 10456-10459
[7] Yamada T, Yamada Y, Nishimura H, Nakaike Y, Wakamiya A, Murata Y and Kanemitsu Y 2016 *Adv. Electron. Mater.* **2** 1500290
[8] Yamada Y, Nakamura T, Endo M, Wakamiya A and Kanemitsu Y 2014 *J. Am. Chem. Soc.* **136** 11610-11613
[9] Yamada Y, Yamada T and KanemitsuY 2017 *Bull. Chem. Soc. Jpn.* **90** 1129-1140
[10] De Wolf S, Holovsky J, Moon S -J, Löper P, Niesen B, Ledinsky M, Haug F -J, Yum J -H and Ballif C 2014 *J. Phys. Chem, Lett.* **5** 1035-1039
[11] Yamada Y, Nakamura T, Endo M, Wakamiya A and Kanemitsu Y 2014 *Appl. Phys. Express* **7** 032302
[12] Kanemitsu Y 2017 *J. Mater. Chem. C* **5** 3427-3437
[13] Yamada Y, Hoyano M, Akashi R, Oto K and Kanemitsu Y 2017 *J. Phys. Chem. Lett.* **8** 5798-5803
[14] Shi D, Adinolfi V, Comin R, Yuan M, Alarousu E, Buin A, Chem Y, Hoogland S, Rothenberger A, Katsiev K, Losovyj Y, Zhang X, Dowben P A, Mohammed O F, Sargent E H and Bakr O M 2015 *Science* **347** 519-522

IOP Conf. Series: Journal of Physics: Conf. Series **1220** (2019) 012053 doi:10.1088/1742-6596/1220/1/012053

Charge carrier migration and hole extraction from $MAPbI_3$

Kęstutis Budzinauskas[1], Simon Ewertowski[2], Selina Olthof[2], Klaus Meerholz[2], Paul H.M. Van Loosdrecht[1]

[1]Institute of Physics 2, University of Cologne, Zülpicher Str. 77, Cologne, Germany
[2]Department of Chemistry, University of Cologne, Luxemburger Str. 116, Cologne, Germany

E-mail: budzinauskas@ph2.uni-koeln.de

Abstract. Linear as well as time resolved absorption measurements were performed on 40 nm and 170 nm thick $MAPbI_3$ films with PEDOT:PSS hole extraction layer, spin-coated on quartz substrate. From linear absorption measurements exciton binding energy of 18 – 19 meV and band gap of 1.60 - 1.62 eV was deduced. Transient absorption spectra after the excitation at 1.77 eV showed a strong difference in carrier recombination dynamics for the two $MAPbI_3$ films of different thicknesses. From the analysis on the decay dynamics, hole population lifetime of 0.3 ns and 3.5 ns for 40 nm and 170 nm films, respectively, are determined. A numerical 1D diffusion model was used to model the carrier relaxation dynamics yielding hole diffusion constants of 0.025 - 0.030 cm^2s^{-1}, which results in a hole mobility of 1 $cm^2(Vs)^{-1}$ in these $MAPbI_3$ films.

1. Introduction

In recent years, hybrid metal halide perovskites gained significant attention in the solar cell community because of a remarkable power conversion efficiency in this material system of more than 20%.[1] Despite the rapid increase in efficiency, the exact origin for the extraordinary photovoltaic properties are still under debate. It is postulated that these are strongly correlated with anomalous $MAPbI_3$ carrier transport properties. Diffusion lengths in disordered $MAPbI_3$ films exceed 1 micrometer but mobilities are very modest, giving bimolecular recombination rates as small as in well studied single-crystalline inorganic semiconductors such as GaAs or Si:H and 5 orders lower than the mobility limited Langevin recombination model.[2] Some authors claim, that charge carriers in $MAPbI_3$ are screened by large polarons which prohibit carrier recombination and are the reason for the long carrier lifetimes observed in $MAPbI_3$.[3] Other arguments for the lifetime are based on the extraordinary defect tolerance due to the unique band structure of this material; both the conduction band and the valence band are formed from antibonding mixed Pb-I orbitals with strong ionic character, and within the band gap only shallow (benign) defects are formed.[4] Up to now, there are wide spread inconsistencies in the estimation of transport properties such as carrier mobility and diffusion constants in hybrid perovskites. Reported mobility values in $MAPbI_3$ films vary from 0.6 $cm^2(Vs)^{-1}$ to 80 $cm^2(Vs)^{-1}$ depending on the both sample preparation method and the measurement technique. [5,6]

In this study, we employ a contactless method to measure carrier transport in $MAPbI_3$ films, using transient absorption spectroscopy; this way enabled us to estimate intrinsic carrier diffusion constants by using a charge extraction layer.

IOP Conf. Series: Journal of Physics: Conf. Series **1220** (2019) 012053 doi:10.1088/1742-6596/1220/1/012053

2. Experimental setup

The sample structure is shown in the inset of the Figure 1a). First, PEDOT:PSS (Clevios 4083, Heraeus) was spin coated onto an ITO covered quartz glass slide. Next, a solution containing PbI_2 (Sigma-Aldrich) and MAI (prepared according to literature)[7] in a molar ratio of 1 : 1 was prepared and 18 mg/ml of the additive NH_4Cl were added. For the thick sample, a concentration of 310 mg/ml in DMF was used, while for the thinner one it was reduced to 80 mg/ml. Films were spin coated inside a nitrogen filled glovebox at 6000 rpm for 50 s and then heated at 110°C for 30 s. Small amounts of toluene were dropped around the spin-coater to create a solvent atmosphere.

Time resolved absorption as well as linear absorption experiments were performed under ambient pressure at room temperature. The time resolved absorption measurements were done using a 150 kHz Yb:KGW laser system with 80 fs resolution. Photoexcitation with the energy of 1.77 eV was used in order to create photo carriers above the band gap. Care was taken to keep the charge carrier density the same in both samples at a value of $2 \cdot 10^{18}\ \mathrm{cm}^{-3}$. This is important to exclude carrier density dependent recombination mechanisms, which would result in different decay times. For carrier density calculations, the carrier concentration was integrated over the sample thickness, while using the Beer-Lamberts law to estimate the excitation depth profile.

3. Experimental results and discussion

The linear absorption Tauc plots of both samples is shown in the Figure 1a), yielding onset for the direct band gap of 1.60 eV and 1.62 eV for the thicker and thinner $MAPbI_3$ film, respectively; the extracted absorption coefficient of both $MAPbI_3$ layers at the 1.75 eV is $2.5 \cdot 10^4\ \mathrm{cm}^{-1}$, which is in good agreement with literature values for spin coated $MAPbI_3$.[8]

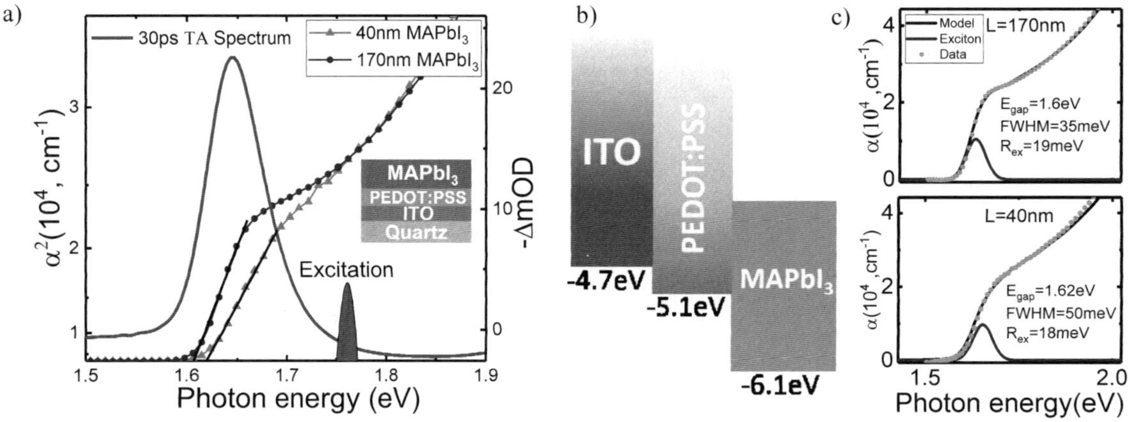

Figure 1 a) Linear absorption Tauc plot of 170 nm and 40 nm thick $MAPbI_3$ layers, together with the transient absorption spectrum taken 30 ps after the 1.77 eV excitation, b) energy diagram of the studied sample, c) modelled absorption spectra of both films using the simplified Elliot formula.

The absorption spectra near the band gap of both 40 nm and 170 nm samples showed a typical exciton offset in the band gap region. To determine its contribution, the absorption spectra were modelled using the simplified Elliot formula [9]:

$$\alpha(\hbar\omega) = \left[A \cdot \theta\left(\hbar\omega - E_g\right) \cdot D_{cv}(\hbar\omega)\right] \cdot \left[\frac{\pi x e^{\pi x}}{\sinh(\pi x)}\right] + A \cdot R_{ex} \sum_{n=1}^{\infty} \frac{4\pi}{n^3} \cdot \delta\left(\hbar\omega - E_g + \frac{R_{ex}}{n^2}\right) \tag{1}$$

where R_{ex} is the exciton binding energy, E_g the band gap, θ a step function, and D_{cv} the density of states of two parabolic bands, $x = R^{\frac{1}{2}}/(\hbar\omega - E_g)^{1/2}$. Convolution with a Gaussian distribution function was applied for incorporating inhomogeneous broadening effects.

Results of this model are shown in a Figure 1c) yielding estimates for the exciton binding energy of $R_{ex} = 19 \pm 3$ meV and 18 ± 3 meV for the two films. The observed slight differences in broadening effects (FWHM of 35 and 50 meV) can be attributed to the different disorder in $MAPbI_3$ films.

Transient absorption (TA) experiments were carried out using above the band gap excitation at 1.77 eV. A typical TA spectrum is shown in Figure 1a); after the excitation, electrons and holes occupy the bottom of the conduction and top of the valence band, resulting in a transient population with a new quasi Fermi levels. This leads to a diminished number of available states for an optical transition resulting in a strong transient absorption peak in the region close to the band gap. The amplitude of state filling peak, assuming a rigid band picture and Pauli blocking only, is directly proportional to the number of photogenerated carriers. This is valid at delay times > 5 ps, because most of electrons and holes have equilibrated and populate the conduction band minima and valence band maxima. By monitoring the amplitude of a state filling one can get insights in to the excited carrier dynamics in the material.

In this study a thin layer of PEDOT:PSS was implemented to extract holes from the $MAPbI_3$ film. The corresponding energy diagram is shown in Figure 1b).

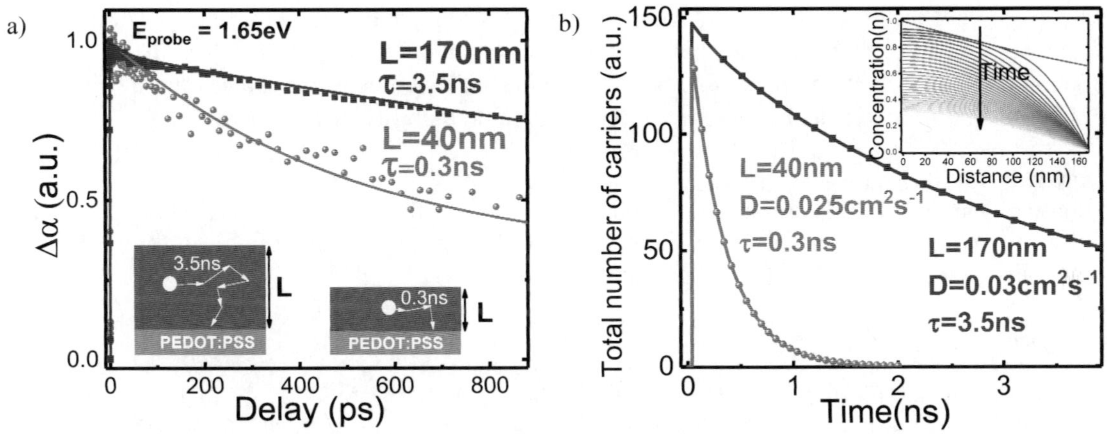

Figure 2 a) Transient absorption dynamics of the 40 nm and 170 nm thick samples after excitation with 1.77 eV, using a probing energy of 1.65 eV, b) results of the numerical simulation based on the data of sub-figure a) using the 1D diffusion model.

The decay dynamics of both films are shown in the Figure 2a). The data was fitted using a single exponential decay model with a non-decaying component C, $A \cdot e^{-t/\tau} + C$, and convolved with a Gaussian instrument response function. For both samples the amplitude of A and C are almost equal, which supports the carrier decay mechanism we describe below. Here we assume that the carrier decay mechanism is dominated by a hole transfer from $MAPbI_3$ to PEDOT:PSS while electrons remain in the film and contribute to the non-decaying component C. This indicates that only half of the carriers decay, which is consistent with the result of $A = C$, where A is proportional to the number of holes and C to electrons. Interestingly, we find one order of magnitude difference in the carrier lifetimes in two films, $\tau = 0.30$ ns for 40 nm and $\tau = 3.5$ ns for 170 nm, even though they are prepared from the same material. Since carrier density was kept identical in both cases, the recombination time differences must originate from differences in carrier drift time, resulting from the differences in layer thickness.

To confirm this story, we numerically solved a 1D diffusion equation $\partial n(x,t)/\partial t = D\, \partial^2 n(x,t)/\partial x^2$ by varying the diffusion constant. The Beer-Lambert law $\exp(-L \cdot \alpha)$ carrier distribution was chosen as the initial condition. The absorption coefficient α was extracted from linear absorption measurements and L corresponds to the sample thickness. In the applied model, one boundary (the surface) is assumed fully reflecting $\partial n/\partial x = 0$ and the other (PEDOT:PSS) fully absorbing, $n = 0$.

Results of the numerical simulation for both samples are shown in the Figure 2b). The carrier concentration at each time step was calculated by integrating over all carriers in the modelled film. The time dependent carrier concentration was then fitted with an exponential decay function. The best match with experimental decay values was found using diffusion constants of $0.025\ \mathrm{cm^2s^{-1}}$ and $0.030\ \mathrm{cm^2s^{-1}}$ for the 40 nm and 170 nm thick films, with estimated errors of $\pm 0.003\ \mathrm{cm^2s^{-1}}$. From these result we are able to extract the hole mobility μ through the Einstein relation: $\mu = \mathrm{De}/\mathrm{k_B T}$, where e is the electron charge $1.6 \cdot 10^{-12}$ C and $\mathrm{k_B} = 1.3810^{-23}\ \mathrm{m^2kgs^{-2}K^{-1}}$. Using this equation, a hole mobility of $1\ \mathrm{cm^2\ (Vs)^{-1}}$ was estimated which again is in good agreement with mobility values, estimated from optical measurements such as photoluminescence decay.[10]

4. Conclusions

In this study, we have shown a simple way to extract the intrinsic carrier diffusion constants and mobilities using transient absorption experiment, combined with a simple 1D carrier diffusion model. By employing a hole extraction layer of PEDOT:PSS we were able to gain insight into the hole transport mechanisms through the excited carrier relaxation time. We found that a difference in $MAPbI_3$ film thickness does not affect the carrier transport properties significantly, meaning that the film quality remains the same. We estimated the hole diffusion constant of $0.025\ \mathrm{cm^2s^{-1}}$ for the 40 nm and $0.030\ \mathrm{cm^2s^{-1}}$ for the 170 nm thick $MAPbI_3$ film. Using diffusion constant values and the Einstein – Smoluchowski relation, resulted in an estimation for the hole mobility of $1\ \mathrm{cm^2 (Vs)^{-1}}$.

Acknowledgements

The authors would like to thank Dr. Jürgen Schelter for the synthesis of the MAI material, Dr. Fumiya Sekiguchi for fruitful discussions and the Ministry of Science of the state of NRW for funding within the PeroBOOST (EFRE) project.

References

[1] Yang W S, Noh J H, Jeon N J, Kim Y C, Ryu S, Seo J and Seok S I 2015 *Science (80).* **348** 1234–7
[2] Wehrenfennig C, Eperon G E, Johnston M B, Snaith H J and Herz L M 2014 *Adv. Mater.* **26** 1584–9
[3] Zhu X Y and Podzorov V 2015 *J. Phys. Chem. Lett.* **6** 4758–61
[4] Brandt R E, Poindexter J R, Gorai P, Kurchin R C, Hoye R L Z, Nienhaus L, Wilson M W B, Polizzotti J A, Sereika R, Žaltauskas R, Lee L C, Macmanus-Driscoll J L, Bawendi M, Stevanović V and Buonassisi T 2017 *Chem. Mater.* **29** 4667–74
[5] Stranks S D, Stranks S D, Eperon G E, Grancini G, Menelaou C, Alcocer M J P, Leijtens T, Herz L M, Petrozza A and Snaith H J 2014 *Science* **342** 341–4
[6] Kim D H, Park J, Li Z, Yang M, Park J S, Park I J, Kim J Y, Berry J J, Rumbles G and Zhu K 2017 *Adv. Mater.* **29** 1–8
[7] Etgar L, Gao P, Xue Z, Peng Q, Chandiran A K, Liu B, Nazeeruddin M K and Grätzel M 2012 *J. Am. Chem. Soc.* **134** 17396–9
[8] Kanemitsu Y 2017 *J. Mater. Chem. C* **5** 3427–37
[9] Yang Y, Ostrowski D P, France R M, Zhu K, van de Lagemaat J, Luther J M and Beard M C 2015 *Nat. Photonics* **10** 53–9
[10] Li Y, Yan W, Li Y, Wang S, Wang W, Bian Z and Xiao L 2015 *Sci. Rep.* **5** 14485

Time-scale dependent Brownian motion of nanoparticles in clusters at a solid-liquid interface by laser trapping

Itsuo Hanasaki[1], Chie Hosokawa[2,3]

[1]Institute of Engineering, Tokyo University of Agriculture and Technology, Naka-cho 2-24-16, Koganei, Tokyo 184-8588, Japan
[2]Biomedical Research Institute, National Institute of Advanced Industrial Science and Technology (AIST), 1-8-31 Midorigaoka, Ikeda, Osaka 563-8577, Japan,
[3]Advanced Photonics and Biosensing Open Innovation Laboratory, AIST, Yamadaoka 2-1, Suita, Osaka 565-0871, Japan

E-mail: hanasaki@cc.tuat.ac.jp

Abstract. Nanoparticles in a cluster trapped by laser-induced force field show Brownian motion at solid-liquid interfaces. The cluster formation means that the particles are highly concentrated. In general, the diffusion coefficients of particles in fluids decrease with substantially high concentration and also in the vicinity of solid walls due to the hydrodynamic effect. The particle trajectory data obtained from the experimental measurements show that the longer time span of observation leads to smaller diffusion coefficient due to the confinement effect. However, they also exhibit higher diffusion coefficient compared to the the bulk condition when evaluated at a sufficiently short time span of the frame interval under the condition of sufficiently high laser powers.

1. Introduction

Laser-induced force field causes colloidal particles in liquid in the vicinity of a solid wall to form clusters. The constituent particles do not necessarily form rigid crystal structures but retains Brownian motion. On the one hand, the position of nanoparticles in the clusters changes from arbitrarily defined initial position in a steady state. On the other hand, these particles are trapped in the laser-induced force field, which means that the particles are confined in a finite domain. When the trapping energy is sufficiently large, diffusion coefficient will vanish in the evaluation with the infinite time span. However, as far as the relative positions of particles in clusters changes in time, a kind of diffusion coefficient can be discussed, paying attention to the time scale of the observation. The combination of Brownian motion and optical force field can give rise to diverse dynamical phenomena. As the thermal fluctuation is the essential characteristics of the system, it has also been proposed to combine with light-induced force to transport nanoparticles [1]. The fluorescence correlation spectroscopy (FCS) [2] and related fluorescence detection approach [3] have been employed to measure the trapping dynamics, sometimes being combined with the independent numerical analysis to compare with and understand the measurement data. A confined motion by optical tweezer has been attempted to be evaluated using a numerical analysis [4]. In this article, we discuss the generalized diffusion characteristics [5] of particles in clusters at a solid-liquid interface by laser trapping. Although the system of interest is rather one of the typical setups in the laser trapping studies, there are complexities with respect to the Brownian motion. The particles are highly crowded because of the trapping, and the particles are in the vicinity of a solid wall, which can slow down the Brownian motion by the hydrodynamic effect [6, 7]. We try to evaluate the dynamics purely from the measurement data without introducing too much assumption, which is realized by the particle tracking analysis from the microscopy movie data [5, 8, 9]. Consequently, we reveal the emergence of enhanced stochastic motion in the cluster of nanoparticles for sufficiently high laser powers. A simple protocol for the detection of multiple modes of stochastic motion plays an important role in this finding [8, 9].

2. Methodological details

2.1. Experimental setup for observation

A 1064-nm near-infrared laser beam from a continuous wave Nd:YVO_4 laser was used for optical trapping. These lasers were introduced to an inverted microscope and focused on the surface of the cover slip that constitutes an 86-μm thick container of nanoparticle suspensions. The objective lens was $\times 60$ magnification, and the images were captured by an EMCCD camera. The exposure time and frame rate were 25 ms and 38.1 frame/s, respectively. All laser experiments were performed at room temperature around 19-22°C. Fluorescent dye-doped polystyrene nanoparticles with a diameter of 500 nm (F8812, Thermo Fisher Scientific) were diluted in distilled water to 3.8×10^9 particles/mL.

2.2. Analysis of the movie data

The irradiated lasers attract the nanoparticles at around the focal point and form nanoparticle clusters. We extracted the particle trajectories data from the obtained microscopy movie using the algorithm of Ref. [10]. Then, we analyzed the set of trajectory data based on the framework of the generalized diffusion coefficient as follows:

$$\langle |\Delta \boldsymbol{r}(t)|^2 \rangle = 2 n_{\text{dim}} \alpha t^{\beta}, \tag{1}$$

where $\langle |\Delta \boldsymbol{r}(t)|^2 \rangle$ is the mean squared displacement, α and β are the coefficient and exponent of the generalized diffusion, and n_{dim} is the dimensionality (i.e, $n_{\text{dim}} = 2$ in this article discussing two dimensional data of $\Delta \boldsymbol{r}(t)$). Note that we used the data in the steady state after the transient period of cluster formation. In the case of normal diffusion with diffusion coefficient D, the exponent is $\beta = 1$, i.e., the MSD evolves linearly as a function of time. Then, $\alpha = D$ and hence

$$D = \frac{1}{2 n_{\text{dim}} t} \langle |\Delta \boldsymbol{r}(t)|^2 \rangle. \tag{2}$$

3. Results and discussion

Fig. 1(a) shows the mean squared displacements (MSDs) of nanoparticles in clusters as a function of time for different laser power conditions. The MSDs increase for a time scale of a second, and the scaling behavior appears to be the same for different conditions of the laser power. Nevertheless, there is a dependence of MSD on the laser power in that the higher power causes larger MSD. Because the higher power for the trapping is likely to cause more tight trapping that can cause the slower diffusion, there must be the driving effect of the laser on the particles. For example, the radiation pressure of the laser is attributed to be part of the mechanism of the enhanced Brownian motion under the concentrated condition.

The MSDs as a function of time can be understood from the framework of the generalized diffusion coefficient represented by Eq. 1. Defining the initial values of fitting parameters $\alpha = \langle |\Delta \boldsymbol{r}(t)|^2 \rangle / (4t)$ and $\beta = 1$ respectively, the nonlinear fitting (based on gnuplot ver.4.6)

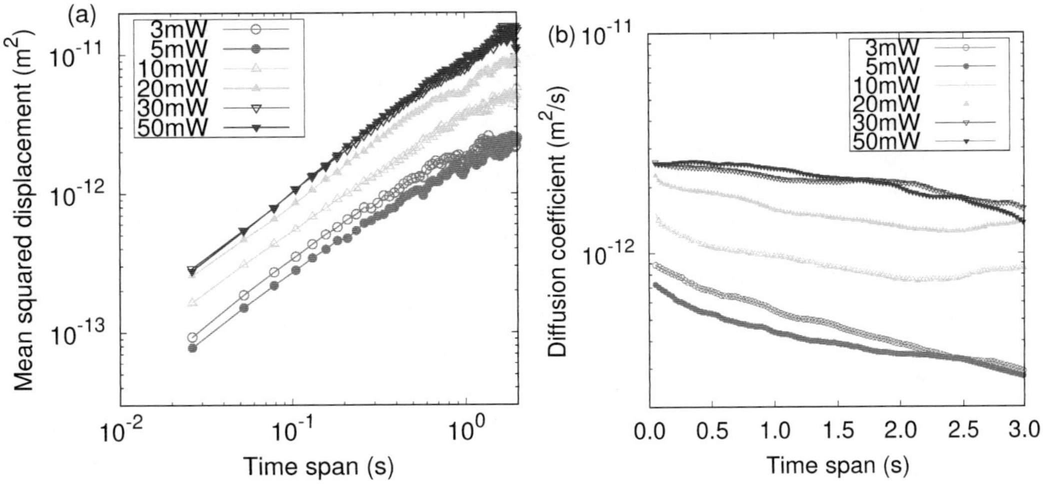

Figure 1. Time evolution of the particle dynamics obtained from the set of trajectory data: (a) mean squared displacements of nanoparticles in clusters as a function of time for different laser power conditions, and (b) diffusion coefficient as a function of time span to observe the Brownian motion. The particles within 2 μm distance from the laser focal point is defined as those that belong to the cluster. The legends show the laser power to trap the nanoparticles.

yields $\beta = 1$ for all cases within the time span of a few seconds. Thus, we use Eq. 2 hereafter. Fig. 1(b) shows the diffusion coefficient as a function of time. There is a common feature that the diffusion coefficients decrease as a function of the time span of observation. This is mainly attributed to the confinement caused by the trapping force fields. On the other hand, the diffusion coefficients depend on the laser power at a fixed time span of observation. The normal diffusion coefficient of 500-nm spherical particles dilutely dispersed in water in bulk can be predicted as 8.5×10^{-13} m^2/s based on the Stokes-Einstein relation. At the smallest time scale available in the experiment, which is more free from the confinement effect, the conditions of laser power less than 10 mW correspond to this value. The higher laser power results in the higher diffusion coefficient. In other words, the sufficiently high laser power enhances the diffusion. This is counterintuitive considering only the conservative force field.

However, the optical force does not really consists of the conservative force field alone. The non-conservative component of the optical force can trigger the enhanced Brownian motion. Fig. 2 shows the distribution of displacements in a unit of diffusion coefficient D at the time span of frame interval $\Delta t = 0.026$ s, displayed by the logarithmic scale. This logarithmic representation reveals the possible multiple modes of Brownian motion [8]. In general, the apparent multiple modes can also originate from the error in the tracking analysis [9]. However, Fig. 2 shows the clear dependence of characteristics on the laser power. The results indicate that the enhanced Brownian motion is triggered by the sufficiently high laser power. Although our set of experimental conditions is not likely to induce thermal convection, the non-conservative force, typically the scattering force, incessantly drives the nanoparticles in the direction that approaches the solid wall. The driven nanoparticles repeatedly collide with the solid wall and adjacent particles. This triggering combined with the stochasticity can induce the enhanced stochastic motion. Substantially higher diffusion coefficients in the systems coexisting with convection compared with those without convection have been reported independently from the context of optical force [11, 12], which is of more relevance in the vicinity of a solid wall [13].

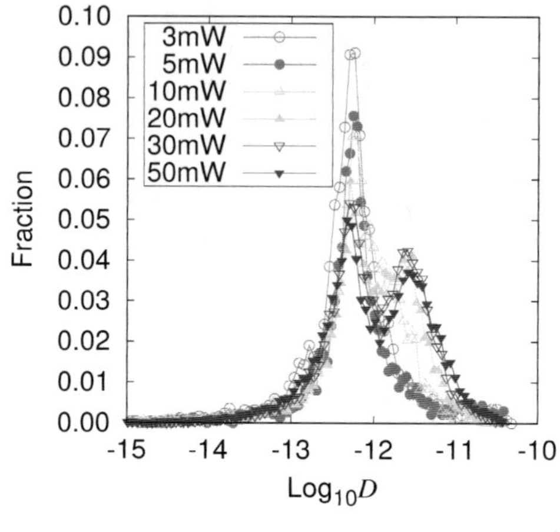

Figure 2. Distribution of displacements in a unit of diffusion coefficient D based on an assumption of normal diffusion at the time span of frame interval $\Delta t = 0.026$ s, displayed by the logarithmic scale. This logarithmic representation reveals the possible multiple modes of Brownian motion [8, 9]. The normal diffusion coefficient of 500 nm spherical particles dilutely dispersed in water in bulk can be predicted as 8.5×10^{-13} m^2/s based on the Stokes-Einstein relation. The legends show the laser power to trap the nanoparticles.

4. Concluding remarks

Nanoparticles in laser-induced clusters at a solid-liquid interface show Brownian motion that is markedly different from the bulk free condition. On the one hand, the confinement effects cause the diffusion coefficients decreasing with the increase of the time scale of observation. On the other hand, in spite of this confinement effect and the location in the vicinity of the solid wall, the Brownian motion is enhanced by the laser. The higher laser power causes the higher diffusion coefficient compared to the bulk free condition at sufficiently short time scale.

Acknowledgement

This work was partly supported by JSPS KAKENHI Grant Numbers 17H05463 and JP16H06504 in Scientific Research on Innovative Areas "Nano-Material Optical-Manipulation", and JP17H01820.

References

[1] Tamura M and Iida T 2012 *Nano Lett.* **12** 5337–5341
[2] Ito S, Toitani N, Yamauchi H and Miyasaka H 2010 *Phys. Rev. E* **81** 061402
[3] Hosokawa C, Yoshikawa H and Masuhara H 2004 *Phys. Rev. E* **70** 061410
[4] Pancorbo M, Rubio M A and Domíngues-García P 2017 *Proc. Comp. Sci.* **108C** 166–174
[5] Motohashi R and Hanasaki I *Nanoscale Advances* Accepted for publication
[6] Brenner H 1961 *Chem. Eng. Sci.* **16** 242–251
[7] Happel J and Brenner H 1983 *Low Reynolds number hydrodynamics* (Kluwer Academic Publishers Group)
[8] Hanasaki I, Uehara S, Arai Y, Nagai T and Kawano S 2015 *Jpn. J. Appl. Phys.* **54** 125601
[9] Motohashi R, Hanasaki I, Ooi Y and Matsuda Y 2017 *Micro Nano Lett.* **12** 506–510
[10] Sbalzarini I and Koumoutsakos P 2005 *J. Struct. Biol.* **151** 182–195
[11] Solomon T H and Gollub J P 1988 *Phys. Rev. A* **38** 6280
[12] Solomon T H, Lee A T and Fogleman M A 2001 *Physica D* **157** 40–53
[13] Zhang J, Fan J and Fei F 2010 *Physics of Fluids* **22** 122005

IOP Conf. Series: Journal of Physics: Conf. Series **1220** (2019) 012055 doi:10.1088/1742-6596/1220/1/012055

Electric-field induced magnetization in YIG observed by Faraday rotation

K Fujimoto[1], T Hasunuma[1] and T Kohmoto[1,2]

[1] Graduate School of Science, Kobe University, Kobe 657-8501, Japan
[2] Molecular Photoscience Research Center, Kobe University, Kobe 657-8501, Japan

E-mail: kohmoto@kobe-u.ac.jp

Abstract. We studied the electric-field induced magnetization caused by the magnetoelectric effect in YIG. The magnetization was observed as the Faraday rotation of a transmitted continuous-wave probe light. From the observed dependence of the Faraday-rotation amplitude on the electric and magnetic fields, it was found that the two components linear and quadratic in the electric field coexist at low temperatures.

1. Introduction

Yttrium iron garnet (YIG, $Y_3Fe_5O_{12}$) is a ferrimagnetic material with a Curie temperature of 560 K and is known to show a large Faraday effect. YIG is also known to show the magnetoelectric (ME) effect. The second-order ME effect (magnetic-field induced electric polarization) at room temperature [1] and the first-order ME effect (electric-field induced magnetization) below 125 K [2,3] have been reported. Recently, the magnetocapacitance effect and the magnetic-field induced electric polarization in YIG were investigated in detail [4,5]. In the present report, we studied the electric-field induced magnetization caused by the magnetoelectric effect in a YIG single crystal by using the Faraday rotation of a probe light.

2. Experiment

A single crystal of YIG, which has a (001) plane of 5 × 5 mm^2 and a thickness of 0.3 mm, was set in a temperature-controlled cryostat. The magnetic and electric fields were applied along the [001] direction. The sample was sandwiched between a transparent electrode and a metal mirror, and a square-wave electric field pulse, whose period is 20 ms, was applied between them. A continuous-wave probe light, whose wavelength is 1064 nm, goes through the electrode and the sample, is reflected by the metal mirror, and goes back through the sample and the electrode. The magnetization induced by the electric field was detected by a polarimeter as the Faraday rotation of the transmitted probe light. The change of the Faraday rotation angle induced by the electric field was obtained as the Faraday-rotation amplitude. Before the measurement, the temperature was decreased to 7 K in an electric field E of 3.3 MV/m and a magnetic field B of 1 T (field cooling).

3. Results and Discussion

3.1. Electric-field dependence

The electric-field dependence of the Faraday-rotation amplitude observed at 50 K and 175 K in the magnetic field of +1 T is shown in figure 1. The observed amplitude can be considered to be a sum of two components linear and quadratic in the electric field. At low temperatures, the amplitude of the linear component is larger than that of the quadratic component, and the inversion of the electric-field direction reverses the sign of the Faraday-rotation signal. The relative amplitude of the quadratic component increases at higher temperatures, and the sign of the signal amplitude at $E = \pm 3.3$ MV/m is the same.

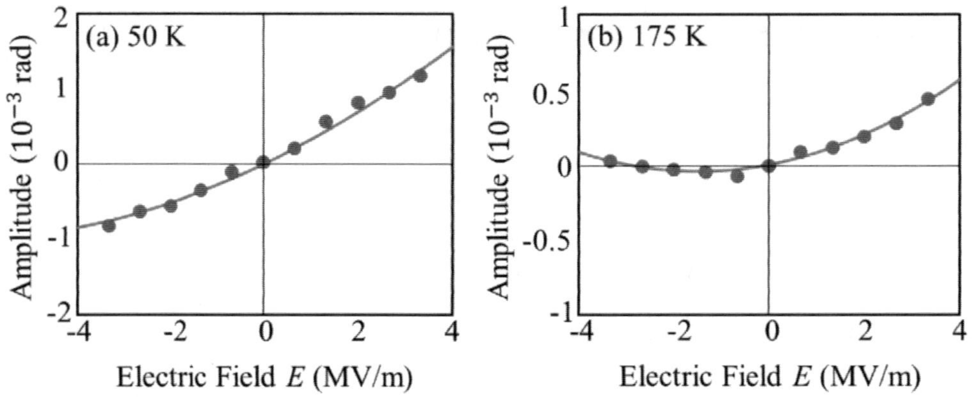

Figure 1. Electric-field dependence of the Faraday-rotation amplitude observed at (a) 50 K and (b) 175 K in the magnetic field of +1 T. The solid curves are the best fit to the observed data using the function of equation (1).

The solid curves in figure 1 are the best fit to the observed data for +1 T using a function of

$$S(E) = \alpha E + \beta E^2 \quad (B = +1\text{ T}), \tag{1}$$

where S is the Faraday-rotation amplitude, and α and β are the coefficients for the linear and quadratic components. From this fitting, the coefficients α and β can be evaluated.

3.2. Magnetic-field dependence

The magnetic-field dependence of the Faraday-rotation amplitude observed for $E = \pm 3.3$ MV/m at 7 K are shown in figure 2. The magnetic field was swept up and down between ± 1 T for $E = +3.3$ and -3.3 MV/m. The inversion of the magnetic-field direction reverses the sign of the Faraday-rotation signal. The absolute value of the amplitude is the same for $B = \pm 1$ T. The large amplitude change in the magnetic field between $\pm$ 0.2 T is considered to be caused by the saturation of magnetization in YIG.

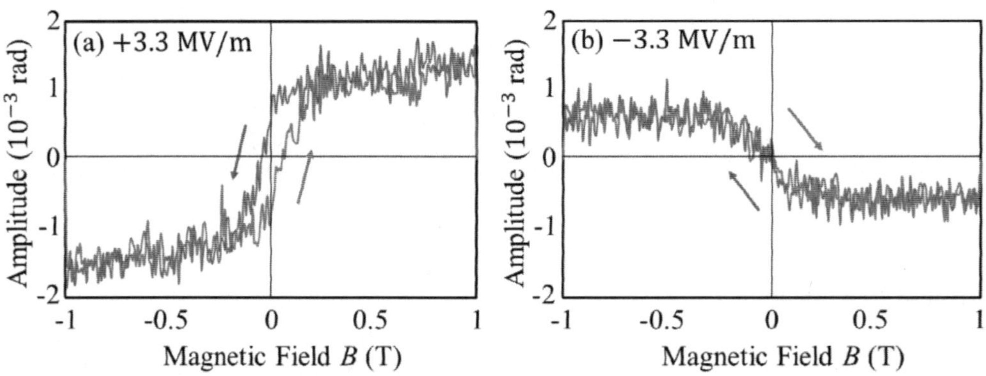

Figure 2. Magnetic-field dependence of the Faraday-rotation amplitude observed at 7 K for the electric field of (a) + 3.3 MV/m and (b) −3.3 MV/m.

Assuming that the same relation is valid for the Faraday-rotation amplitude for $E = \pm 3.3$ MV/m and $B = +1$ T in the measurements of electric-field dependence and magnetic-field dependence, equation (1) for the observed amplitude at $B = +1$ T in the measurement of magnetic-field dependence yields

$$\begin{cases} S(+3.3\ \mathrm{MV/m}) = \alpha(3.3\ \mathrm{MV/m}) + \beta(3.3\ \mathrm{MV/m})^2\,, \\ S(-3.3\ \mathrm{MV/m}) = -\alpha(3.3\ \mathrm{MV/m}) + \beta(3.3\ \mathrm{MV/m})^2\,. \end{cases} \quad (2)$$

The value of the Faraday rotation amplitude $S(E)$ in the saturated magnetization at $B = +1$ T is expected to be $\alpha|E| + \beta E^2$ for positive E and $-\alpha|E| + \beta E^2$ for negative E. The coefficients α and β for the linear and quadratic components can be evaluated from equations (2).

3.3. Temperature dependence of α and β

Figure 3 shows the temperature dependence of α, β, and α/β, where both of those values evaluated from the measurements of electric-field dependence and magnetic-field dependence are plotted. The agreement of the values of α and β evaluated from the two measurements is not good. In the magnetoelectric effect in YIG, oxygen vacancies were suggested to play an important role [4,5], and significant sample dependence was reported [6]. The disagreement may be because that the efficiency of the electric-field induced magnetization is not the same for different spots in the sample and/or for each field-cooling process. However, the value of ratio α/β obtained from the two measurements is in good agreement.

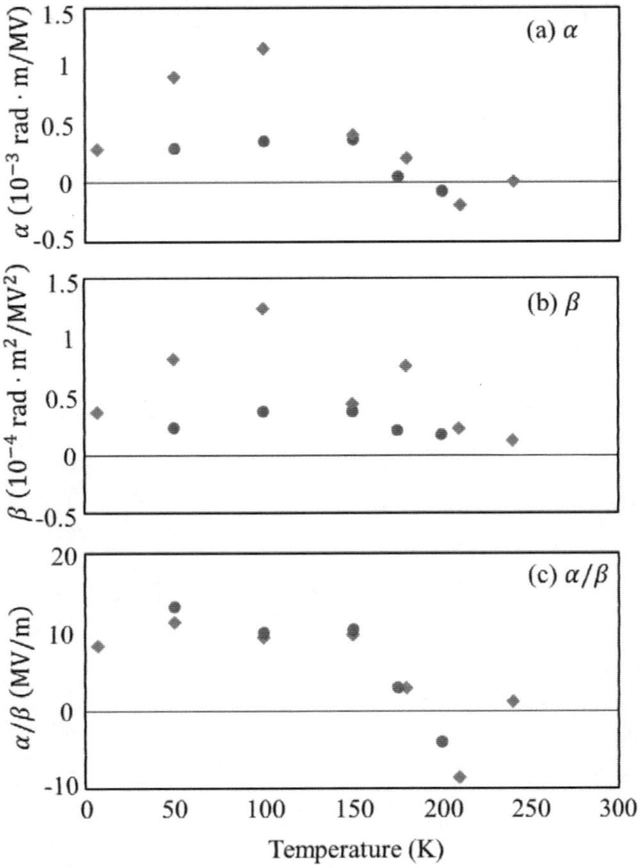

Figure 3. Temperature dependence of the values of α, β, and α/β evaluated from the measurements of electric-field dependence (circle) and magnetic-field dependence (diamond).

Figure 3 shows the coexistence of linear and quadratic components in the electric-field induced magnetization in YIG. The value of α/β begins to decrease at about 150 K, its sign is changed at about 180 K, and the Faraday-rotation signal disappears above 250 K. The sign change in α/β is caused by that in α, which suggests the direction inversion of the induced magnetization for the linear component.

4. Summary

We observed the electric-field dependence and the magnetic-field dependence of the electric-filed induced magnetization in YIG by Faraday rotation. The Faraday-rotation amplitude can be given by a sum of two components linear and quadratic in the electric field, and we evaluated the coefficients for the two components from the two measurements independently. It was found that the linear and quadratic components coexist below 250 K in the electric-field induced magnetization in YIG. The linear component is dominant below 150 K, the ratio of the linear component to the quadratic one decreases around 180 K, and the sign of its coefficient is inverted above 180 K.

References

[1] O'Dell T H 1967 *Phil. Mag.* **16** 487
[2] Ogawa H, Kita E, Mochida Y, Kohn K, Kimura S, Tasaki A and Siratori K 1987 *J. Phys. Soc. Jpn.* **56** 452
[3] Kita E, Takeno S, Tasaki A, Shiratori K, Kohn K and Kimura S 1988 *J. Appl. Phys.* **64** 5659
[4] Yamasaki Y, Kohara Y and Tokura Y 2009 *Phys. Rev. B* **80** 140412(R)
[5] Kohara Y, Yamasaki Y, Onose Y and Tokura Y 2010 *Phys. Rev. B* **82** 104419
[6] Kita E, Takano S, Kohn K, Siratori K, Kimura S and Tasaki A 1992 *J. Magn. Magn. Matter.* **104-107** 449

IOP Conf. Series: Journal of Physics: Conf. Series **1220** (2019) 012056 doi:10.1088/1742-6596/1220/1/012056

Optical selection rule of monolayer transition metal dichalcogenide by an optical vortex

Shodai Ishii[1], **Nobuhiko Yokoshi**[1], **Hajime Ishihara**[1,2]
[1]Department of Physics and Electronics, Osaka Prefecture University, Sakai, Osaka, Japan
[2]Department of Materials Engineering Science, Osaka University, Toyonaka, Osaka, Japan

E-mail: ishii-2@pe.osakafu-u.ac.jp

Abstract. We have derived the optical selection rule of monolayer transition metal dichalcogenide (TMD) for various optical vortices. Monolayer TMD, which is two-dimensional direct semiconductor, has a twofold valley degree of freedom. The valley degree of freedom can be controlled by circular polarized light. However, it is not clear how the orbital angular momentum of light relates to the valley degree of freedom. Here we clarified that the orbital angular momentum of light modifies the optical selection rule at valley points and the selection rule reflects the three-fold rotation symmetry of monolayer TMD. We expect that this modified selection rule broadens the research field of two-dimensional layered materials and spin-valleytronics.

1. Introduction

Since the famous discovery of graphene[1], two-dimensional layered materials have attracted significant interest because of their excellent electronic and physical properties. Monolayer transition metal dichalcogenide (TMD) such as MoS_2 has direct bandgaps at the two corners of the Brillouin zone $\pm$ K points (see Fig. 1). Thus, in addition to charge and spin, carriers in monolayer TMD have an extra freedom, which is called "valley degree of freedom". The valley degree of freedom can be selectively operated using circularly polarized light due to optical selection rule[2]. Furthermore, the valence bands at $\pm$ K-points have large valley-contrasting spin splitting ($0.1 \sim 0.5$eV) due to strong spin-orbit coupling and broken inversion symmetry. Therefore, monolayer TMD is expected to a proper platform for spin-valleytronics[3, 4].

The operation of spin and valley mentioned above uses the spin angular momentum (SAM) of light. Here, the SAM corresponds to circular polarization of light. The SAM of light has been applied in a wide range of fields and we have often considered only SAM as the angular momentum of light. However, light with finite orbital angular momentum (OAM) has attracted much attention in the field of optics[5]. Such a light has spiral equiphase surface and corkscrew potential (see Fig. 2), so it is often called "optical vortex". Applications of optical vortex are actively researched, e.g., nano chiral laser ablation[6] and optical manipulation[7]. However, it has not been clear how the OAM of light affects carriers in material.

In this work, we focus on an interaction between the OAM of light and carriers of monolayer TMD. To investigate the interaction, we aim to clarify the optical selection rule for various optical vortices. In general, it is believed that the optical selection rule is changed by the presence of OAM of light[8, 9]. However, the relation between the OAM of light and the valley degree of freedom is still elusive.

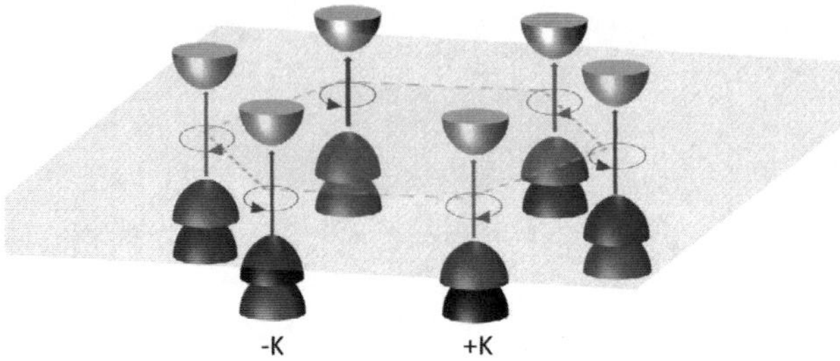

Figure 1. A schematic of the band structure of monolayer TMD. The band edges at the corners (±K) of 2D hexagonal Brillouin zone. The inequivalent valleys can be operated by circular polarized light.

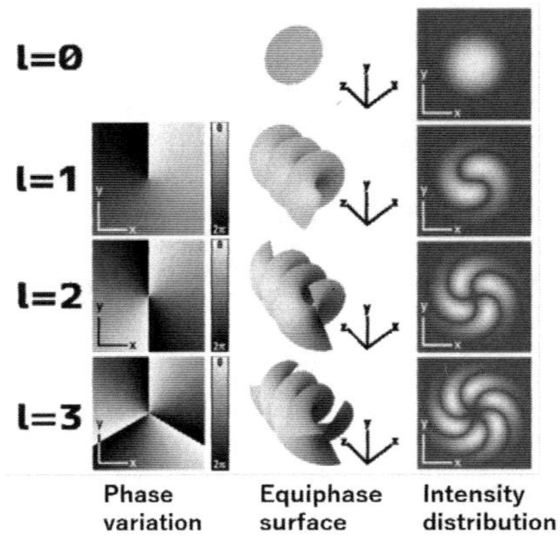

Figure 2. A schematic of optical vortex beam. This picture shows phase variation, equiphase surface and intensity distribution about $l = 0$ to $l = 3$. This beam have an azimuthal phase dependence of the type $\exp(il\phi)$. In the case of $l \neq 0$, optical vortex has phase singularity and helical equiphase surface. The beam intensity on the propagation axis is zero.

2. Method

We have derived the optical selection rule of monolayer TMD by various optical vortices. Early theoretical studies reveals that the band edges at K and −K are mainly dominated by transition metal d-orbitals in monolayer TMD[10]. The trigonal prismatic coordination classifies the transition metal d-orbitals into three groups with irreducible representations: $A\,(d_{z^2})\,, E(d_{xy}, d_{x^2-y^2})$ and $E'\,(d_{xz}, d_{yz})$ of the C_{3h} point group. The wave functions of conduction and valence band at K and −K are

$$|\Psi_c\rangle = |d_{z^2}\rangle, \qquad |\Psi_v^\tau\rangle = \frac{1}{\sqrt{2}}\left(|d_{x^2-y^2}\rangle + i\tau|d_{xy}\rangle\right), \tag{1}$$

where the index $c\,(v)$ means conduction (valence) band and $\tau = \pm 1$ is the valley subscript.

In order to clarify inter-band transition at K and −K, we calculated the optical transition amplitude between the d-orbitals mentioned above. The calculated transition matrix is

$$\langle\Psi_f|\mathbf{p}\cdot\mathbf{A}^l + \mathbf{A}^l\cdot\mathbf{p}|\Psi_i\rangle \sim \frac{2im}{\hbar}\,(\Delta E)\,\langle\Psi_f|\mathbf{r}\cdot\mathbf{A}^l|\Psi_i\rangle, \tag{2}$$

where $|\Psi_i\rangle$ indicates the initial state, $|\Psi_f\rangle$ is the final state, ΔE means the energy difference between them. Here, we assume that the optical vortex is Laguerre-Gaussian beam. Its vector potential is described as

$$\mathbf{A}^l(r,\phi,z) = \mathbf{e_A}\sqrt{\frac{2p!}{\pi(p+|l|)!}\frac{1}{\omega_0}}\left[\frac{r\sqrt{2}}{\omega_0}\right]^{|l|} L_p^{|l|}\left(\frac{2r^2}{\omega_0^2}\right)\exp\left(-\frac{r^2}{\omega_0^2}\right)\exp\left(il\phi\right), \tag{3}$$

where $l \in \mathbf{Z}$ characterizes the OAM, $p \in \{0,1,2,\cdots\}$ indicates the radial index, ω_0 is the beam waist and $\mathbf{e_A}$ is the polarization vector

$$\mathbf{e_A} = \begin{cases} (1/\sqrt{2}, i/\sqrt{2}, 0) & \text{(right-handed circular polarization)} \\ (1/\sqrt{2}, -i/\sqrt{2}, 0) & \text{(left-handed circular polarization).} \end{cases} \tag{4}$$

Here, we have considered only the case for $p = 0$ because the radial component of the beam does not affect the selection rule. In this calculation, we have not considered the spatial spread of the d-orbital wave function. The ratio of the spread and the beam waist ω_0 affects the optical transition intensity. However, the selection rule itself does not depend on the ratio.

We calculated the transition matrix mentioned for the case of $l = -4$ to $l = +4$. When the transition matrix is zero, the transition is forbidden. When the matrix is finite, the transition is allowed. In calculating the transition matrix, we took into account the three-fold rotation symmetry of monolayer TMD.

3. Result

The resultant selection rules are summarized in the Table 1 and Fig. 3. These table and figure indicate the transition from the valence band to the conduction band at ±K. For the circularly polarized plane wave ($l = 0$), the selection rule is the well-known one[2]. One can see that the OAM of light modifies the optical selection rule. Furthermore, we clarified that the selection rules are the same for every 3 of l, e.g., the selection rules coincide for the cases of $l = 0$ and $l = \pm 3$. This result reflects the three-fold rotation symmetry of monolayer TMD.

Table 1. Optical selection rule at ±K. l characterizes the orbital angular momentum of light ($l = -4 \sim +4$). RCP and LCP mean right-handed circular polarization and left-handed circular polarization respectively. Check mark indicates that the corresponding transition is allowed, and cross mark indicates that the transition is forbidden.

l		−4	−3	−2	−1	0	+1	+2	+3	+4
K	RCP	×	×	✓	×	×	✓	×	×	✓
	LCP	×	✓	×	×	✓	×	×	✓	×
−K	RCP	×	✓	×	×	✓	×	×	✓	×
	LCP	✓	×	×	✓	×	×	✓	×	×

4. Summary

We have investigated selection rules of monolayer TMD for various optical vortices. We clarified that OAM of light modifies the optical selection rule at ± K points and the result reflects the three-fold rotation symmetry of monolayer TMD. The modified selection rules have the potential

IOP Conf. Series: Journal of Physics: Conf. Series **1220** (2019) 012056 doi:10.1088/1742-6596/1220/1/012056

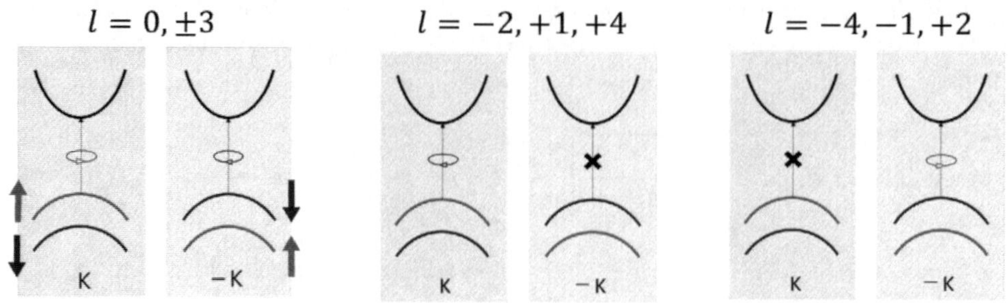

Figure 3. Inter-band transition by various optical vortices. Red circle arrow indicates left-handed circular polarized light and blew circle arrow is right-handed circular polarized light. The valence bands have large valley-contrasting spin splitting ($0.1 \sim 0.5$eV).

of new physics and application. This achievement broadens the research fields of two-dimensional layered materials and spin-valley physics.

It is considered that the optical vortex radiation to semiconductors causes new light-matter interactions. In future, we will reveal an interaction between the OAM of light and the carrier spin of monolayer TMDs. We expect that the new interaction makes a nontrivial spin-texture of the itinerant electrons, like skyrmion[11], without strong electron-electron correlations. Furthermore, the spin-texture may reflect the valley information because of strong spin-valley coupling. Finally, we want to create the new field of spin-valleytronics by optical vortex.

Acknowledgments
This work was supported by JSPS KAKENHI Grant No. JP16H06504 in Scientific Research on Innovative Areas " Nano-Material Optical-Manipulation ".

Reference

[1] Novoselov K S, Geim A K, Morozov S V, Jiang D, Zhang Y, Dubonos S V, Grigorieva I V and Firsov A A 2004 *Science* **306** 666
[2] Cao T, Wang G, Han W, Ye H, Zhu C, Shi J, Niu Q, Tan P, Wang E, Liu B and Feng J 2012 *Nat. Commn.* **3** 887
[3] Xiao D, Liu G B, Feng W, Xu X and Yao W 2012 *Phys. Rev. Lett.* **108** 196802
[4] Xiaodong X, Wang Y, Di X and Tony F. H 2014 *Nature Physics.* **10**, 343-350
[5] Allen L, Beijersbergen M W, Spreeuw R J C, and Woerdman J P 1992 *Phys. Rev. A.* **45**, 8185
[6] Toyoda K, Takahashi F, Takizawa S, Tokizane Y, Miyamoto K, Morita R and Omatsu T 2013 *Phys. Rev. Lett.* **110** 143603
[7] Garces-Chavez V, McGloin D, Padgett M J, Dultz W, Schmitzer H and Dholakia K 2003 *Phys. Rev. Lett* **91** 093602
[8] Picon A, Benseny A, Mompart J, Vazquez de Aldana J R, Plaja L, Calvo G F. and Roso L 2010 *New Journal of Physics* **12**, 083053
[9] Schmieglow C T, Schulz J, Kaufmann H, Ruster T, Poschinger U G and Schmidt-Kaler F 2016 *Nat. Commn.* **7** 12998
[10] Coehoorn R, Haas C and de Groot R A 1987 *Phys. Rev. B* **35**, 6203
[11] Yu X Z, Onoue Y, Kanazawa N, Park J H, Han J H, Matsui Y, Nagaosa N. and Tokura Y 2010 *Nature* **465**, 901-904

IOP Conf. Series: Journal of Physics: Conf. Series **1220** (2019) 012057 doi:10.1088/1742-6596/1220/1/012057

Optical characterization of MoS_2 sputtered thin films

N Hasuike[1], S Yamauchi[1], K Seki[1], S Kamoi[2], K Nishio[1] and K Kisoda[3]

[1] Department of Electronics, Kyoto Institute of Technology, Matsugasaki, Sakyo-ku, Kyoto, 606-8585, Japan
[2] Kyoto Prefectural Technology Center, Chudoji Minami-cho, Shimogyo-ku, Kyoto, 600-8813, Japan
[3] Department of Physics, Wakayama University, Sakaedani, Wakayama, 640-8510, Japan

Corresponding author e-mail: hasuike@kit.ac.jp

Abstract. We studied on the crystallinity of as-sputtered and annealed MoS_2 thin films by Raman scattering. The samples were prepared by RF magnetron sputtering, and the thermal annealing was carried out under sulfurous atmosphere. Although as-sputtered MoS_2 thin films clearly showed the deterioration of the lattice ordering, it was drastically improved by the thermal annealing due to the sulfurization of the sample. And since the sulfurization occurred remarkably on the top surface of MoS_2 sputtered thin films, it was expected to be an effective method to realize a few-layer MoS_2 sputtered thin films with high crystallinity.

1. Introduction
After the discovery of graphene, two-dimensional layered materials have played a major role in materials science. Members of two-dimensional materials are still growing from insulators, topological insulators, semiconductors, semimetals, to superconductors. Among them, transition metal dichalcogenides (TMDs) such as molybdenum disulfides (MoS_2) demonstrated unexpected photoluminescence [1-4] and other favorable physical properties over those of the bulk toward real devices when they were thinned to monolayers [5-11]. Therefore, family of TMDs has been expected to substitute for silicon as new electronic devices. To realize such devices based on MoS_2 and other TMDs, appropriate synthetic methods must be established. In this study, we studied on MoS_2 thin films fabricated by sputtering technique, which has been expected as a new method to produce the large area MoS_2 thin films.

2. Experimental
MoS_2 thin films were prepared on SiO_2/Si substrates by RF magnetron sputtering method. MoS_2 ceramics target with purity of 99 % was used, and the growth temperature was varied from 100 °C to 520 °C. The samples were prepared by following procedures. First, the chamber was evacuated to the order of 10^{-3} Pa by using the combination of diffusion and rotary pumps. And then, Ar gas was introduced to the chamber and the pressure was adjusted to be about 0.5 Pa. The substrate was heated to desired temperature and RF electric field with the power of 100 W was applied to the target for the excitation of Ar ion plasma. To clean the surface of the target, sputtering was carried out for several seconds with the shutter closed. After that, the shutter was opened and MoS_2 thin film was deposited for 5 minutes. Furthermore, we carried out the thermal annealing for the samples under sulfurous

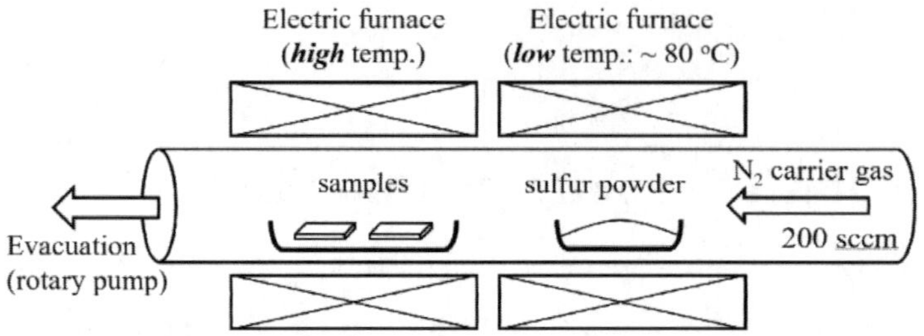

Figure 1. Schematic diagram of the equipment for the thermal annealing under sulfurous atmosphere.

atmosphere with different duration. For the process, we newly prepared the hand-made annealing equipment as shown in figure 1. It was consisted of two electric furnaces connected in series and the quartz tube was installed through furnaces. Sulfur powder was placed on the upstream side and vaporized by heating at 80 °C, and the sulfur vapor was carried to the downstream side by flowing N_2 carrier gas of 200 sccm. As-sputtered MoS_2 thin films were placed on the downstream side and annealed at 600 °C. The crystallinity of the samples was evaluated by Raman scattering technique. Hexagonal MoS_2 belongs to the space group D_{6h} ($P6_3/mmc$), and a group theoretical analysis gives 3 Raman-active modes of E_{1g}, E_{2g} and A_{1g} at the Γ-point in the Brillouin zone [12-16]. Raman spectra were obtained by using an Ar-ion laser at 514.5 nm with incident laser power of a few milliwatt.

3. Results and discussion

3.1. Growth temperature dependence

The film thicknesses were evaluated by cross-sectional scanning electron microscopy. All the samples had the thickness of several tens of nano-meters or more. Figure 2(a) shows Raman spectra of as-sputtered samples prepared with different growth temperature. For comparison, the spectrum taken from the bulk MoS_2 is shown in the bottom of the figure. No Raman signals were observed in the sample grown at room temperature, indicating that MoS_2 lattice could not be structured well and it might be amorphous. On the other hand, broad Raman peaks were appeared at 376 and 411 cm^{-1} in samples grown above 100 °C. Although they could be assigned as E_{2g} and A_{1g} mode in terms of their frequencies, they appeared at different peak positions compared to those of bulk. It is well-known that the peak frequency depends on the number of MoS_2 layer up to a few layers[17]. However, the samples could be regarded as bulk MoS_2 from the viewpoint of Raman scattering, because they have the thickness which corresponds to several tens of MoS_2 layers. Therefore, the observed peak shift is not due to the thickness of MoS_2 film. In addition, observed peaks showed an extremely broad shape compared to that of bulk, which suggests that the observed spectra showed a shape similar to the phonon density of states (p-DOS)

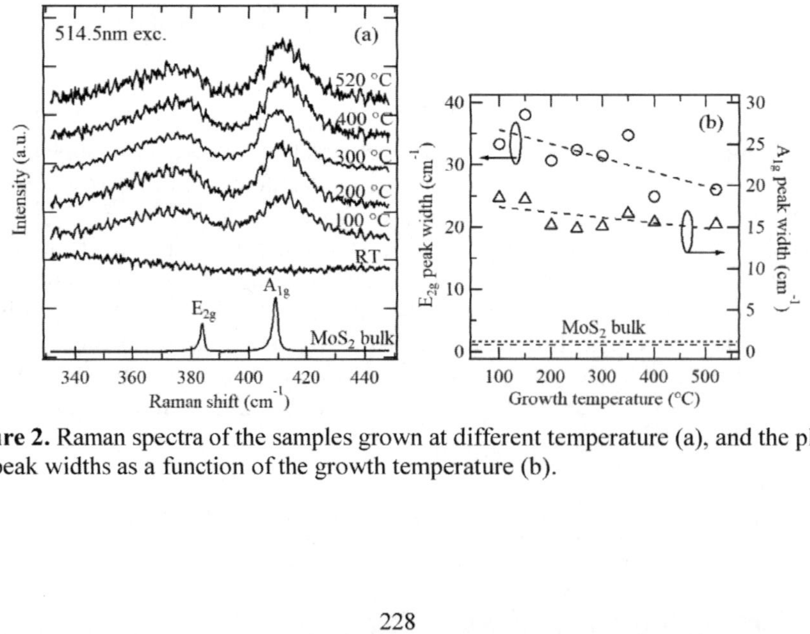

Figure 2. Raman spectra of the samples grown at different temperature (a), and the plot of the peak widths as a function of the growth temperature (b).

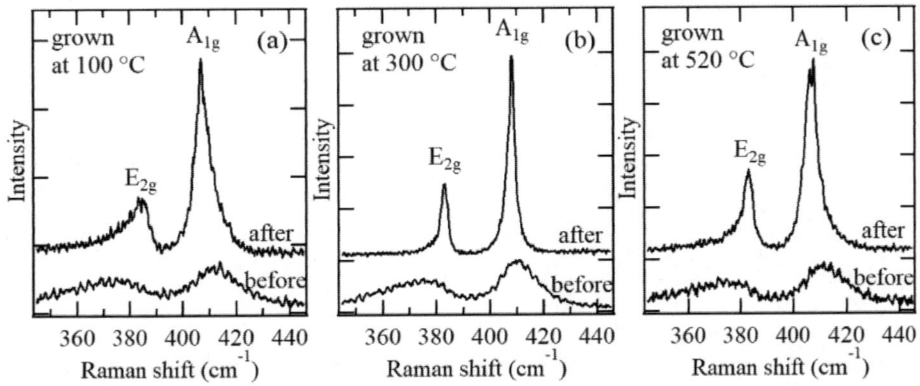

Figure 3. Raman spectra for the samples grown at 100 °C (a), 300 °C (b) and 520 °C (c) before/after thermal annealing under sulfurous atmosphere.

of MoS_2[15]. It is also well-known that the Raman spectrum taken from defective materials shows a spectral shape which well-reflects the shape of p-DOS due to the breaking of Raman selection rule. In our samples, it is obvious that the samples contain a lot of defects due to its broad Raman features. And then, these peaks slightly narrowed with the increase in growth temperature as shown in figure 2(b). This shows that the crystallinity could be improved slightly by growing with higher temperature. However, it was not significant, because both peaks show still broad features compared with those of bulk. This might be caused by sulfur vacancies in MoS_2 lattice, which could not be suppressed by its high vapor pressure at higher temperature.

3.2. The effect of the thermal annealing under sulfurous atmosphere

We carried out the thermal annealing for MoS_2 sputtered thin films under sulfurous atmosphere by using the annealing equipment shown in fig. 1. The thermal annealing was performed at 600 °C for 2 hours for the samples grown at 100 °C, 300 °C and 520°C. Figure 3 shows Raman spectra of the annealed samples, and both spectra taken from the sample before/after annealing were shown to compare the effect of thermal annealing. Raman signals of MoS_2 become obviously sharp in all samples after the thermal annealing, indicating the thermal annealing has remarkable effect on the improvement of the crystallinity due to the sulfurization of the sample. Furthermore, it was also observed the difference in the peak shapes between each annealed sample. Among them, the sample grown at 300 °C showed the sharpest peaks of 3.0 and 3.3 cm^{-1} for E_{2g} and A_{1g} modes with high symmetry shape, respectively. These values were close to those of bulk MoS_2 (E_{2g}: 1.6 cm^{-1} and A_{1g}: 2.0 cm^{-1}). On the other hand, the annealed samples grown at 100 °C and 520 °C showed the broader peaks and asymmetric shape. E_{2g} and A_{1g} peaks of the sample grown at 100 °C had the widths of 10 and 7.2 cm^{-1} and showed asymmetry shape tailed to lower and higher frequency side, respectively. And, both peaks became narrower to 4.4 and 5.8 cm^{-1} in the annealed sample grown at 520 °C, but these were broader than those of the annealed sample grown at 300 °C. This suggests that the annealing effect depends on the sputtering condition of the samples, that is, it is necessary to prepare the as-sputtered samples fabricated by optimal growth temperature.

3.3. The thermal annealing time dependence

We prepared the samples annealed for different duration. The duration was varied from 30 minutes to 120 minutes. Raman spectra of the samples were shown in figure 4, and it contains the spectra of as-sputtered and bulk MoS_2 for comparison. Although all the spectra showed very similar shapes and almost overlapped, the spectrum taken from the sample annealed for 30 minutes had slight broader peaks in both peaks of E_{2g} and A_{1g} mode than others. From the result, the sulfurization process occurs at the initial stage of the annealing and sufficient effect can not be obtained even if the duration of the thermal annealing is prolonged. And, it was found that the broad bands tailed to lower and higher frequency side of E_{2g} and A_{1g} peaks correspond to those of as-sputtered sample. Roughly speaking, the annealed

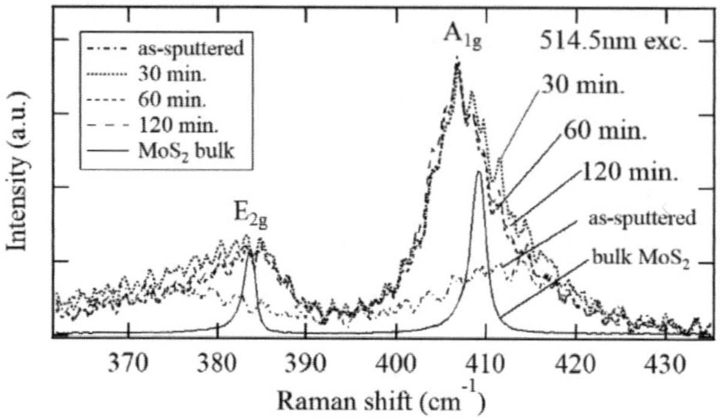

Figure 4. Raman spectra of the samples annealed at 600 °C for different duration.

samples may consist of the two components, in which the one has the MoS_2 lattice with comparatively long-range lattice ordering and the other does not. Here, we consider the process of sulfurization. Since sulfurization occurs from the top surface toward to the inside, sulfur atoms need to diffuse into the inside of the films to be sulfurized for entire the film. However, the top surface become covered with high-quality two-dimensional MoS_2 layer in the initial stage of sulfurization process, and it might make the incorporation of sulfur atoms from the atmosphere more difficult. And, this becomes one of the reasons why sufficient improvement could not be obtained even if the annealing is carried out for long time. Therefore, the thermal annealing under sulfurous atmosphere will become effective method to improve the crystallinity for a few-layer MoS_2 thin films.

4. Conclusion

We investigated the influence of the thermal annealing for MoS_2 sputtered thin films by Raman scattering. As-sputtered MoS_2 thin films showed broad Raman signals due to the deterioration of the crystallinity in MoS_2 lattice. And in MoS_2 sputtered thin films after thermal annealing under sulfurous atmosphere, Raman signals became sharp drastically because the crystallinity was improved by the incorporation of sulfur atoms from the atmosphere and the formation of the long-range lattice ordering by thermal annealing. Since these phenomena occurred remarkably on the top surface of MoS_2 sputtered thin films, the thermal annealing under sulfurous atmosphere has a great effect on the improvement of the crystallinity for a few-layer MoS_2 thin films.

References

[1] Splendiani A, Sun L, Zhang Y, Li T, Kim J, Chim C Y, Gualli G and Wang F 2010 *Nano Lett.* **10** 1271
[2] Eda G, Yamaguchi H, Voiry D, Fujita T, Chen M and Chhowalla M 2011 *Nano Lett.* **11** 5111
[3] Buscema M, Steele G A, van der Zant H S J and Castellanos-Gomez A 2014 *Nano Res.* **7(4)** 561
[4] Özden A, Sar H, Yeltik A, Madenoğlu B, Sevik C, Ay F and Perkgöz N K 2016 *Phys. Stat. Sol. RRL* **10** 792
[5] Radisavljevic B, Whitwick M B and Kis A 2011 *ACS NANO* **5** 9934
[6] Bae H, Kim C K and Choi Y K 2017 *AIP ADVANCES* **7** 075304
[7] Piatti E, Chen Q and Ye J 2017 *Appl. Phys. Lett.* **111** 013106
[8] Das S, Chen H Y, Penumatcha A V and Appenzeller J 2013 *Nano Lett.* **13** 100
[9] Baugher B W H, Churchill H O H, Yang Y and Herrero P J 2013 *Nano Lett.* **13** 4212
[10] Kufer D and Konstantatos G 2015 *Nano Lett.* **15** 7307
[11] Xu E Z, Liu H M, Park K, Li Z, Losovyj Y, Starr M, Werbianskyj M, Fertig H A and Zhang S X 2017 *Nanoscale* **9** 3576
[12] Wieting T J and Verble J L 1971 *Phys. Rev. B* **3** 4286
[13] Sandoval S J, Yang D, Frindt R F and Irwin J C 1991 *Phys. Rev B* **44** 3955
[14] Frey G L, Tenne R, Matthews M J, Dresselhaus M S and Dresselhaus G 1999 *Phys. Rev. B* **60** 2883
[15] Molina-Sa'nchez A and Wirtz L 2011 *Phys. Rev. B* **84** 155413
[16] Verble J L and Wieting T J 1970 *Phys. Rev. Lett.* **25** 362
[17] Chakraborty B, Matte R, Sood A K, Rao C N R 2013 *J. Raman Spectrosc.* **44** 92

Complex-angle analysis of electromagnetic waves on interfaces

Daigo Oue[1]
[1]Division of Frontier Materials Science, Osaka University, 1-3 Machikaneyama, Toyonaka, Osaka, Japan 560-0871

E-mail: daigo.oue@gmail.com

Abstract. Electromagnetic wave is reflected and refracted at interfaces, satisfying Fresnel-Snell law which is required by conservations of energy and momentum. If the incident angle is lower than the critical angle, we can use this Fresnel-Snell law, but the Fresnel-Snell law is modified in the case of existence of dissipation ($\tilde{n} = n + i\kappa, \kappa > 0$) or in the condition of total internal reflection. In the cases, we have to extend the angle of refraction from real number to complex number ($\theta \to \tilde{\theta} = \theta + i\psi, \psi \neq 0$). In this paper, by using complex-angle approach, we analyse the behaviour of electromagnetic waves in various kind of interfaces: dielectric - dielectric system, dissipative dielectric - dielectric system, and metal - dielectric system. We show that iso-frequency curves in wavenumber space is opened in the case where $n, \kappa > 0$, and closed either when $n \to 0$ or when $\kappa \to 0$ ('Lifshitz transition' of electromagnetic waves). Excess momentum (wavevector) and anomalous circular polarisation emerging with the transition are also discussed.

1. Introduction

1.1. *Fresnel-Snell law*

Snell's law (1) is one of the fundermental laws which we use to analyse the behaviours of electromagnetic waves on interfaces, and which is required by momentum conservation on interfaces [1].

$$n_i \sin\theta_i = n \sin\theta \tag{1}$$

One of the most important thing about Snell's law is that once we obtain angle of refraction from Snell's law, which gives wavevector as will be explained in the subsection 1.2, we can obtain the amplitude of reflected and refracted waves by using Fresnel coefficients [1].

1.2. *Rotation for calculation of reflected and refracted waves*

To give explicit forms of reflected and refracted waves, we need to take the propagation directions into consideration. One method to do so is to rotate the wave vector and the field vector:

$$\boldsymbol{k} = \begin{bmatrix} 0 \\ 0 \\ k_0 \end{bmatrix} \to \boldsymbol{k} = R(\Theta) \begin{bmatrix} 0 \\ 0 \\ nk_0 \end{bmatrix}, \boldsymbol{E} = \begin{bmatrix} E_p \\ E_s \\ 0 \end{bmatrix} \to \boldsymbol{E} = R(\Theta) \begin{bmatrix} t_pE_p \\ t_sE_s \\ 0 \end{bmatrix} \tag{2}$$

IOP Conf. Series: Journal of Physics: Conf. Series **1220** (2019) 012058 doi:10.1088/1742-6596/1220/1/012058

by using a rotation matrix:

$$R(\Theta) = \begin{bmatrix} \cos\Theta & 0 & \sin\Theta \\ 0 & 1 & 0 \\ -\sin\Theta & 0 & \cos\Theta \end{bmatrix}. \quad (3)$$

Here, for reflected waves $\Theta = \pi - \theta_i$, whereas for refracted waves $\Theta = \theta$. Then, we can get the explicit forms by substituting them into $\boldsymbol{E}\exp(i\boldsymbol{k}\cdot\boldsymbol{r})$.

2. Complex angle

2.1. Requirement of complex number for angle of refraction

We need to introduce imaginary part in the angle of refraction ($\theta \to \tilde{\theta} = \theta + i\psi$) due to satisfying momentum conservation on interfaces when the refracted waves have decay structure such as the presence of dissipation (or total internal reflection which is discussed in [2, 3]).

2.2. Complex-angle Snell's law and complex-angle rotation

With the complex-angle ($\tilde{\theta} = \theta + i\psi$) and complex-index of refraction ($\tilde{n} = n + i\kappa$), Snell's law are modified as

$$n_i \sin\theta_i = \tilde{n}\sin\tilde{\theta}. \quad (4)$$

Note that from the imarginary part of (4) the imaginary part of the complex-angle is a function of the real part ($\psi = \psi(\theta)$).

Complex-angle rotation matrix is given by

$$R\left(\tilde{\theta} = \theta + i\psi\right) = \cosh\psi \begin{bmatrix} \cos\theta & 0 & \sin\theta \\ 0 & 1/\cosh\psi & 0 \\ -\sin\theta & 0 & \cos\theta \end{bmatrix} - i\sinh\psi \begin{bmatrix} \sin\theta & 0 & -\cos\theta \\ 0 & 0 & 0 \\ \cos\theta & 0 & \sin\theta \end{bmatrix}. \quad (5)$$

Complex-angle rotation yields wavevector of refracted wave:

$$\tilde{\boldsymbol{k}} = R\left(\tilde{\theta}\right) \begin{bmatrix} 0 \\ 0 \\ \tilde{n}k_0 \end{bmatrix} = \boldsymbol{k} + i\boldsymbol{\eta} \quad (6)$$

where

$$\boldsymbol{k} = k_0\bar{\boldsymbol{k}} \equiv k_0 \begin{bmatrix} n\sin\theta\cosh\psi - \kappa\cos\theta\sinh\psi \\ 0 \\ n\cos\theta\cosh\psi + \kappa\sin\theta\sinh\psi \end{bmatrix}, \boldsymbol{\eta} = k_0\bar{\boldsymbol{\eta}} \equiv k_0 \begin{bmatrix} n\cos\theta\sinh\psi + \kappa\sin\theta\cosh\psi \\ 0 \\ -n\sin\theta\sinh\psi + \kappa\cos\theta\cosh\psi \end{bmatrix}. \quad (7)$$

Likewise, we obtain the field vector of refracted wave:

$$\boldsymbol{E}_0 = R\left(\tilde{\theta}\right) \begin{bmatrix} t_pE_p \\ t_sE_s \\ 0 \end{bmatrix} = \begin{bmatrix} t_pE_p\left(\cos\theta\cosh\psi - i\sin\theta\sinh\psi\right) \\ t_sE_s \\ -t_pE_p\left(\sin\theta\cosh\psi + i\cos\theta\sinh\psi\right) \end{bmatrix} \quad (8)$$

3. 'Lifshitz transition' of electromagnetic waves

3.1. Discontinuous transition of iso-frequency curve

The wavevector (6) which we obtain by complex-angle rotation satisfies the complex dispersion relation (9), which holds in the dissipative media.

$$\tilde{\boldsymbol{k}}\cdot\tilde{\boldsymbol{k}} = \left(\frac{\omega}{c}\tilde{n}\right)^2 \quad (9)$$

Figure 1 shows iso-frequency curve drawn by wavevector discontinuously changes its shape (open↔close transition). We call this 'Lifshitz transition' of electromagnetic waves.

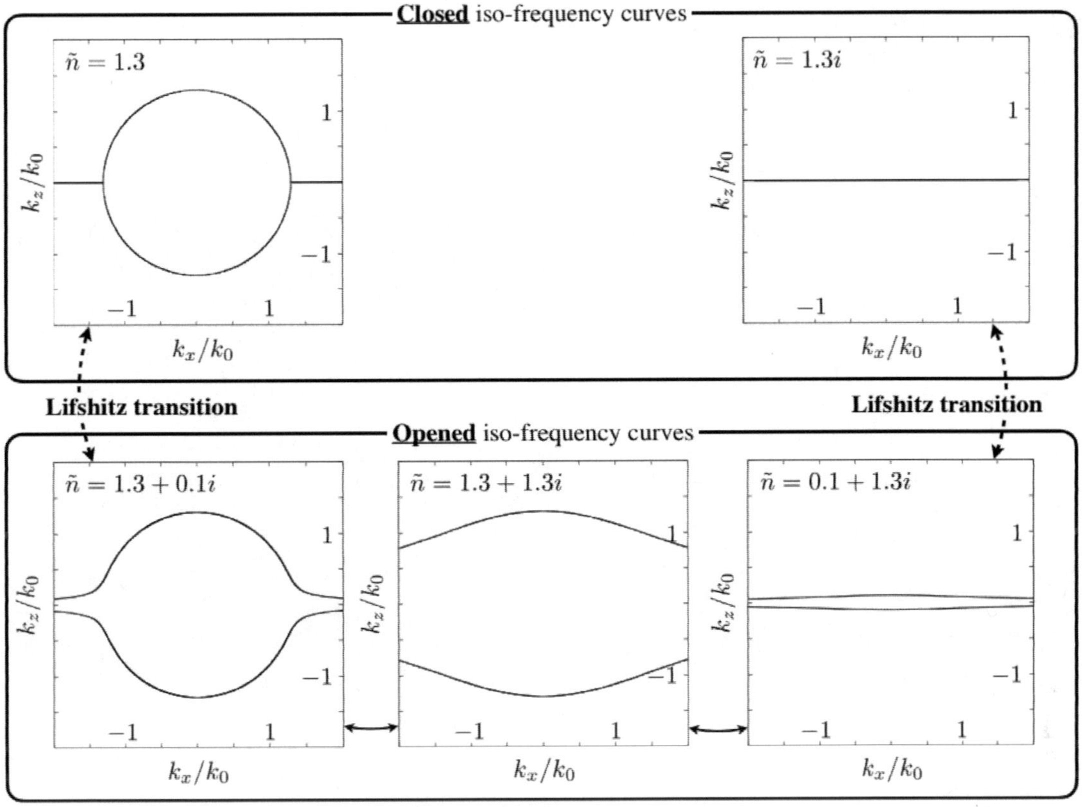

Figure 1. Iso-frequency curves drawn by wavevector for interfaces between various complex refractive indeces $\tilde{n}$ and vacuum. We assume that the interface is at $z = z_0$ and that we have vacuum in the region of $z < z_0$ and dissipative media in $z > z_0$. The iso-frequency curve is closed either when real part vanishes ($\mathrm{Re}(\tilde{n}) = n \to 0$) or when imaginary part ($\mathrm{Im}(\tilde{n}) = \kappa \to 0$) vanishes, while it is opened when both real and imaginary part are non-zero ($\mathrm{Re}(\tilde{n}) \times \mathrm{Im}(\tilde{n}) = n\kappa \neq 0$).

3.2. Surface mode does not exist

Once iso-frequency curves are open, surface mode cannot exist on the interfaces (observe that there is no condition of $k_z = 0$ in the bottom graphs in Figure 1). Surface modes have pure imarginary wavenumber perpendicular to the interfaces, but by dissipation this pure imaginary component couples out to real (propagating) component, and the modes become reaky (For the rigorous proof, see [4]).

3.3. Excess momentum and anomalous circular polarisation

In Figure 2, The black curve is the iso-frequency curve with the presence of dissipation ($\tilde{n} = 1.3{+}1.3i$), while the grey circle is the iso-frequency curve without dissipation ($\tilde{n} = 1.3{+}0i$). The black curve covers the grey circle. This means that the electromagnetic waves near interfaces with dissipation have larger wavenumber (momentum) than the waves without dissipation:

$$\left|\mathrm{Re}\left(\tilde{\boldsymbol{k}}\right)\right| = |\boldsymbol{k}| = k_0\sqrt{\left(\bar{k}_x\right)^2 + \left(\bar{k}_z\right)^2} > nk_0, \tag{10}$$

IOP Conf. Series: Journal of Physics: Conf. Series **1220** (2019) 012058 doi:10.1088/1742-6596/1220/1/012058

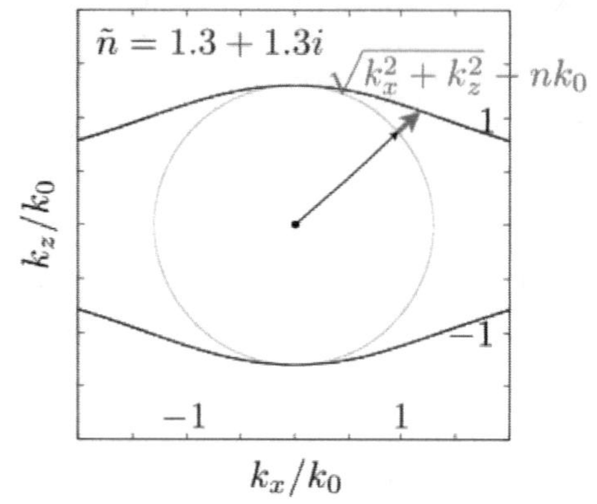

Figure 2. Excess momentum production by dissipation. The solid curve is the iso-frequency curve with the presence of dissipation ($\tilde{n} = 1.3+1.3i$), while the grey circle is the iso-frequency curve without the presence of dissipation ($\tilde{n} = 1.3 + 0i$).

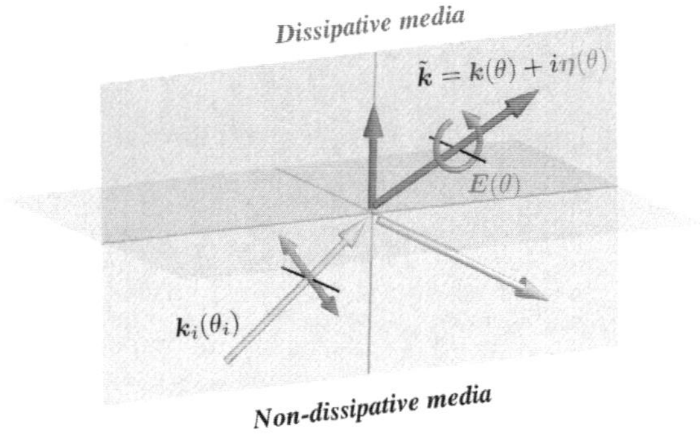

Figure 3. Anomalous circular polarisation

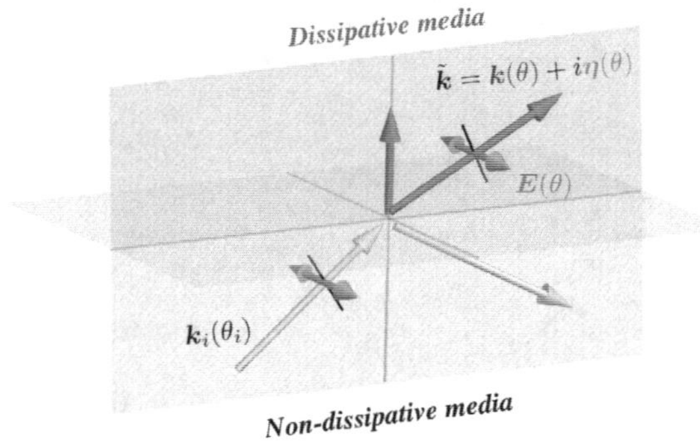

Figure 4. Ordinary linear polarisation

although dissipation generally changes not the wavenumber but only the intensity profile of waves. This curious phenomena is caused by the presence both of dissipation and interface (cannot happen in bulk).

Define *circular-polarisation vector* of electromagnetic waves as $\boldsymbol{s} \equiv \frac{g}{4\omega}\mathrm{Im}\,(\boldsymbol{E}^{\star} \times \boldsymbol{E})$. Here, $g = 1/4\pi$ is a Gaussian-unit factor. This circular-polarisation vector is generally parallel to the wavevector because of Maxwell transversality condition. Near the dissipative interfaces, however, electromagnetic wave has the longitudinal component, and the circular-polarisation vector can be non-parallel to the wavevector. With p-polarisation incidence, anomalous circular polarisation ($s_y \neq 0$) is generated, while no circular polarisation ($s = 0$) with s-polarisation (compare Figure 3 and 4).

4. Conclusion

By utilizing complex-angle approach, we have shown that discontinuous transition of iso-frequency curve ('Lifshitz transition' of electromagnetic waves) occurs when either real part or imaginary part of the complex-index of refraction vanish ($\mathrm{Re}(\tilde{n}) \times \mathrm{Im}(\tilde{n}) = n\kappa = 0$). Once the transition occurs, surface mode does not exist on the interfaces. Also, extraordinary behaviours of electromagnetic waves: excess momentum and anomalous circular polarisation emerge.

References

[1] Novotny L and Hecht B 2012 *Principles of nano-optics* (Cambridge university press)
[2] Bekshaev A Y, Bliokh K Y and Nori F 2013 *Optics express* **21** 7082–7095
[3] Bliokh K Y, Bekshaev A Y and Nori F 2014 *Nature communications* **5** 3300
[4] See the supplementary materials.

Second harmonic generation from complementary triangular Au metamaterials

Y B Habibullah, K Iwata, and T Ishihara*

Department of Physics, Graduate School of Science, Tohoku University, Sendai 980-8578, Japan

*E-mail: t-ishihara@tohoku.ac.jp

Abstract. Sub-wavelength square array of a triangular Au platelet and its complementary structure (i.e. that of a triangular hole in a square film of Au) are compared in terms of second harmonic generation efficiency for fundamental light in the near infrared and visible region of spectrum for normal incidence. Electric field strength around the convex corners of a triangular particle is at least 10 times larger than the one around the concave corners of triangular hole in the complementary structure. Nevertheless the SHG intensity at the respective resonant frequency is found to be comparable, which are numerically estimated by an overlap integral of nonlinear polarization and electric field at the SHG frequency in the nonlinear optical scattering theory originally proposed by Roke et al. (Phys. Rev. B 70, 115106 (2004)). The reason is due to the large electric field strength at the sides of the triangular hole at the resonance frequency, which compensates the suppressed electric field at the concave corners in the overlap integral.

1. Introduction

Metamaterials (MMs) are artificial periodic or non-periodic structure, whose electromagnetic properties are due to both the subwavelength cellular architecture and chemical composition, which are not readily found in nature. The emergence of MMs has changed the spirit of optics, such that today, researchers are not just restricted to variety of natural materials, but rather, they now have the freedom to nanostructure metallic surface with any geometry to tailor optical properties at will. By nanostructuring a metallic surface with geometries that breaks inversion symmetry, not only that new features such as strong local near field enhancement and surface plasmon resonance start to emerge but also second harmonic (SH) response is achievable from such surface[1].

In recent time, there has been growing interest in understanding the relationship between the strong near field enhancement in MMs and their Second Harmonic (SH) response. So far SH generation from MM are mostly from Split Ring Resonators (SRRs) of different geometries, which are usually characterized with mixed corners (i.e. convex and concave corners)[2]. It is a known fact that the local field enhancement around convex corners of metal can be very large compare to other areas with zero curvature, which is due to the affinity of electric field lines to confine around geometric corners[1]. Depending on whether the corners are convex or concave, the level of enhancement can be different, but since SRRs is characterized with mixed corners, it is difficult to separate the effect of local field enhancement around each corners on SH response.

IOP Conf. Series: Journal of Physics: Conf. Series **1220** (2019) 012059 doi:10.1088/1742-6596/1220/1/012059

In this study, we separate the effect of local field enhancement at the convex and concave corners, by designing a complementary Au MM with triangular resonators (i.e. Au MM with triangular particle and Au MM with triangular hole), where the Au MM with triangular particle is characterized with three convex corners while Au MM with triangular hole is characterized with three concave corners. In this design the triangular particle is simply replaced by air in metal to have a complementary structure. The aim for this design is to confirm how the difference in local field enhancement around convex/concave corners in complementary Au MM with triangular resonators translates into high SH emission intensity, in an attempt to understand the optical origin of nonlinearity. To achieve this objective, we will employ numerical simulation based on the nonlinear scattering theory, which is a numerical method that has proven to be efficient in predicting nonlinear emission from non-centrosymmetric medium using microscopic linear response[3].

2. Methods

The linear responses of our complementary Au MMs with isosceles triangular resonators were numerically calculated using the CST microwave electromagnetic solver. A plane wave (polarized either x- or y- direction) propagating along the z- direction is considered as an excitation and the symmetry axis of the triangle is fixed in x- direction throughout this computation.

In order to estimate SHG from the artificial structure, the nonlinear optical scattering method originally proposed by Roke et al [4] and later applied for MMs by O'Brien et al [3], is used in this work. The SH electric field is evaluated in MATLAB by the following surface integral:

$$E_{ijk}^{\mathrm{SHG}} \propto \iint E_{i,\mathrm{n}}(2\omega)\chi_{\mathrm{nnn}}^{(2)} E_{j,\mathrm{n}}(\omega)E_{k,\mathrm{n}}(\omega)dS \qquad (1)$$

where E_{ijk}^{SHG} is the second order non-linear emission, $\chi_{\mathrm{nnn}}^{(2)}$ is the local nonlinear susceptibility at the surface, and $E_{j,\mathrm{n}}(\omega)$ and $E_{k,\mathrm{n}}(\omega)$ are the near field of the excitation by fundamental wave and $E_{i,\mathrm{n}}(2\omega)$ is the mode at second harmonic frequency normal to the surface of the nanostructure. The notation i, j and k correspond to the direction of second harmonic emission and two fundamental excitations, respectively. In this work, we focus on the case where $j = k$. Then the integral vanishes if $i = y$ when the second harmonic mode is odd upon y-inversion for isosceles triangles. In general, the integral can vanish depending on the symmetry, which gives the selection rule. If the symmetry allows, the SHG intensity is determined by the overlap of SH electric field and nonlinear polarization at the surface. In order to make our calculation simple, only the component $\chi_{\mathrm{nnn}}^{(2)}$ is considered, where n indicates the normal component to the surface, which is justified by a recent experimental result showing the dominance of this component for a nano-structured metal [3]. The overlap integral estimates the efficiency of the mesoscopic nonlinear polarization distribution to emit SHG field at the far field. The SH emission intensity is evaluated by using the relation

$$I_{ijk}^{\mathrm{SHG}} \propto \left|E_{ijk}^{\mathrm{SHG}}\right|^2 \qquad (2)$$

3. Results and Discussion

The structure of considered complementary Au MMs with triangular resonators is defined by the base (b) and height (h) of the triangle and a unit cell size (L) as in Figure 1. In all cases, the thickness of the complementary Au MMs is 25 nm. In this work we consider the cases for $b = h = K$; $K = 150$, 200, 250, 300 nm and $L = 400$ nm. To mimic the experimentally fabricated samples we blend the vertex of a triangle with radius of 30 nm in all cases. In the following we call a MM with triangular platelets a "particle" and its complementary structure with triangular holes a "hole" for simplicity. Figure 1 shows transmission spectra of the complementary structures for different resonator sizes (K). The curves are shifted vertically for better visibility. The resonant wavelength of linear response is identified by evaluating the transmission as a function wavelength,

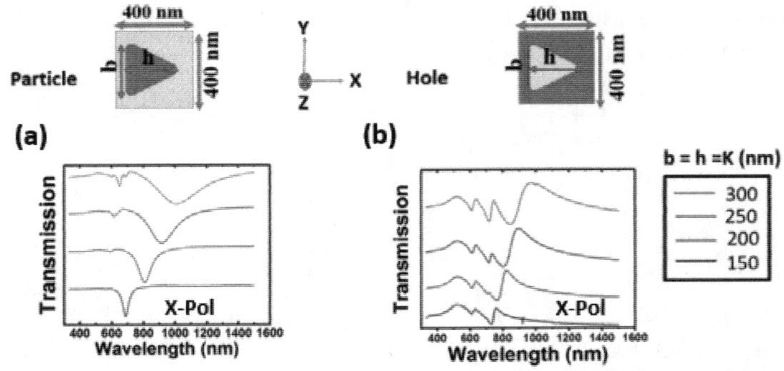

Figure 1. Transmission spectra as a function of wavelength for different sizes of (a) Particle (b) Hole.

the resonant wavelength for different sizes of complementary particle and hole upon *x*-polarization excitation is as shown in Figure (1a and b), where we have the resonant wavelength of particle and hole shifting to longer wavelength with increase in size of the particle and hole. The same trend was observed upon *y*-polarization excitation (not shown). For larger particle size with $K = 250$ nm and 300 nm, two resonance dips (i.e. narrow and wide) emerged, which might be due to multipolar effect coming from the increment in size of the particle. More interestingly at every wavelength for all K where we have wide resonant dip in particle, we always have a corresponding wide peak rather than a dip in hole at approximate wavelength, which should be due to the complementary nature of the two structures. Based on the identified resonance wavelength from the linear response of complementary particle and hole as shown in Figure 1, the optimum SH emission intensity was evaluated for *x*- and *y*-directions upon *x* and *y* polarization excitation at normal incidence using nonlinear scattering theory as defined in Equation (1).

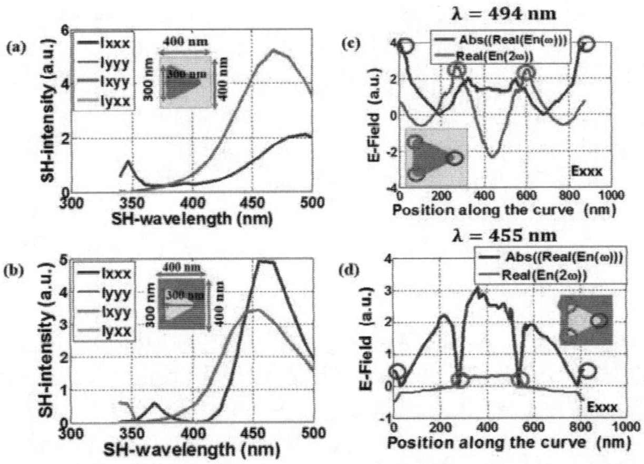

Figure 2. SH intensity as a function of SH wavelength with $K = 300$ nm for (a) Particle (b) Hole. Local field enhancement at optimum SH resonant wavelength as a function of position along the curve surrounding the surface upon *x*-polarization excitation for (c) particle (d) hole.

Figure 2 (a and b) shows the SH intensity as a function of SH wavelength for particle and hole respectively with the dimension as specified in the figure. Due to the inversion symmetry that is broken along x-direction of the particle and hole, SH intensity is only observed in the *x*-direction upon *x* and *y* polarization excitation as indicated by term *Ixxx* and *Ixyy* in Fig. 2 (a and b), with optimum SH wavelength for particle at approximately 467 nm for *x* and *y* polarization while the optimum SH wavelength for hole upon *x* and *y* polarization is 455 nm. In order to investigate local field enhancement, we plot the field amplitude along the surface of the particle and hole 5 nm away from the surface in Figure 2 (c and d). As is seen in the figure, the local field enhancement around the convex corners are at least ten times more than that of concave corners. The same trend was observed for other sizes of the complementary particle and hole.

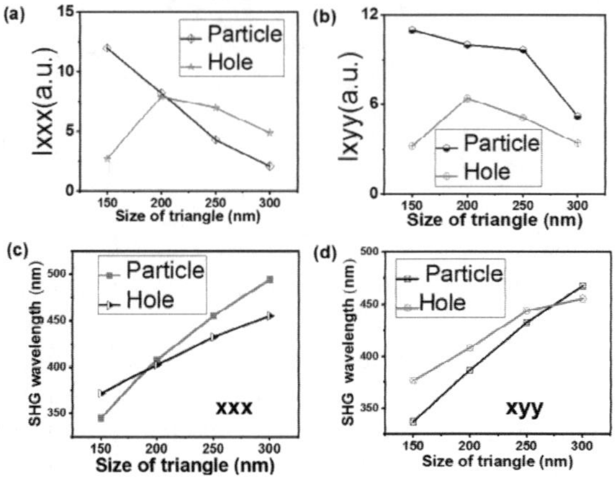

Figure 3. Optimum SH intensity for different sizes of resonators (a) in the *x*-direction upon *x*-polarization excitation. (b) in the *x*-direction upon *y*-polarization excitation. Optimum SH wavelength for different sizes of resonators (c) upon *x*-polarization excitation (d) upon *y*-polarization excitation.

Despite the large difference in enhancement around convex corners of a particle as compared to concave corners hole, the optimum SH intensity for particle and hole are found to be comparable for *x* and *y* polarization excitation at normal incidence as depicted in Figure 2 (a and b). A comparable optimum SH intensity is also observed for the other sizes of complementary particle and hole for *x* and *y* polarization excitation at normal incidence as depicted in Figure 3 (a and b) respectively. The optimum SH wavelength at which the comparable optimum SH intensity were recorded for the different sizes of particle and hole upon *x* and *y* polarization excitation is as shown in Figure 3(c and d) respectively. By comparing the optimum SH resonant wavelength for different sizes of particle and hole, we can obviously notice how close they are for both *x* and *y* polarization excitation. This closeness of optimum SH resonant wavelength can be related to complementary nature of the particle and hole structures.

4. Conclusion

In summary, we compared SHG from a Au MM with triangular particle and its complementary Au MM with triangular hole, with each having a convex and concave corners respectively. Evaluation of the local field enhancement at optimum SH resonant wavelength for the complementary structure confirms that local field enhancement around convex corners are far larger than the enhancement around concave corners. Despite the large difference in local field enhancement around corners of the complementary structure, their respective SH intensity are comparable upon *x* and *y* polarization excitation at normal incidence, which implies that near field enhancement at the corners are not always the major determinant of SH emission intensity from metamaterial, the major contributor to SH efficiency can be other part of the geometrical structure in concave cornered nanostructures.

References

[1] Valev V K 2012 *Langmuir* **28** 15454
[2] Ciracì C, Poutrina E, Scalora M, Smith D R 2012 *Phys. Rev.* B **85** 201403
[3] O'Brien K , Suchowski H, Rho J, Salandrino A, Kante B, Yin X, Zhang X, 2015 *Nat. Matr* **14** 379
[4] Roke S, Bonn M, Petukhov A V 2004 *Phys. Rev. B* **70**, 115106

AUTHOR INDEX

Institute of Physics
Dirac House, Temple Back
Bristol BS1 6BE UK

ISSN: 1742-6588
ISBN 978-1-5108-8820-3

CURRAN ASSOCIATES INC.
proceedings
.com